Modern Medicines
from Plants

Royal College
of Physicians

The beds of the *Pharmacopoea Londinensis* of 1618 at the Royal College of Physicians

Modern Medicines from Plants

Botanical histories of some of modern medicine's most important drugs

From the

Garden of Medicinal Plants at the Royal College of Physicians

11 St Andrews Place, Regent's Park

London NW1 4LE

Editor: Dr Henry Oakeley

Authors: Dr Susan Burge, Dr Timothy Cutler, Professor Anthony Dayan, Professor Michael de Swiet, Dr Arjun Devanesan, Professor Graham Foster, Jane Knowles, Professor John Newton, Dr Henry Oakeley, Dr Noel Snell, Angela Tunstall.

2024

CRC Press is an imprint of the
Taylor & Francis Group, an **informa** business

Layout and typesetting: Smallfish Designs Ltd. email info@smallfish.tv

First published by CRC Press 2024

6000 Broken Sound Parkway NW, Suite 300, Boca Raton, FL 33487-2742

and by CRC Press

4 Park Square, Milton Park, Abingdon, Oxon, OX14 4RN

CRC Press is an imprint of Taylor & Francis Group, LLC

on behalf of the Royal College of Physicians

11 St Andrews Place, Regent's Park, London NW1 4LE, United Kingdom

British Library Cataloguing-in-Publication Data
A catalogue record for this book is available from the British Library

ISBN: 9781032534343 (hbk)
ISBN: 9781032536170 (pbk)
ISBN: 9781003413004 (ebk)

Typeset in Baskerville

Printed in the UK by Severn, Gloucester on responsibly sourced paper

https://garden.rcplondon.ac.uk
for information on the Garden of Medicinal Plants at the Royal College of Physicians, London

Contents

Preface   10

Introduction   11

Chapter 1   ***Ammi majus*** – ammi, false Queen Anne's lace, bullwort, bishop's weed, herb William – methoxsalen (8-methoxypsoralen)   14

Chapter 2   ***Artemisia annua*** – sweet wormwood, annual wormwood, Qing-Hao – artemisinin and derivatives   24

Chapter 3   ***Atropa belladonna*** – deadly nightshade; ***Datura stramonium*** – Jimson weed; and ***Mandragora***, ***Hyoscyamus***, ***Brugmansia***, ***Scopolia*** – atropine and hyoscine   30

Chapter 4   ***Betula pendula*** – silver birch – ß-sitosterol   44

Chapter 5   ***Camellia sinensis*** – tea; ***Coffea arabica*** – coffee; ***Theobroma cacao*** – chocolate – caffeine, theophylline   50

Chapter 6   ***Camptotheca acuminata*** – camptotheca – camptothecin, topotecan and irinotecan   62

Chapter 7   ***Capsicum annuum*** – chilli pepper – capsaicin   68

Chapter 8   ***Catharanthus roseus*** – Madagascar periwinkle, Cayenne jasmine, old maid, rosy periwinkle – vincristine, vinblastine   76

Chapter 9   ***Cephalotaxus harringtonia*** – Chinese plum yew – harringtonine   82

Chapter 10   ***Chondrodendron tomentosum*** – curare vine – tubocurarine   86

Chapter 11   ***Cinchona species*** – quinine tree, Jesuit's bark, Peruvian bark – quinine, quinidine   92

Chapter 12   ***Citrus x limon*** – lemon – vitamin C   102

Chapter 13   ***Colchicum autumnale*** – meadow saffron, autumn crocus, naked ladies – colchicine   112

Chapter 14   ***Digitalis purpurea*** – purple foxglove – digitoxin; ***Digitalis lanata*** – woolly foxglove – digoxin   120

Chapter 15   ***Dioscorea polystachya*** – yam; ***Glycine max*** – soybean – steroids   128

Chapter 16   ***Ephedra sinica*** – joint pine, Mormon tea – ephedrine, ecstasy   136

Chapter 17   ***Erythroxylum coca*** – coca bush – cocaine   144

Chapter 18   ***Euphorbia peplus*** – common spurge – ingenol mebutate   150

Chapter 19 **Galanthus nivalis** – snowdrop – galantamine   158

Chapter 20 **Galega officinalis** – goat's rue – phenformin, metformin   166

Chapter 21 **Glycyrrhiza glabra** – liquorice – carbenoxolone, glycyrrhizic acid   174

Chapter 22 **Guaiacum officinale** – roughbark lignum-vitae, guaiacwood –
alpha-guaiaconic acid   180

Chapter 23 **Hordeum jubatum** – foxtail barley; **Arundo donax** – giant reed –
lignocaine (lidocaine) and derivatives   186

Chapter 24 **Hordeum vulgare, Claviceps purpurea** – common barley, ergot –
ergometrine, ergotamine   194

Chapter 25 **Hydrangea febrifuga** – Asian hydrangea, Chinese quinine – febrifugine,
methaqualone, halofuginone   200

Chapter 26 **Illicium verum** – Chinese star anise; **I. anisatum** – Japanese star
anise – oseltamivir   206

Chapter 27 **Inula helenium** – elecampane, enula, horseheal, scabwort, wild
sunflower – inulin   212

Chapter 28 **Melilotus officinalis** – melilot, yellow sweet clover, king's clover,
yellow melilot – warfarin   218

Chapter 29 **Morus alba** – white mulberry – miglustat, migalastat, miglitol   224

Chapter 30 **Nicotiana tabacum** – tobacco – nicotine, with a note on **Lobelia** –
lobeline, and **Laburnum anagyroides** – cytisine   228

Chapter 31 **Papaver rhoeas** – corn or Flanders poppy – rhoeadine, thebaine,
oxycodone, etorphine and other derivatives   242

Chapter 32 **Papaver somniferum** – opium poppy – morphine, codeine,
noscapine, protopine   248

Chapter 33 **Physostigma venenosum** – Calabar bean – physostigmine   256

Chapter 34 **Pilocarpus microphyllus** – jaborandi – pilocarpine   262

Chapter 35 **Podophyllum peltatum** – mayapple, American mandrake, ground
lemon; **Podophyllum hexandrum** – Himalayan mayapple –
podophyllotoxin, etoposide, teniposide   264

Chapter 36 **Rauvolfia serpentina** – snake root – reserpine   272

Chapter 37 **Salix alba** – willow; **Filipendula ulmaria** – meadowsweet; **Gaultheria
procumbens** – wintergreen – aspirin, salicylic acid, methyl salicylate   280

Chapter 38 **Silybum marianum** – milk thistle – silymarin, Legalon-SIL   292

Chapter 39 **Tanacetum cinerariifolium** – pyrethrum, Dalmatian chrysanthemum –
pyrethrins   298

Chapter 40 **Taxus baccata** – European yew; **Taxus brevifolia** – Pacific yew –
paclitaxel and derivatives   306

Chapter 41   *Valeriana officinalis* – valerian – sodium valproate   314

Chapter 42   *Veratrum album*, *V. nigrum* – false hellebores – protoveratrine –
V. californicum – cyclopamine, sonidegib   320

Chapter 43   *Visnaga daucoides* – khella – nifedipine, amiodarone, sodium
cromoglicate, nedocromil sodium   328

Chapter 44   **Excipients and Solvents**   336

Chapter 45   **Vitamins**   345

Appendices

Appendix 1   **Historical references before 1700 and Linnaeus**   362

Appendix 2   *Dramatis personae*. The main historical personages   367

Appendix 3   **Glossary**   372

Index   **Chemicals and Drugs**   379

Index   **Plant Names**   384

The Authors   390

Preface
Michael de Swiet

This book is about the plants which have given rise to modern prescription medicines or have proved useful in medicine, such as elecampane, *Inula helenium* and guaiac, *Guaiacum officinale*. Most of these plants grow in the Garden of Medicinal Plants at the Royal College of Physicians in central London. A few that cannot be grown in our Garden such as *Chondrodendron tomentosum*, the source of curare which is the original muscle relaxant, and *Physostigma venenosum*, the source of physostigmine, are included because of their huge importance in medicine.

These accounts of about fifty plants, the history of their use in the past two thousand years, the 'Prescription Only' medicines (POMs) that are derived from them, and their modern uses, have been compiled by the Garden Fellows, all senior physicians at the Royal College of Physicians of London, with horticultural notes written by the College Gardeners. Photographs, and woodcuts from the 16th and 17th centuries, provide a continuity with the illustrated medicinal herbals of the past.

A section is included about the more important plant-sourced excipients, the other substances essential in producing a useful and practical therapy. Plants have been essential in the discovery of vitamins, and they are also discussed.

References are given that provide more detailed information about the plants, their history, the ancient physicians who wrote about them, the important chemicals they contain and their effects. For ease of reading, the historical and more scientific sections have been separated graphically as have the 'Plant profiles' regarding their habitats and cultivation.

This is a unique book about the historical medicinal uses of plants and the effective medicines that have been derived from them.

Introduction
Henry Oakeley

Plants have been used to provide us with medicines for millennia. Pedanius Dioscorides [40-90 CE] , a Greek physician from Cecilia, a region of southern Turkey, recorded the uses of 600 plants in 70 CE. Claudius Galen [129-c. 200/216 CE] a Greek physician and surgeon in Rome, set out the Humoral concept of disease and how medicines worked in 200 CE. Their observations and beliefs dominated: they were repeated and elaborated on by Avicenna [980-1037 CE] and the other great physicians of the Golden Age of Islam. However, they stultified and held back the practice of medicine until the 19th century, and, with important exceptions, the beliefs of our predecessors have little relationship to the modern medicinal uses of plants. This book describes plant-sourced treatments from Classical to modern times, documenting the transition from 'then' to 'now' and providing the information and primary sources for knowing the story of a plant and its role in producing medicines and treating disease.

There are about 7000 names of 'Prescription Only' medicines (POMs) in Britain of which about 1000 are in regular use and many of these are minor modifications of others. They are the powerful orthodox medicines, which can only be prescribed by a medical practitioner or dentist; they are listed in the official British National Formulary (accessible on-line at https://bnf.nice. org.uk). It may be a surprise that the chemical structure of about one third of them, including many of the most modern, are related to chemicals found in plants (Newman & Cragg, 2016). Such a statistic may be used to suggest that over 2000 plants are currently used to make our POMs or have been the original sources of the chemical structures from which the medicinal compound has been derived but, in truth, of the 400,000 different plant species on the planet, the number which have been used in the production of modern 'Prescription Only' medicines in the West is nearer 50. With no attempt to identify all the medicines derived from our 50 plants we have mentioned over 200.

Most of these medicinal plants are from the world's temperate zones, only a handful from the tropics, and most are grown in the Garden of Medicinal Plants at the Royal College of Physicians in London, opposite the southeast corner of Regent's Park. With its collection of over 1000 plants, it is a living museum of the history of medicines, old and new, open to the public and free for all to visit.

A plant-sourced chemical that is a POM is rarely found to exist as a safe therapeutic substance in the plant but has usually been altered to make it more effective or safer. Hundreds of potentially medicinal chemicals with varied uses may be obtained from the original natural molecule and study of them provides leads to the creation of purely synthetic chemicals. This book introduces the rapidly expanding world of the chemistry of plants, of diseases and how medicines work.

Many different plants may contain the same medicinally useful chemical but for various reasons are not used as a source for producing medicines: this book is about those plants which are used, with occasional notes on the ones that are not.

There was no evolutionary benefit for any of the 400,000 known plant species, whose ancestors came on the planet at some time in the last 400 million years, to provide medicines for humans whose predecessors came on the planet less than five million years ago. Plants that became poisonous or immediately repulsive to predators survived; humans who developed taste genes to detect poisonous alkaloids as bitter, or simply vomited, also survived. Children who ate 'greens' such as aconite, foxglove, yew leaves, and most wayside plants not found in the vegetable aisle of the supermarket, did not survive until lunchtime let alone long enough to continue the evolutionary process, so it is not surprising that many children have to be taught to eat edible vegetables. It is of note that from the vast number of plants on the planet, only about 30 vegetables (excluding fruit) can be found in the vegetable aisles of our supermarkets in the West. The inherently poisonous inedibility of most of the plant kingdom is nowhere more clearly shown. Except for vitamin C, without which we die horribly from scurvy, all our medicines share this property and are poisonous if taken in excess.

We have used the poisons that have evolved in plants to deter their parasites and their other predators, to make our medicines. The lethal yew tree, *Taxus baccata*, through a symbiotic fungus, produces chemicals, to deter other fungal attack; these chemicals can be extracted and converted into potent anti-cancer drugs for us. No animal eats the equally poisonous autumn crocus, *Colchicum autumnale*, at least not twice. It contains colchicine, whose toxic action on dividing cells we exploit to treat several serious medical conditions. This book unravels the complexity of making medicines and the diligence of pharmacists in their search to make them safer and more effective.

All modern medicines, including those sourced from plants, are carefully tested by double-blind trials and with long-term follow-up, to ensure that they are effective and not toxic in short- or long-term use. It is no longer enough to believe that 'if it is natural, it is good for you', nor that a plant which has been 'used for thousands of years' is effective or even safe.

This book recounts the successes and failures of many drugs, discovered from mouldy cattle fodder, ordeal and arrow poisons, one-eyed lambs, an unloved chilli pepper, infected barley, serendipity, hard work, garden plants, ancient Chinese medicinal texts, etc. etc., and our unstoppable drive to know more.

Note:

We have primarily used the Latin names of plants, as the local names from different countries differ so much, and Latin names are standard across the world. Information about visiting the Medicinal Garden at the Royal College of Physicians in London can be found at https://garden.rcplondon.ac.uk

Warning:

The plants, and the medicines that have been derived from them, are discussed here for information only. It is dangerous to attempt to use them for self-medication. If you need treatment for a medical condition the advice of a doctor should be sought. Never make your own medicines.

References

Newman DJ and Cragg GM. Natural Products as Sources of New Drugs from 1981 to 2014. *J Natural Products* 2016;79:629-661

Acknowledgments

The authors thank the Funding and Awards Management Committee of the Royal College of Physicians for their generous grant towards the production of this book and to the Officers and staff, particularly Natalie Wilder, Head of Corporate Communications and Publishing, for their unfailing support.

Special thanks are due to Christopher Oakeley for reading and commenting on the early draft, to Sara Oakeley for copy editing the entire work and to Ms Randy Brehm, Senior Editor, Life Sciences and Nutrition with CRC Press in the Taylor and Francis Group LLC for guiding us through the process to publication.

We are very grateful also to Sophia and Marcel Kral of Smallfish Designs, for layout and typesetting, readying the book for print.

We express a particular thank you to Dr Sofia Labbouz for her research on the 13th century Arabian physician, Ibn al-Bayṭār, with respect to the use of *Ammi* for treating vitiligo.

Photographs are by Henry Oakeley unless otherwise stated.

CHAPTER 1 – SUSAN BURGE

Ammi majus
The source of 8-methoxypsoralen (methoxsalen)

Ammi majus, known vernacularly as ammi or bullwort, is a plant in the Apiaceae (cow parsley) family with a long history of dermatological use which continues to be important to this day.

Introduction

The seeds of *Ammi majus* are a major source of a furanocoumarin called xanthotoxin, better known as 8-methoxypsoralen or methoxsalen, a photosensitising agent that is widely cultivated in some parts of the world, especially India, for the furanocoumarins which are used in the treatment of skin conditions. In a process known as photochemotherapy it is used to treat psoriasis (an inflammatory skin disease characterised by raised areas of red, flaky skin), some forms of dermatitis, vitiligo (white patches of skin that have lost the pigment melanin) and cutaneous T-cell lymphoma (a rare very slow-growing cancer of the lymphatic system which presents in the skin with dry, red, scaly patches or lumps).

Additionally, 8-methoxypsoralen has also been extracted from other members of the Apiaceae including *Heracleum mantegazzianum* (giant hogweed) and *H. sphondylium*, (hogweed). Laboratory cultures of *A. majus* cells may be induced to produce furanocoumarins when exposed to fungal infections (Trease *et al*, 2009; Hamerski *et al*, 1988) but this does not appear to be used commercially.

Plant profile

Ammi majus is a tall and elegant hardy annual with the characteristic umbel-shaped flowerheads of the Apiaceae family. The delicate lace-like appearance of their white flowers and divided leaves belies their robust nature. It is this combination which makes it popular as a garden plant and in the cut flower trade. Native to Macaronesia and the Mediterranean region through to Iran and the Arabian Peninsula, it is found in scrub and wasteland. The plant is in flower from early summer to autumn and the seeds ripen from July onwards. These can be sown *in situ* in the spring but if they seed themselves in the autumn they will flower earlier and produce stronger plants.

Common names

Ammi majus has a variety of common names, most of uncertain etymology. Culpeper (1653) refers to bullwort, bishop's weed, herb William, and – giving far earlier names – Ethiopian cumin-seed and cummin royal that originate with Dioscorides [70 CE] and Hippocrates [4th C BCE]. The link with cumin is straightforward – cumin (*Cuminum cyminum*) is a similar-looking flowering plant in the Apiaceae enjoyed for its spicy, flavoursome seeds. 'Wort' is derived from the Old English word 'wyrt' meaning plant, herb or root. The derivation of 'bull' is far from certain, but 'bull' might be an abbreviation of bulla, a fluid-filled blister, referring to the effect the sap has on skin – the 'blister plant'. The links to Bishops or indeed William remain a tantalising mystery. False Queen Anne's lace, another name, refers to the lacy appearance of the flowers and the similarity to wild carrot, *Daucus carota*, which is known as Queen Anne's lace. None of these names is in common usage. Many gardeners refer to the plant quite simply as ammi, derived from the Classical Greek name, *Ami*, according to Pliny [70 CE].

History

Ammi appears to have been used traditionally for treating a wide range of diseases for the first 1700 years of the current era, and possibly much earlier, but then disappeared from the literature until the 20th century. While it is certain that the plant for treating skin conditions is now regarded as being *A. majus*, it is more likely that it was *Ammi visnaga* (now called *Visnaga daucoides*) which was the original *Ammi* of Dioscorides [70 CE]. The painting of *Ammi* in the *Juliana Anicia Codex* (514 CE) which illustrates the plants described by Dioscorides is clearly of *Visnaga daucoides*.

Modern writers have suggested that Egyptian physicians since at least 2000 BCE used the remarkable capacity of *Ammi* to cause hyperpigmentation to treat conditions such as vitiligo in which melanin is lost from the skin and the story has been repeated in numerous articles (McGovern & Berkley, 1998). We cannot identify any of the diseases or indeed most of the plants and drugs referred to in the Ebers papyrus (1500 BCE) with certainty and *Ammi* is not mentioned in it despite innumerable articles stating that it is (Hartmann 2016). Apparently, the Ebers papyrus does describe a condition with flat areas of colour loss that was regarded as 'treatable' (Ebbell, 1937), but the origin of this condition is debatable. Amongst the numerous causes of acquired loss of skin colour are infections such as leprosy (referred to as *baras* in ancient Arabic texts), nutritional deficiencies, vitiligo, morphoea and toxins such as arsenic. However, the commonest cause of loss of colour is post-inflammatory hypopigmentation which occurs after an injury to the skin or after inflammatory skin diseases such as psoriasis or eczema. Post-inflammatory hypopigmentation slowly resolves spontaneously so it is just as likely that the author of the Ebers papyrus was referring to this 'treatable' condition as to any other cause of hypopigmentation.

Dioscorides [70 CE] recommends *Ammi* topically, mixed with honey, for taking away '*sugillata*' [Latin = bruises, 'black and blue spots'] and states that taken either as a drink or smeared on it makes skin paler. Like all the other uses, this is unlikely to be effective as *Ammi* would have darkened rather than lightened sun-exposed skin. However, Dioscorides additionally reports that the seed of *Ammi* cures severe abdominal colic and painful urination, which equates to renal colic from renal and bladder calculi, for which *V. daucoides* (*q.v.*) is still used. The latter causes photosensitivity in the same way as *A. majus*, and it is possible that *A. majus* came to be used because of its close similarity to *V. daucoides*.

Galen [200 CE] said that the seed (as opposed to the leaves or roots) of *Ammi* was the most useful medicinally. The Arabic physician Avicenna [980-1037] wrote that *Ammi* when applied as a liniment or drunk, mixed with honey, treated *albaras* [Arabic for 'white leprosy'] and morphea

Ammi majus – herbarium specimen from the Pharmaceutical Society's herbarium at the Royal College of Physicians

(in the modern era morphoea refers to a condition with pale thickened patches of skin which may resemble vitiligo). In its other actions he follows Dioscorides, in its treating abdominal pain, difficulty in urination, and renal stones. It is also recorded in the Latin compilation of earlier authors, the *Pandectarum* of Matthaeus Sylvaticus written in 1317, that it heals the passages of the kidney and bladder and expels [bladder] stones, and as a liniment *Ammi* cures and cleans morphoea and *baras* when mixed with honey. One to two thousand years ago the ability even to tell leprosy from vitiligo would have been difficult, but nevertheless it is clear that Avicenna was advising *Ammi* for the treatment of such a skin condition. The historic Arabic names were *nachana* according to Sylvaticus [14th century], *anazue* according to Serapion [11th century], and both as well as *Hanochach* (probably a misprint for *nanochach*) according to Mattioli (1581).

The 13th century Arabian physician Ibn al-Bayṭār is said to have described the treatment of leukoderma or vitiligo with *Aatrilla* (regarded now as *Ammi majus*) and sunlight in his book *Mofradat El-Adwiya* [*Compendium on Simple Medicaments and Foods*], which was essentially a compilation of the works of Avicenna and Galen. However, the German translation of this work (Sontheimer, 1840) does not confirm this. Apparently, the Ben Shoeib, a Berber tribe from Northwest Africa, were aware of this remedy (Marquis, 1985; Pathak and Fitzpatrick, 1992).

16th and 17th century European pharmacopoeias and herbals mention *Ammi* as a treatment for blue marks on the skin, but its medicinal use for anything disappears thereafter – partly because it was imported as the seed from Egypt and India and its identity was lost or confused with other Apiaceae. There is no mention of *Ammi* for treating morphea or leprosy in Bernard de Gordon's *Opus Lilium medicinae* (1550), written in 1305, and for the next 500 years *Ammi* had no role in treating skin diseases except in Egypt until the isolation of 8-methoxypsoralen from *A. majus* in the mid-20th century.

Medicinal

The root, sap and seeds of many plants in the Apiaceae family contain furanocoumarins (also known as furocoumarins) which have evolved to protect the plant from mammal or insect predators as well as some fungal and bacterial diseases. They cause skin to become photosensitive.

Chemistry

Ammi majus contains two furanocoumarins, bergapten (5-methoxypsoralen) and xanthotoxin (better known as 8-methoxypsoralen or methoxsalen), both of which are activated by ultraviolet A radiation (UVA; 320-400 nm). Activated furanocoumarins react with nucleobases in DNA to cause cutaneous inflammation and blistering. Unlike allergy, which only occurs in previously sensitised individuals, everyone getting furanocoumarins on skin that is subsequently exposed to sunlight will experience phototoxicity.

Egyptians would surely have observed the effects that sap from a bruised leaf or broken stem of this common Nile Valley weed had on skin later exposed to sun. Over the next two to three days the skin that had been in contact with the sap would have become uncomfortable, red and swollen and finally might even have blistered. Blisters would have healed slowly over several days but one or two weeks later the affected individual would have noticed persistent dark brown streaks at the sites of the reaction – the end result of what is called a phytophotodermatitis (phyto = plant; photo = light; dermatitis = skin inflammation). The hyperpigmentation of phytophotodermatitis can take months or years to fade, or may persist indefinitely.

Chemistry

This hyperpigmentation is produced by melanin, the pigment that gives skin its normal colour and produces a tan after sun exposure. Melanin is produced by melanocytes, dendritic cells (cells with long processes) that reside in the basal layer of the epidermis (the outermost layer of skin). The epidermis is composed of keratinocytes arranged like bricks in a wall. Melanocytes transfer small packages of melanin along their dendrites into the keratinocytes, where the melanin provides the nuclear DNA of the keratinocytes with some protection from sun damage (and subsequent skin cancer). Furanocoumarins that have been activated by UVA radiation increase the pigmentation through a variety of mechanisms: melanocytes increase in number; they make more melanin; and melanin is transferred into the keratinocytes in more and larger packages than normal (McGovern & Berkley, 1998).

Vitiligo is a common, long-lasting condition in which pigment-producing skin cells, the melanocytes, are either destroyed or stop producing melanin leading to disfiguring and distressing white patches that may have considerable psychosocial impact. Vitiligo is not life-threatening or contagious, but it was and still is confused with leprosy, a mycobacterial infection that also may cause discrete pale skin patches. Leprosy is a stigmatised disease that even now, although it is curable, may be synonymous with abandonment and social isolation. Leprosy would have been prevalent in Egypt, so it would not be surprising if Egyptians explored methods of inducing repigmentation using extracts of *Ammi*, which were either applied to the skin or taken orally before exposing the skin to sunlight. Egyptian herbalists still sell Aatrillal, a yellowish-brown powder made from *Ammi* seeds, for the treatment of vitiligo.

Modern use in psoriasis

In the 1940s, scientists at the University of Cairo, reportedly stimulated by the writings of Ibn al-Bayṭār, investigated the folklore surrounding *Ammi majus* and isolated xanthotoxin (8-methoxypsoralen) in addition to other furanocoumarins from its seeds (Fahmy *et al*, 1947; Schönberg *et al*, 1948).

Subsequently el-Mofty, an Egyptian dermatologist, demonstrated that crystalline 8-methoxypsoralen applied to the skin followed by exposure to sunlight could be helpful in managing vitiligo (el-Mofty, 1948; el-Mofty *et al*, 1994). This work led to the development of modern photochemotherapy – the combination of a drug with light – to treat skin disease.

Patients prescribed photochemotherapy either take 8-methoxypsoralen tablets or apply a solution of 8-methoxypsoralen to the affected skin. Then the skin is exposed to an artificial source of UVA radiation for a limited time to activate the 8-methoxypsoralen. This treatment is known colloquially as PUVA – Psoralen plus UltraViolet-A radiation. In countries such as India, with strong natural sunlight, psoralen plus sunlight (PUVAsol), is a cost-effective and practical alternative to PUVA.

Treating vitiligo with PUVA is only moderately successful – treatments must be repeated over many months to induce acceptable repigmentation. Normal skin darkens, accentuating the patchy loss of colour, and repigmentation is variable.

The pigmentation produced by PUVA has been used for very different ends. In 1959 the white journalist, John Howard Griffin, wished to explore and document the difficulties of living with skin of colour under racial segregation in southern states of America. Griffin (with the help of a dermatologist) underwent a prolonged course of PUVA so that eventually his skin colour resembled that of an African American (a temporary change). Griffin and a photographer travelled for six weeks through Louisiana, Mississippi, Alabama, Arkansas and Georgia. He published an account of his disturbing experiences in *Black Like Me* (Griffin, 1961).

PUVA is still used to treat severe psoriasis and can be effective in other inflammatory skin diseases, including hand dermatitis, cutaneous T-cell lymphoma, a rare slowly progressive, non-Hodgkin lymphoma in which there are abnormal growths of T-lymphocytes in the skin. (Pathak & Fitzpatrick, 1992; Singer & Berneburg, 2018).

Chemistry

PUVA for psoriasis

Photoactivated 8-methoxypsoralen does much more than stimulate pigmentation. PUVA has anti-proliferative, anti-inflammatory and immunosuppressive effects. In the 1970s, American dermatologists at Harvard Medical School showed that PUVA was a remarkably effective treatment for severe psoriasis, a chronic inflammatory skin disease that is driven by the T-lymphocytes of the immune system (Parrish *et al*, 1974). The disease is characterised by sustained inflammation and hyperproliferation of keratinocytes. The action of PUVA in psoriasis is complex and incompletely understood. Treatment both alters the immune response and reduces cell proliferation. Damage to DNA leads to cell-cycle arrest with programmed cell death (apoptosis), especially of lymphocytes. PUVA also alters the secretion of cytokines – the signalling molecules that mediate and regulate immunity (Singer & Berneburg, 2018).

Other uses of 8-methoxypsoralen

8-methoxypsoralen is also used to modify white blood cells when they have been removed from the patient's blood, a cell-based immunomodulatory therapy called extra-corporeal photopheresis (ECP). White blood cells are separated from the patient's blood, mixed with 8-methoxypsoralen and exposed to UVA radiation before being returned to the patient. The modes of action of ECP are far from clear but the wide-ranging effects include modification of the release of cytokines and activation of some T-cell lines that down-regulate immune reactions. Indications for ECP include advanced cutaneous T-cell lymphoma, severe cutaneous scleroderma and chronic graft versus host disease, a complication of stem cell transplantation (Cho *et al*, 2018).

Toxicity

Excess 8-methoxypsoralen or overexposure to UVA during treatment with PUVA causes severe phototoxicity (sunburn) with pain, swelling, redness and blisters: a serious reaction may even lead to hospital admission. Patients who take 8-methoxypsoralen by mouth must protect their eyes (to prevent cataracts) both during treatment with UVA radiation and for the rest of that day by wearing UVA-absorbing, wraparound sunglasses. The safe lifetime dose of UVA is limited, and the cumulative dose of UVA must be monitored carefully in every patient because 8-methoxypsoralen in combination with UVA is carcinogenic. Too much PUVA causes sun damage and, eventually, skin cancers such as squamous cell carcinoma and malignant melanoma. Psoralens such as 8-methoxypsoralen were even added to tanning lotions but this practice was banned in the 1990s when the increased risk of skin cancer was appreciated.

Phytophotodermatitis can be a problem in gardeners who have been wearing inadequate protective clothing while weeding or strimming members of the Apiaceae in sunny weather, or children who have been playing in the sunshine with plants such as the notorious giant hogweed (*Heracleum mantegazzianum*). Pickers of celery (*Apium graveolens*) working outside are also at risk, particularly if the celery is infected with the fungus *Sclerotina sclerotiorium* or 'pink rot', as the plant responds to fungal infection (and other stresses) by increasing the production of furanocoumarins. Plants in the Rutaceae family, including rue (*Ruta graveolens*) and the peel of citrus fruit such as lime (Mojito drinkers take care), also contain furanocoumarins that cause similar problems if sap or fruit juice splashes onto sun-exposed skin.

Conclusion

Exploration of the folklore surrounding *Ammi* led to the isolation of 8-methoxypsoralen for the treatment of vitiligo and the development of photochemotherapy in 20th and 21st centuries. This story demonstrates the importance of conserving plants and reviewing the practices of physicians and herbalists from long ago.

References

Al-Ismail D, Edwards C, Anstey AV. PUVA through the ages: Ibn al-Bitar to El-Mofty. *Br J Dermatol* 2012;167 suppl1:161-2

Cho A, Jantschitsch C, Knobler R. Extracorporeal Photopheresis – An Overview. *Front Med (Lausanne)* 2018;5:236. doi: 10.3389/fmed.2018.00236

Ibn al-Bayṭār [Ebn Baithaar, Ibn al-Baitar] *Grosse Zusammenstellung über die Kräfte der bekannten einfachen Heil- und Nahrungsmittel*, [Compendium on Simple Medicines and Foods], translated by JV Sontheimer. Stuttgart, Hallbergersche Verlagshandlung, 1840

Ebbell B. *The Papyrus*. Ebers. Levin and Munksgard, Copenhagen, 1937 el-Mofty AM. A preliminary clinical report on the treatment of leukoderma with Ammi majus linn. *J R Egypt Med Assn* 1948;31:651-65

el-Mofty AM, el-Sawalhy H, el-Mofty M. Clinical study of a new preparation of 8-methoxypsoralen in photochemotherapy. *Int J Dermatol* 1994;8:588-92

Fahmy IR, Abu-Shady H. Ammi Majus Linn. pharmacognostical study and isolation of a crystalline constituent, ammoidin. *Q J Pharm Pharmacol* 1947;20:281-91

Griffin JH. *Black Like Me*. Boston, Houghton Mifflin, 1961

Hamerski D, Matern U. Elicitor-induced biosynthesis of psoralens in Ammi majus L. suspension cultures. Microsomal conversion of demethylsuberosin into (+)marmesin and psoralen. *Eur J Biochem* 1988;171:369-75

Hartmann A. Back to the roots – dermatology in ancient Egyptian medicine. *J Dtsch Dermatol Ges* 2016;14:389-96 doi: 10.1111/ddg.12947

Hossain MA, Al Touby S. Ammi majus an endemic medicinal plant: a review of the medicinal uses, pharmacological and phytochemicals. *Ann Toxicol* 2020;2:9-14

McGovern TW, Barkley TM. Botanical dermatology. *Int J Dermatol* 1998;37:321-34

Marquis L. Arabian contributors to dermatology. *Int J Dermatol* 1985;24:60-4

Parrish JA, Fitzpatrick TB, Tanenbaum L, *et al*. Photochemotherapy of psoriasis with oral methoxsalen and long wave ultraviolet light. *N Engl J Med* 1974;291:1207-11

Pathak MA, Fitzpatrick TB. The evolution of photochemotherapy with psoralens and UVA (PUVA): 2000 BC to 1992 AD. *J Photochem Photobiol B* 1992;14:3-22

Schönberg A, Sina A. Xanthotoxin from the fruits of Ammi majus L. *Nature* 1948;161:481-2

Singer S, Berneburg M. Phototherapy. *J Dtsch Dermatol Ges* 2018;16:1120-9

Trease WC, Evans D, Trease GE. *Trease and Evans Pharmacognosy*, 16th edn. Saunders/Elsevier, 2009

Title page of Avicenna's *Liber Canonis*, written in c. 1000CE, translated into Latin by Gerard of Cremona, 1522. It was from this work by the great Arabic physician that Ibn al-Bayṭār would have learnt of the uses of *Ammi*

CHAPTER 2 – MICHAEL DE SWIET

Artemisia annua
The source of artemisinin

Artemisia annua, also known as sweet wormwood, in the family Asteraceae, is the source of artemisinin, a very important treatment for malaria.

Plant profile

Artemisia annua, also known as *Qing-Hao*, is an annual, sweetly aromatic herb, native to many countries throughout Asia and North Africa, but it has been naturalised in many other countries. Its natural habitat is scrub and light woodland on rocky slopes. It grows quickly up to 2m in height on thin, stiff, brownish stems laced with soft, finely divided, pinnate leaves of greyish green. The tiny greenish-yellow flowers appear in July on loose drooping panicles. In a garden it can look rather unwieldy and is less decorative than other members of the genus. It will grow stronger, for longer and with more aromatic foliage in poorer soils but it has no tolerance for drought or waterlogging. It can be sown directly outdoors and once established it may self-seed. It does come up slightly later in the season than other hardy annuals. Due to a lifecycle which is complete in just six months, *A. annua* is considered a weed in many areas of the world. It is pollinated by insects.

History

Artemisia annua does not feature specifically in European herbals, but *A. absinthium* was rubbed on the body with oil to drive away mosquitos and, taken as a tea, killed intestinal parasites (an anti-helminthic), according to Dioscorides [70 CE]. A wine 'called absinthitis' was made from the plant and to this day it is used to flavour spirits, as absinthe and vermouth. Its use as an anti-helminthic was recorded in the 6th century by Alexander of Tralles (1570). *Artemisia* is known as wormwood 'because it killeth [intestinal] worms' according to Turner (1551) who notes that its bitterness is a defence against it being eaten by animals – the plants are also remarkably free of parasites, insects and caterpillars. This bitterness is recorded in the Old Testament 'bitter as wormwood' (Proverbs 5:4). It was still used for treating intestinal worms in the 18th century (Quincy, 1718; Woodville, 1792) and 19th century (Lindley, 1838; Bentley, 1861). Woodville also reported it being used for malaria ('intermittent fevers'). It is serendipitous that the chemicals produced by *Artemisia* for its own survival have been harnessed so effectively to treat human parasites.

In the Far East, *Qing-Hao* has long been known as a treatment for malaria (*q.v. Cinchona* for further details about malaria). Identified in a Chinese tomb book, it has been used as an anti-inflammatory drug for the treatment of fever and specifically for the treatment of malaria from the 4th century CE. It was also used for the treatment of haemorrhoids.

Introduction and discovery

During the Vietnam War (1955-75) malaria was a major problem. The *Plasmodium* parasite that causes malaria was becoming resistant to the then conventional drugs, particularly chloroquine. Communist North Vietnam was fighting a guerrilla war against South Vietnam and the United States. The North Vietnamese leader, Ho Chi Minh, asked Chinese Premier Zhou Enlai for help in developing an effective new treatment for malaria. Since malaria was also a major problem in China, Zhou Enlai was able to persuade Chairman Mao Zedong to help. He set up a top-secret military scheme named Project 523 in 1967. Tu Youyou, a Chinese pharmaceutical chemist and malariologist was appointed head in 1969. Scientists worldwide had screened over 214,000 substances up to 1972 without success, but Tu had the idea to go back to the original Chinese traditional medicine sources.

Tu looked at 640 prescriptions in ancient Chinese manuscripts. Her team made 380 herbal extracts, from some 200 herbs, which were tested on mice. One extract was effective. It was named *Qinghaosu* and came from *A. annua* which, as noted above, had been used for over 1000 years for intermittent fevers and specifically for malaria. The active medicinal chemical, artemisinin, is inactivated in hot water but Tu remembered that the original Chinese medicine source stated that leaves of the plant should be steeped in cold water. She developed a cold extraction procedure and produced artemisinin which was shown to be effective in both mice and monkeys infected with malaria. By 1973 she had determined the chemical structure.

In 2015 Tu was awarded the Nobel Prize in Physiology or Medicine for this outstanding work (Philips, 2015; Campbell *et al*, 2015).

Treatment of malaria, synthesis of derivatives

Even though artemisinin was effective, artemisinin preparations did not perform well due to the drug's poor solubility and poor absorption when swallowed in tablet form. Tu discovered dihydroartemisinin when she altered the structure of artemisinin in a structure-activity relationship study. The resulting compound dihydroartemisinin was ten times as potent as artemisinin. Other currently recommended artemisinin derivatives are artemether which can be given intramuscularly (i.m.), and artesunate which can be given i.m., intravenously (i.v.) and rectally as a suppository, but all are metabolised in the body to the active metabolite, dihydroartemisinin.

The current World Health Organisation (WHO) recommendations for the treatment of uncomplicated malaria are to use one of five different artemisinin-based combination therapies consisting of dihydroartemisinin or another artemisinin derivative, together with another antimalarial such as amodiaquine or mefloquine (WHO, 2019). Resistance of the malaria parasite to artemisinin is already a problem in some parts of Southeast Asia, so artemisinin derivatives should not be used on their own.

Simultaneous use of two antimalarials makes the development of a resistant strain less likely because the *Plasmodium* parasite would have to become resistant to both agents. Furthermore, artemisinin derivatives are relatively short acting, being rapidly eliminated from patients. Each of the second agents in the five WHO recommended artemisinin-based combination therapies is a long-acting antimalarial. Therefore, the theory is that any artemisinin-resistant organisms that have escaped the short-acting artemisinin derivative will be killed by the other long-acting antimalarial.

Severe life-threatening malaria requires parenteral (i.v. or i.m.) treatment (or rectal suppositories in very sick young babies and children). The patient should be treated with injections of artesunate which may be given intravenously or intramuscularly. Artemether intramuscularly is a less preferable alternative. *In extremis* and if neither is available, quinine injections should still be used. Where facilities do not exist for parenteral treatment, rectal suppositories are used. Randomised controlled trials in severe *Plasmodium falciparum* malaria have shown that artesunate reduces the mortality rate by about 30% more than does quinine. Parenteral treatment should be continued for at least 24 hours and until the patient can take oral therapy (WHO, 2019).

Because artemisinin derivatives are so short acting they should not be used for malaria prophylaxis. In addition, their indiscriminate use has increased the prevalence of malarial parasites that are resistant to them.

Chemistry

Mode of action

All current therapeutic artemisinins are precursors of dihydroartemisinin. Several mechanisms have been proposed for the action of dihydroartemisinin, including a process called oxidative stress. At the schizont (multiplication) stage in its complicated lifecycle, the malaria parasite consumes haemoglobin within the victim's red blood cells to liberate haem. The iron of the haem reacts with dihydroartemisinin to produce a highly toxic compound that kills the parasite by oxidative stress, so the parasite is especially vulnerable to dihydroartemisinin when it is present in the red blood cells. However, dihydroartemisinin has been shown to act against a large number of protein targets including those present in the mitochondria (which are responsible for cell function) of the *Plasmodium* so it is also vulnerable to dihydroartemisinin at all stages of its lifecycle within the human host (German, 2008).

Artemisinin and its derivatives are also effective anti-helminthics in other protozoan parasitic infections such as schistosomiasis and liver flukes (Keiser & Utzinger, 2007). Although artemisinin derivative therapy has been shown to be effective in schistosomiasis, it has been used mostly in sheep liver fluke disease. Artemisinin also has treatment possibilities in autoimmune diseases such as lupus erythematosus and in cancer treatment.

Artemisia annua plants still provide artemisinin for therapeutic purposes, but a culture method growing *A. annua* cells in giant fermenters is being developed for easier production. It is extracted from the leaves with a solvent. Much work has been done to develop particularly high-yielding strains of *A. annua* that contain up to 20 times the amount of artemisinin compared with the wild type.

Comment

The work of Tu Youyou that led to the production of therapeutic artemisinins making use of the innate properties of the plant's defence mechanisms is one of the great achievements of the past century. The statistic that over 200,000 compounds had previously been tested for antimalarial activity without success, including 640 by Tu Youyou, emphasizes the rarity of finding novel therapies from plants, and the single-minded dedication of scientists.

References

Campbell WC, Satoshi Omura, Tu Youyou. Tu Youyou Biographical Tu Youyou – Biographical. https://www.nobelprize.org/prizes/medicine/2015/tu/biographical/ accessed 6 December 2011

German PI, Aweeka FT. Clinical pharmacology of artemisinin-based combination therapies. *Clinical Pharmacokinetics* 2008;47:91-102

Keiser J, Utzinger J. Artemisinins and synthetic trioxolanes in the treatment of helminth infections. *Current Opinion in Infectious Diseases* 2007;20:605-12

Philips T, Tu Youyou. How Mao's challenge to malaria pioneer led to Nobel prize. *The Guardian* 2015. https://www.theguardian.com/science/2015/oct/05/youyou-tu-how-maos-challenge-to-malaria-pioneer-led-to-nobel-prize accessed 28 May 2021

Tu Youyou. Artemisinin – A Gift from Traditional Chinese Medicine to the World. Nobel Lecture 2015 https://www.nobelprize.org/prizes/medicine/2015/tu/lecture/#:~:text=Youyou%20Tu%20delivered%20her%20Nobel,Committee%20for%20Physiology%20or%20Medicine accessed 26 May 2021

World Health Organisation. Compendium of WHO malaria guidance – prevention, diagnosis, treatment, surveillance 2019 https://apps.who.int/iris/bitstream/handle/10665/312082/WHO-CDS-GMP-2019.03-eng.pdf accessed 26 May 2021

Synthesis

Artemisinin synthesis can be performed from basic organic reagents but at present this is too expensive for commercial use. While cell cultures of *A. annua* are a satisfactory source, more promising is synthesis in genetically-modified organisms. International cooperation, initially funded by the Bill and Melinda Gates Foundation, has produced a genetically-engineered yeast that produces artemisinin, and this process is now producing tonnage quantities. In addition, transgenic specimens of *Nicotiana tabacum* plants have been bio-engineered to manufacture artemisinic acid, the precursor of artemisinin.

CHAPTER 3 – NOEL SNELL

Atropa belladonna
Datura stramonium
and other plants containing
antimuscarinic tropane alkaloids

Introduction

Several members of the potato family, Solanaceae, contain chemical compounds known as antimuscarinic tropane alkaloids, which include atropine, hyoscine (the l-isomer of which is known as scopolamine) and hyoscyamine. These are pharmacologically active (and toxic in large doses).

The principal plants that produce antimuscarinic tropane alkaloids are: *Atropa belladonna* (deadly nightshade), *Datura stramonium* (thorn apple), *D. ferox, D. innoxia, Scopolia carniolica, Brugmansia suaveolens* (angels' trumpets), *Mandragora officinarum* (mandrake), *Hyoscyamus albus* and *H. niger* (henbanes), *Latua pubiflora* (sorcerers' tree), and *Duboisia leichhardtii* and *D. myoporoides* (corkwood trees). All contain varying amounts of atropine, hyoscine and hyoscyamine depending on the season and the part of the plant. Before their alkaloids were in use medicinally, the unrefined extracts were given their plant names: belladonna, stramonium and henbane.

Plant profiles

Atropa belladonna is native to West and Central Europe, the Mediterranean through to North Iran. It is a vigorous hardy perennial growing to 2m high found in woodland on damp, shady sites and alkaline soils. It has chunky stems and large green simple leaves. The small, purple, bell-like flowers are tinged with green and appear in summer followed by luscious-looking, shiny black berries. These may look delicious but, in common with the rest of the plant, they are in fact poisonous.

Flowers of *Atropa belladonna*

Datura stramonium

Mandragora officinarum

Scopolia carniolica

Datura stramonium, or Jimson weed, is an annual from the Americas which can grow up to 2m high but rarely does so in the Royal College of Physicians' Garden. It has showy white or purple, bell-shaped flowers, but the leaves are much greener, less felty looking and more lobed than those of *D. innoxia*. It has very distinctive 50mm seed pods: egg-shaped with spikes. Hence its other common name, the thorn apple.

Mandragora officinarum is a hardy herbaceous perennial which emerges early in the spring from thick, sturdy tap roots. The star-shaped white or mauve flowers appear first, held in tight rosettes sitting close to the ground and looking a bit like little cauliflowers. Next come the large coarse leaves which fall loosely and wide apart. The fruits resemble golf balls and are held on short stalks close to the centre of the plant, turning from green to yellow-brown as they ripen. By mid-summer the plant is returning to dormancy.

The plant has a narrow distribution, occurring naturally in northern Italy and along coastal areas of former Yugoslavia, but it is easily cultivated in any ordinary garden soil. Once established it will happily seed itself around, proving a source of interest rather than beauty.

Scopolia carniolica is a low-growing perennial originally from East, Central, and Southeast Europe to the West Caucasus where it is found in hilly areas on damp, stony sites. Clumps of thin, bright green, membranous leaves develop from the fleshy, underground rhizomes and, in early spring, solitary bell-shaped flowers appear. These pretty flowers hang from long stems and are brown/purple to vivid red with a yellowish colour on the inside. The fruit containing the seed is a small, lidded capsule. Plant in a light, fertile, well-drained soil in a shady position. *S. carniolica* is pollinated by insects.

Brugmansia suaveolens is a semi-woody shrub or small tree native to Brazil where its original distribution was the forested Atlantic coastal strip of eastern Brazil, from northern Rio Grande do Sul to southern Bahia. It can reach up to 3m high and has large, bright green leaves and stunning, trumpet-shaped, white, yellow, orange or pink flowers which are night-scented. A mature specimen can display as many as 80-100 blooms at any one time throughout the summer months and so, unsurprisingly, it has been widely introduced as an ornamental. Being tropical, the plant is not hardy and in temperate regions will need the winter protection of a warm greenhouse or conservatory. At the Royal College of Physicians' Garden it is grown outside in very sheltered conditions and is wrapped in fleece from October until April. It is very easy to grow from cuttings.

Hyoscyamus albus is a small, hardy annual or biennial originating in Macaronesia, the Mediterranean to North Iraq and the Arabian Peninsula. It has erect, branching, woolly stems and sticky, serrated leaves. The pretty flowers are tubular and white, tinged with green or yellow, and with pale yellow stamens. In the wild the plant is found on dry, uncultivated ground in walls and on field margins. Older plants resent their roots being disturbed.

Hyoscyamus niger is a similar but smaller annual or biennial from temperate Eurasia and Northwest Africa. It has velvety leaves with toothed edges and distinctive olive-green/dull yellow and purple-veined flowers.

Brugmansia suaveolens

Hyoscyamus albus

Hyoscyamus niger

Chemistry

Antimuscarinic tropane alkaloids

They are called antimuscarinic tropane alkaloids as they act by blocking muscarinic receptors in the parasympathetic nervous system (part of the autonomic nervous system). In the final step of their biosynthesis, hyoscyamine is converted to hyoscine. The hyoscyamine molecule can take two different 'mirror-image' forms, known as D- and L-hyoscyamine; atropine is simply a mixture of these two forms.

Antimuscarinic tropane alkaloids – physiological action

By blocking the action of the autonomic nervous system, these alkaloids relax smooth muscle in the bronchial tubes and the gut, reduce secretions, increase the heart rate, and dilate the pupils.

They are absorbed into the bloodstream after ingesting them, from the skin, or from the lungs (if smoked or inhaled) and readily pass into the brain where they can cause both sedation and stimulation and, in large doses, delirium and hallucinations followed by amnesia. Historically, they have been used by different cultures for their psychogenic effects, and medicinally for the treatment of pain, inflammation, asthma and colic.

Other tropane alkaloids

Tropane alkaloids with different mechanisms of action are produced by other plants, for example cocaine from *Erythroxylum coca (q.v.)*.

History

It is clear that the medicinal and toxic effects of these plants have been recognised for millennia. From Roman times in Europe to the late medieval period in England, numerous recipes can be found for mixtures used to induce surgical anaesthesia. Common ingredients included opium, henbane, and mandrake – the last two containing hyoscyamine and hyoscine, which would have hypnotic and sedative effects; together with the analgesic and sedative properties of opium this would have been an effective anaesthetic agent (Carter, 1999)

Mandrake, *Scopolia carniolica*, and other plants from the Solanaceae containing tropane alkaloids, were used in the Middle Ages and later as ingredients of 'love potions' despite their inherent dangers. They were all constituents of the 'flying ointments' said to be used by witches before mounting their broomsticks and flying off to the Sabbat. Sufficient hyoscine and hyoscyamine would have been absorbed through the skin and mucous membranes to induce hallucinations of flying and attending satanic orgies (Mann, 1992; Fatur, 2020), rather than any magical property of levitation. Anyone who used such a potion or ointment and recounted her experience (and it was almost invariably a woman) would be regarded as a 'witch' with magical powers, and not as a victim of intoxication. Drinking a 'tea' made from the leaves, roots or fruits would have had the same effect. The disinhibition induced would have been an aid to orgies, but the literature appears to be based on individual accounts of the hallucinations and not on witnessed 'satanic orgies'.

Legend has it that Macbeth used a brew of deadly nightshade to stupefy and overcome the army of Harold Harefoot invading Scotland from England in the 11th century.

Hyoscyamus

Dioscorides [70 CE] wrote of henbane that it was scarcely useable as it caused frenzies and sleep, but that the yellow henbane (*Hyoscyamus alba*) was safer than the black (*H. niger*). He reports that orally it was good for coughs, although too much caused 'a disturbance of the senses' and while good topically for inflammation and as an analgesic, an enema of it for anal pain would also cause confusion.

Mandragora

Dioscorides gives much the same properties to mandrake as he does to henbane, with the addition of its use to induce anaesthesia before surgery or cautery to a wound – a practice still recommended in the 16th century.

Scopolia carniolica – woodcut from Mattioli's *Discorsi*, 1568

Hyoscyamus niger – woodcut from Mattioli's *Discorsi*, 1568

MANDRAGORA MASCHIO.

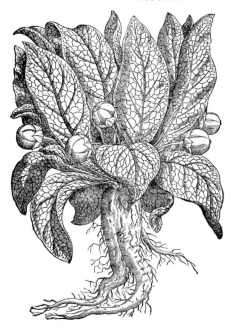

Mandragora officinarum – woodcut from Mattioli's
Discorsi, 1568

Stramonicum minus five perogrinum simplex & duplex.
Single and double small Thorny Apples.

Datura stramonium – woodcut from Parkinson's
Theatrum Botanicum, 1640

Datura

Datura stramonium originates from Mexico and
Central America although it is now found in
most parts of the world. In the early colonial era
it was known as Jimson weed after the town of
Jamestown in Virginia. British troops sent there in
1676 to put down a rebellion of colonists led by
Nathaniel Bacon collected and ate it as a 'boiled
salad' and hallucinated, danced, dozed and read
poetry. When they eventually recovered after
11 days, they must have found that their weapons
had been removed and would have left without
fighting. Linnaeus (1749, 1782) documented its
stupefying, hallucinatory properties.

While *D. stramonium* is alleged to have been described
by Dioscorides (Gunther, 1959; Beck, 2005), he
describes the fruits as soft, in clusters of 10 to 12,
and like the seeds of ivy (which are 5mm across
and smooth unlike the 3cm spherical spiky fruits of
Datura). This is another plant – the story that *Datura*
was used in India and Arabia, and described by
Ibn al-Bayṭār in the 12th century, conflicts with its
Mesoamerican origins and the lack of evidence of
plant introductions from there before the 'discovery'
of the Americas in the 15th century. It appears likely
that another sedative hallucinogenic plant was called
Dhatura in India and the name later transferred to
Datura metel. Thorn apple is mentioned as a purgative
by Hildegard von Bingen [fl 1200 CE] which may
have been *Strychnos nux-vomica* (historically *Nux metalla*
and *Strychnon*), another candidate for the early identity
of the plant from the Graeco-Roman period.

The powdered root of *D. ferox* plant is smoked
as a treatment for asthma in traditional Indian
Ayurvedic medicine, but we can be confident that
it was a different *Dhatura* in earlier times.

Atropa

Deadly nightshade was named *Atropa*, by
Linnaeus, after Atropos, the eldest of the three
Fates from Greek mythology. Her sisters Clotho
and Lachesis spun and measured the thread
of life and Atropos defined our end by cutting
it. Its historical names were sometimes benign
e.g. Solanum majus [large] or *majora* [greater],
but more often reflected its toxic properties –
Solanum (or *Solatrum/Solatum*) *manicum, somnificum,*
dormitorium, lethale and *furiosum*, and *Strychnos* – a
corruption of *Trychnon* – the name used by Galen
[200 CE]. Pliny [70 CE] calls it *Dorycnion* and says
it was used as an arrow poison. He details the
disinhibition that results from taking it internally.
In the 16th century in England it was known as
Dwale, from the Old Norse name of the plant.

It has been called Bella Donna (Italian for a
beautiful lady) and *Bella Dona* (the plural of
Bellum + Donum, Latin for 'wars' + 'gifts') and
is a corruption of Bellonaria, the priestesses
of Bellona, the Goddess of War, who used it
as an arrow poison (Bodaeus, 1644; Apuleius,
6th century). Wecker's *Antidotarium* (1608) calls it
'Solano, which the common people call Bellam
Donnam'. The belief that it was called 'beautiful
lady' as Italian ladies used the juice of the berries
(which contain atropine) to dilate their pupils to
enhance their charms, is a myth. Its ability (by
blocking the autonomic nervous system) to dilate
pupils (mydriasis) was first discovered by John
Ray (1686) and no Italian lady would have used it
twice as its effect lasts for 7-14 days – a very long
time to have dilated pupils, to be so short-sighted
as to be almost blind, and to be dazzled by even
the slightest light.

Belladona.

Atropa belladona – woodcut from Bodaeus's *Theophrasti
Eresii*, 1644

EXT.
BELLAD.
LIQ.

Early 20th century bottle of belladonna extract

Mattioli (1586), calling it *Solatro maius* or *bella donna*, says that the dark purple juice of the fruits was used by Italian ladies to darken their hair, which is more likely. He reported that four to six berries eaten caused delirium, relieved visceral inflammation (*i.e.* colic) and, on the skin, treated erysipelas.

Brugmansia, Latua

There is no record of use of *Brugmansia* and *Latua pubiflora* from Latin America for respiratory disease, but they were employed in Shamanistic ceremonies and as intoxicants. In the present day, *D. stramonium* (known locally as 'zombie cucumber') has been implicated in the creation of zombies in Haiti (Davis, 1983). The story that powdered *Brugmansia* leaves are blown into the faces of tourists in Colombia to induce stupor and facilitate street robberies is a widely held popular belief.

Modern medicinal uses – development

Datura

The psychotropic properties of *Datura* led to it being tested in 1762 and afterwards as a treatment for various neuropsychiatric disorders including epilepsy, mania and depression (Chapman 1999). *Datura* inhalations for asthma were introduced to Europe from India; a Dr Anderson of Madras had found smoking the powdered roots of *D. ferox* (first described in 1753 by Johan Christian Höjer in presenting his dissertation on the 1450 plants growing in the Uppsala Botanic Garden to Linnaeus) beneficial for his asthma and recommended it to a General William Gent in the Madras army, who on his return to Britain in 1802 introduced the remedy to a Dr Sims, who tried it on some of his patients, with success (Sims, 1802). It was soon discovered that *D. stramonium* (which grew wild in England) was just as effective in relaxing the bronchi (Sigmond,1836). By the mid-19th century a variety of *D. stramonium* – containing smoking powders, cigarettes and cigars were available for domestic use. These usually contained other ingredients including tobacco, lobelia, cannabis, henbane, and even arsenic; the popular and long-lasting 'Potter's asthma remedy' contained *D. stramonium*, *Lobelia*, coltsfoot (*Tussilago farfara*), aniseed (*Pimpinella anisum*), and potassium nitrate. Tobacco and *Lobelia* are discussed elsewhere – the other botanical ingredients are not known to have any medicinal action – but potassium nitrate is a bronchodilator.

Atropa belladonna

Linnaeus (1749) noted its narcotic, hallucinogenic, anti-inflammatory and analgesic properties, as well as its usefulness in dysentery. As 'belladonna' it was used as an external liniment on superficial tumours, and as 'belladonna bandages' as a treatment for neuralgia and musculoskeletal aches and pains (Flückiger & Hanbury, 1874), and taken internally (in small doses) to relieve colic. Although previously noted by John Ray (1686), its effect on dilating the pupil (his own, following accidentally getting sap in his eye, and then of a cat) was reported in his academic thesis by Peter Daries (1776). Consequently, Professor Reimarus (1729-1814) in Germany started using it to dilate the pupil to facilitate cataract surgery (Reimarus, 1797; Earles, 1961).

Its use for asthma, taken as a tincture, was endorsed by the English physician Hyde Salter, who considered its main mode of action to be that of a sedative, though also to diminish reflex irritability (Salter, 1869). Like stramonium, it also became a component of some asthma cigarettes and smoking mixtures. By the end of the 19th century it was recommended for asthma by the great physician, Sir William Osler (1849-1919): 'The sedative anti-spasmodics, such as belladonna, henbane, stramonium, and lobelia, may be given in solution or used in the form of cigarettes' (Osler, 1895). In the 20th century the effects of asthma cigarettes containing atropine or stramonium were studied in subjects with mild airway obstruction (asthma or emphysema) and clinically relevant improvements in lung function were confirmed (Herxheimer, 1959; Charpin, 1979).

Chemistry

Pharmacology and Drug Development

In a series of key animal experiments reported in 1840, the eminent chest physician Charles Williams showed that extracts of both belladonna and stramonium almost abolished the airway contraction induced by stimulation of the vagus nerve (hence establishing their antimuscarinic properties) (Lotvall, 1994).

Atropine

Atropine, the active principle in *A. belladonna*, was first isolated in 1831 by the German pharmacist Heinrich Mein and synthesised in 1901. As mentioned above, atropine is a racemic mixture of D- and L-hyoscyamine. In 1867, Albert von Bezold showed that it blocked the slowing of the heart rate induced by vagal nerve stimulation. In 1880, Albert Ladenburg synthesised an atropine analogue, homatropine, which acted more quickly on dilating the pupil of the eye (mydriasis) and had the great advantage of a much shorter duration of action, so was used when an ophthalmologist wishes to examine the retina.

Analogues of atropine are still used to treat bradycardia (slow heart rate), and as a protection against nerve gas poisons such as sarin and organophosphorus pesticides. Soldiers in the Gulf war in Iraq carried a loaded syringe of an atropine analogue to self-administer in case of a nerve gas attack. It is lifesaving in aconite poisoning where it reverses the progressive bradycardia which otherwise terminates in death.

Hyoscine and scopolamine

Ladenburg was also the first, in 1881, to isolate hyoscine and show it to be chemically similar to atropine.

Chemistry

It was subsequently discovered that in solution, hyoscine also forms a racemic mixture of L- and D-isomers, and that the D-isomer had a stimulant effect on the CNS while the L-isomer, scopolamine, was sedative, explaining the unpredictable effects of the mixture.

Scopolamine was then developed as part of a premedication for surgery together with a formulation of opium called Omnopon, a mixture of morphine, papaverine and codeine, known colloquially as 'Om and Scop', utilising the sedative properties of both drugs together with the secretion-drying effect of scopolamine and the analgesic effects of morphine and codeine. The combination became widely used in obstetric practice in North America, where it was known as 'twilight sleep' (Anon, 1915) and did not fall out of favour until the 1960s.

Safer analogues developed

More recently, it was found that quaternary salts of atropine and hyoscine penetrated much less easily into the central nervous system (CNS), thus avoiding side effects such as hallucinations, but kept their useful effects on the peripheral nervous system. The German company Bayer had introduced such a compound, atropine methonitrate, as a mydriatic, in 1902, and it was subsequently developed as a nebulised bronchodilator. Hyoscine butylbromide was patented in 1950 and introduced as a treatment for colic and a preventative against motion sickness (as a pill or a transdermal patch). Ipratropium bromide was successfully introduced as an inhaled bronchodilator in the 1970s and a longer-acting agent, tiotropium bromide, was marketed in 2005; several more long-acting antimuscarinic bronchodilators have since been developed.

Chemistry

New developments that treat other conditions – new fields of pharmacology

During the late 20th century it became clear that that there were actually five subtypes of the muscarinic receptor, with differing distribution and activities. For example, M1 and M4 receptors are found mainly in the CNS, where they are involved in memory and learning, and pain processes, respectively; M2 receptors are found in the heart and lungs; M3 receptors in the respiratory tract where they (and to a lesser extent M2 receptors) are involved in bronchoconstriction. M5 receptors may have a role in regulating the brain microcirculation. This opened a new field of pharmacology, and novel compounds which selectively blocked or stimulated individual receptors were soon synthesised. One of the first was pirenzepine, which selectively blocks M1 receptors and reduces gastric acid secretion and was used in the management of peptic ulcers. Selective M1 agonists have been tried in the treatment of dementia. The M1 antagonist benztropine is a treatment for Parkinson's disease (M4 antagonists have also been studied in this indication). M3 antagonists such as darifenacin are used to treat urinary incontinence (Eglen, 2012; Kohnen-Johannsen, 2019).

Although hyoscyamine and hyoscine can be synthesised chemically, most of the world's commercial supply is still obtained from plants, principally from *Duboisia myoporoides* and *D. leichhardtii* and their hybrids, which are native to Australia (Pearn, 2005). Atropine is no longer extracted from plants and is made synthetically.

Toxicology of the antimuscarinic tropane alkaloids

The effects of atropine poisoning have been summarised as 'blind as a bat, mad as a hatter, red as a beet, hot as a hare, dry as a bone'. Subjects are flushed, with dry mouth and skin, and may become hyperthermic; their vision is blurred due to pupillary dilatation; they have a tachycardia and may develop delirium and hallucinations before proceeding to seizures or respiratory arrest. Potter's asthma remedy, referred to above, became a drug of abuse among teenagers in the UK in the 1970s, and quite a number of subjects required hospital admission as a result of overdosing (Barnett, 1977). It was eventually taken off the market in 1988.

Treatment of poisoning is generally supportive, but in severe cases physostigmine (an alkaloid which occurs naturally in the Calabar bean, *Physostigma venenosum, q.v.*) is administered.

Chemistry

Physostigmine is a reversible inhibitor of acetylcholinesterase, which breaks down acetylcholine, the transmitter which stimulates the muscarinic receptor. Increasing the acetylcholine levels can at least partially overcome the antimuscarinic effects of the overdose (Krenzelok, 2010).

Chronic use of some antimuscarinic agents is associated with an increased risk of developing dementia; this is not seen with the antimuscarinic bronchodilators which are used at low doses (Coupland, 2019).

Summary

The properties of extracts of plants that contain atropine, hyoscine and hyoscyamine have been known for millennia. The identification and extraction of these alkaloids has firmly established them in the pharmacopoeia. Modification of their chemical structures has made them safer, with fewer side effects, and they are now used as therapies for a different range of illnesses. That they are both poisons and lifesaving is one of the many paradoxes of medicinal substances from plants.

References

Anonymous Editorial. Twilight sleep: the Dammerschlaf of the Germans. *Canadian Medical Association Journal* 1915;5(9):805-8

Barnett A, Jones F, Williams E. Acute poisoning with Potter's asthma remedy. *BMJ* 1977;2:1635

Carter A. Dwale: an anaesthetic from old England. *BMJ* 1999;319:1623-6

Chapman K. *Historical aspects of medicinal uses of antimuscarinics. In: Spector S (ed) Anticholinergic agents in the upper and lower airways* (Lung biology in health and disease vol 134). New York, Marcel Dekker Inc, pp155-169, 1999

Charpin D, Orehek J, Velardocchio J. Bronchodilator effects of antiasthmatic cigarette smoke (*Datura stramonium*). *Thorax* 1979;34:259-61

Coupland C, Hill T, Dening T, *et al*. Anticholinergic drug exposure and the risk of dementia. *JAMA Internal Medicine* 2019;179:1084-93 doi: 10.1001/jamainternmed.2019.0677

Daries PJA. *De Atropa Belladonna*. Lipsiae, 1776

Davis E. The ethnobiology of the Haitian zombie. *Journal of Ethnopharmacology* 1983;9:85-104

Earles MP. *Studies in the development of experimental pharmacology in the eighteenth and early nineteenth centuries*. PhD thesis, University College, University of London, 1961 https://discovery.ucl.ac.uk/id/eprint/1317887/1/296133.pdf accessed 1 August 2021

Eglen R. *Overview of muscarinic receptor subtypes*. In: Fryer A, Christopoulos I, Nathanson N (eds), *Muscarinic receptors* (*Handbook of experimental pharmacology vol 28*). Heidelberg, Springer, pp3-28, 2012. doi: 10.1007/978-3-642-23274-9_1

Fatur K. 'Hexing herbs' in ethnobotanical perspective: a historical review of the uses of anticholinergic Solanaceae plants in Europe. *Economic Botany* 2020;74(2):140-58

Flückiger F, Hanbury D. *Pharmacographia. A History of the Principal Drugs of vegetable origin met with in Great Britain and British India*. London, Macmillan and Co., 1874

Herxheimer H. Atropine cigarettes in asthma and emphysema. *BMJ* 1959;2:169-71

Kohnen-Johannsen K, Kayser O. Tropane alkaloids: chemistry, pharmacology, biosynthesis and production. *Molecules* 2019;24:796. doi: 10.3390/molecules24040796

Krenzelok E. Aspects of datura poisoning and treatment. *Clinical Toxicology* 2010;48:104-10

Ladenburg A. Die natürlich vorkommenden mydriatisch wirkenden Alkaloïde *Annalen der Chemie* 1881;206(3):274-307

Höjer, JC in Linnaeus C. *Demonstrationes Plantarum in Horto Upsaliensi 1753* (note re *Datura ferox* in a dissertation to Linnaeus from Johan Christian Höjer). Upsalla, 1753

Lotvall J. Contractility of lungs and air-tubes: experiments performed in 1840 by Charles B Williams. *European Respiratory Journal* 1994;7:592-5

Mann J. *Murder, Magic, and Medicine.* Oxford; Oxford University Press, pp76-82, 1992

Osler W. *The Principles and Practice of Medicine*, 2nd edn. Edinburgh & London, Young J Pentland, p535, 1895

Pearn J. Medical ethnobotany of Australia past and present. *The Linnean* 2005;2(4):16-24

Ray J. *Historia Plantarum* London, Henricum Faithorne & Joannum Kersey. vol 1, p680, 1686

Reimarus JAH. *Memoires de la Societé Philomathique* (reported in the *Edinburgh Medical and Surgical Journal*, vol 9. 1813), 1797

Salter H. On the treatment of asthma by Belladonna. *The Lancet* 1869;i:152-3

Sigmond G. Dr Sigmond on stramonium and its use for smoking. *The Lancet* 1836;ii:392-7

Sims (letter of 1802) Anon. Communications relative to the Datura stramonium, or Thorn-apple as a cure for the relief of asthma. *Edinburgh Medical & Surgical J* 1812;8:364-7

In the College Garden one of the Garden Fellows taking a succession of guided tours over a weekend, would rub the leaves of *Brugmansia* and invite the visitors to smell the chemical aroma on his fingers.

Curiously, his finger tips became partially numb, and remained so for eight years – a phenomenon not previously reported, and unexplained.

CHAPTER 4 – ANTHONY DAYAN

Betula pendula
The source of β-sitosterol

The genus *Betula* belongs to the Betulaceae family of trees, which includes the hazels, alders and hornbeams, all of which flourish in cooler areas of the world. These birches are an ancient and prominent symbol of fertility and rejuvenation in Celtic and Norse mythology.

Introduction

The plant sterol β-sitosterol is extracted from the wood pulp of the European white birch, *B. pendula*, and used as a food supplement to reduce blood cholesterol and symptoms in benign prostatic enlargement. It is not a 'Prescription Only' medicine and is often incorporated into margarine and yoghurts. It is also the source of a novel, prescription, wound healing preparation.

Plant profile

The graceful deciduous *B. pendula*, known as European white birch, silver birch or lady of the woods, has grown in open woodland and on heaths throughout Britain since the end of the Ice Age. Reaching up to 30m high with a drooping, swaying habit, it is easily identified by its trunk of peeling, silvery-white bark which develops black fissures with age. The plant is monoecious (with male and female flowers on the same plant). The male flowers are long catkins which release clouds of pollen in April before the delicate, ovate green leaves emerge. After pollination the shorter female flowers thicken and change colour to dark red or crimson; later they produce masses of winged seeds which are dispersed by the autumn wind.

Betula pendula is known as a pioneer species due to its ability to colonise open spaces swiftly. It is therefore very useful in the development of woodland areas where it helps to protect the slower-growing species like oaks by providing shelter and improving the soil environment. Fast-growing but with an average lifespan of just seventy years, it then cedes the ground to longer-lived species.

It is a good tree for wildlife as it provides both food and habitat for more than 300 insect species, and forms associations with specific fungi including the fly agaric, *Amanita muscaria*. Birds such as woodpeckers nest in the trunk and many others feed on the copious seed.

Betula pendula, the European white birch (photo by Susan Burge)

The ornamental hybrids which are now available are relatively easy to grow. They are suitable for many soil types, even quite acid ones, but they prefer a sunny site and will become bent over if exposed to strong winds.

Unfortunately, *Betula* species are notably susceptible to honey fungus.

History

It has no history of being used medicinally in Europe or the Middle East in the past 3000 years. Branches were antidotes to witches, but tying a knot in a birch tree twig after a short incantation was reported as being used as a cure for gout in Marburg, Germany, in the 19th century (Fraser, 1933).

Chemistry

The bark and the concentrate in the distillate contain a number of phenols, which contribute to its characteristic pine-like odour and many of its biological activities.

The outer part of the bark contains up to 20% betulin. The main components in the essential oil of the buds are α-copaene (~10%), germacrene D (~15%) and δ-cadinene (~13%). Also present in the bark are other triterpene substances which have been used in laboratory research to identify possible biological properties.

Betula – woodcut from Mattioli's *Discorsi*, 1568

Uses – non-medical

Extracts of birch bark and tars made from them by distillation have been used in Europe since at least the Neolithic period (about 12,000-1900 BCE) as water-proofing agents, adhesives and perfumes, and as a healing and anti-infective dressing for skin damage (Stacey *et al*, 2019).

It long had many uses before the recent agro-industrial and current medical interest in it. Bundles of smaller branches are typically used to make besom brooms, and many insects rely on it for food and as somewhere to lay eggs. The ecological and agronomic importance of the large forests of silver birches in Finland have made it the Finnish national tree.

Betula pendula, and other birches, have two particular uses in Scandinavia today – as a source of lumber and the other, very large-scale use, is as a source of wood pulp from which paper and similar materials are made. In rural Finland and nearby areas of Scandinavia and Slavic and Baltic countries, the sap is collected by tapping the tree at the winter-spring boundary and drunk fresh or after fermentation and even distillation to make an alcoholic drink (Svanberg *et al*, 2012).

Uses – medical

All birches are medically interesting for several reasons. Birch pollen is the predominant tree pollen in northern Europe that causes asthma and allergic rhinitis and is also associated with certain types of food allergy (Biedermann *et al*, 2019).

The principal medical-pharmaceutical interest in *Betula* comes from the fact that their wood pulp is naturally rich in plant sterols and related stanols. They are natural components of the cell walls of all plants. Sitosterols and the closely related stanols have a molecular structure resembling cholesterol, which is a component of the membranes of animal cells more than in plant cells. The plant sterols are not synthesised by animals, including humans, and they are not normally absorbed from the diet. In the intestines several plant sterols, particularly β-sitosterol and the similar stanols, are taken up by transporter mechanisms in the intestinal lining and inhibit the absorption of cholesterol from the diet. This can reduce the circulating level of cholesterol in the blood without causing any harmful effects.

Chemistry

Cholesterol – intestinal absorption reduced by β-sitosterol

A major cause of human ill-health and death in all developed and most developing countries is vascular disease due to deposition of excessive amounts of cholesterol in the walls of blood vessels as atheromatous plaques resulting in impairment or even blockage of the blood supply to the heart (and so to angina and heart attacks), or to the brain where the result may be a stroke, or to the kidneys, which can end in renal failure and high blood pressure. Although some cholesterol is naturally synthesised in the body, much of the dangerous atheromatous excess comes from the diet, especially from fat-rich dairy products such as cream and butter. Amongst the methods to treat and prevent atheroma is prevention of the absorption of dietary cholesterol.

In the 1950s it was discovered, initially in animals and subsequently in humans, that many plant stanols, a broad chemical class that includes sterols, could reduce the absorption of cholesterol from the intestines; β-sitosterol was particularly effective (Gylling *et al*, 2014). An initial difficulty about exploiting this as a preventative treatment was the difficulty of obtaining a sufficient supply at an affordable cost. That was solved when a method was developed to extract it from the wood pulp left after paper had been made from the fibres (Vats, 2017). Starting in Finland in 1993, and subsequently throughout the world, a number of margarine-like fat spreads, oils, milks and yoghurts have been developed that contain sufficient β-sitosterol or a related stanol to produce a small but useful reduction in cholesterol absorption from the diet. Coupled with other dietary measures and medicines, these products have been found to be clinically useful. They have been registered and are available in many countries as 'Functional Foods' that can help to prevent and control vascular diseases (EU, 2003).

β-sitosterol and the stanols can reduce blood cholesterol levels by up to 10% but they may interact adversely with some other medical treatments for high cholesterol so people taking cholesterol-lowering drugs should consult their doctor before taking it. While lowering cholesterol is known to be beneficial to health, there is no scientific proof that taking β-sitosterol reduces cardiovascular disease (Genser *et al*, 2012). Toxicity studies indicate it should not be taken by pregnant or breastfeeding women.

Prostatic enlargement and wound healing

Use of β-sitosterol for mild benign prostatic enlargement in older men improves urinary flow but has no effect on the size of the prostate (Wilt *et al*, 1999).

Recent research led to European approval in 2016 of Episalvan as a prescription medicine to speed the healing of skin wounds; it is an oleogel extract of bark from *B. pendula* and *B. pubescens* (EMA).

References

Biedermann T, Winther L, Till SJ, *et al*. Birch pollen allergy in Europe. *Allergy* 2018;74,1237-48

EU 2003. *Opinion of the Scientific Committee on Food on an application from MultiBene for approval of Plant-Sterol enriched Foods*. https://food.ec.europa.eu/system/files/2020-12/sci-com_scf_out191_en.pdf accessed 25 January 2021

EMA European Medicines Agency re Episalvan https://www.ema.europa.eu/en/medicines/human/EPAR/episalvan accessed 25 January 2021

Fraser JG. *The Golden Bough, A study in Magic and Religion. Part VI The Scapegoat*, 3rd edn. London, Macmillan and Co. Ltd. p56, 1933

Genser B, Silbernagel G, De Backer D, *et al*. Plant sterols and cardiovascular disease: A systematic review and meta-analysis. *European Heart Journal* 2012;33(4):444-51

Gylling H, Plat J, Turley S, *et al*. Plant sterols and plant stanols in the management of dyslipidaemia and prevention of cardiovascular disease. *Atherosclerosis* 2014;203:346-60

Stacey RJ, Dunne J, Brunning S, *et al.* Birch bark tar in early Medieval England – Continuity of tradition or technological revival? *J Archaeol Sci. Reports* 2020;29 102118

Svanberg I, Sõukand R, Łucsaj L, *et al.* Uses of tree saps in northern and eastern parts of Europe. *Acta Bot Soc Pol* 2012;81:343-57

Vats S. *Methods for Extractions of Value-Added Nutraceuticals From Lignocellulosic Wastes and Their Health Application.* In: Grumuzescu AM, Holban AM (eds) *Handbook of Food Bioengineering.* London, Elsevier, pp1-64, 2017

Wilt TJ, MacDonald R, Ishani A. Beta-sitosterol for treatment of benign prostatic hyperplasia: a systemic review. *BJU Int* 1999;83(9):976-83

> The plant sterols are not synthesised by animals, including humans, and they are not normally absorbed from the diet. In the intestines several plant sterols, particularly β-sitosterol (from *Betula pendula*) and the similar stanols, are taken up by transporter mechanisms in the intestinal lining and inhibit the absorption of cholesterol from the diet. This can reduce the circulating level of cholesterol in the blood without causing any harmful effects.

CHAPTER 5 – NOEL SNELL

Camellia sinensis
Coffea arabica
Theobroma cacao
Caffeine-containing plants –
the source of many useful medicines

This chapter is about caffeine containing plants including *Camelia sinensis*, *Coffea arabica* and *Theobroma cacao*, the medicinal chemicals obtained from them, and their properties.

Introduction

There are numerous caffeine-containing plant species, the most commercially important being the coffee shrubs, particularly *Coffea arabica* and *C. robusta*; tea plants, *Camellia sinensis* and its relatives, and *Theobroma cacao*, the tree from which cocoa and chocolate are made. All have been used medicinally and as beverages by local peoples; brews made from these plants have been recognised for centuries to possess useful pharmacological activities including warding off fatigue and hunger, and having diuretic properties. Caffeine itself has proven therapeutic benefit, particularly for its central nervous system effects and in stimulating respiration in newborn infants. About 8% of ingested caffeine is metabolised in the body to theophylline; this chemical, manufactured by the pharmaceutical industry, was used as a diuretic and coronary vasodilator and is still in use as a bronchodilator for patients with asthma and chronic obstructive pulmonary disease. Synthetic derivatives have found uses in heart failure (milrinone), peripheral vascular disorders (pentoxyfylline), for erectile dysfunction and pulmonary hypertension (sildenafil/ Viagra), and in the management of obstructive airways disease (roflumilast).

Plant profile

Camellia sinensis is a tender evergreen shrub in the Theaceae family, native to South-Central and Southeast China where it grows in the damp shade of deciduous forests. It has smallish, elliptic, glossy, leathery leaves, lighter green on the undersides. The small, white, cup-shaped flowers are scented and have showy golden-yellow stamens. They appear in autumn and winter. In the plant's natural habitat summers are warm and wet and winters are dry and frost free. It needs an acidic soil which is fertile and moist, so it is advisable to incorporate plenty of organic matter when planting.

Coffea arabica is a shrub or small tree in the Rubiaceae family. Widely cultivated through the Americas, Africa and Asia, it is actually native to northeast tropical Africa where it grows at 950-1950m above sea level. The plant has evergreen, glossy, dark green, ovate leaves and clusters of scented white flowers in its leaf axils. These are followed by small red (sometimes yellow or purple) fruit which contain two seeds each: the coffee beans. As this is a tropical plant it needs the protection of a warm glasshouse with temperatures set between 18-21°C to thrive. Lower temperatures and sudden changes may slow its growth and cause the fruit and leaves to drop. It requires a slightly acidic, fertile, well-drained compost and careful watering so that the plant doesn't become either waterlogged or too dry.

Theobroma cacao is an evergreen tree in the Malvaceae family, native to Mexico, Central America and northern South America where it forms part of the understorey of the evergreen tropical rainforest. Often growing in clumps along riverbanks, its roots may be flooded for long periods of the year. Cocoa grows at low elevations, usually below 300m above sea level, in areas with a high yearly rainfall. Its leaves are dark green and leathery, and the small, yellowish-white to pale pink flowers are clustered and grow directly on the trunk. In the wild, cocoa trees are pollinated by midges and, after pollination, their tiny flowers (which are produced throughout the year) develop into massive fruits, the cocoa pods. The 'pod' is an egg-shaped, red to brown berry 15-25cm long, with a knobbly surface and lines from top to bottom. Each fruit produces 30 to 40 seeds which after drying become the brownish-red cocoa beans. We do not grow the plant at the Royal College of Physicians as we cannot provide the right conditions for it to thrive.

The fact that numerous generally unrelated species of plant contain caffeine is fascinating; apart from the tea plants and coffee shrubs, and *Theobroma cacao*, caffeine is also found in *Ilex* and *Paullinia* species from South America, *Cola acuminata* from Africa and several *Citrus* species. Recent research has shown that the ability to synthesise caffeine arose separately in these unrelated genera, using several different synthetic pathways, an example of convergent evolution (Huang *et al*, 2016). This implies a benefit to the plant from the ability to make caffeine; one likely effect of value is that caffeine is toxic to a variety of insects and other pests (including molluscs such as slugs and snails). In coffee, tea, and cacao plants, caffeine is synthesised from a chemical called xanthosine via the intermediaries 7-methylxanthosine and theobromine. In man, caffeine is produced by metabolism of theophylline. Theophylline, caffeine and theobromine all belong to the chemical class known as methylxanthines.

Several different cultures discovered the value of caffeine-containing beverages as psychostimulants. In Asia tea was first drunk for medicinal purposes and later as a recreational drink. Coffee was employed in the Horn of Africa for its alerting, invigorating, and appetite-suppressing effects; an early form of chocolate was drunk in Mexico and Central America, and a brew from *Ilex paraguariensis* (called Yerba Mate) was drunk in South America. A drink made from another holly species, *Ilex vomitoria*, was used by the North American Indians to brew a very strong tea to provoke vomiting. Another, known as Guaraná, made from the seeds of *Paullinia* species, was used by the inhabitants of the Amazon region for its invigorating effect. In tropical Africa the nuts of the kola tree (*Cola acuminata*) were chewed as a stimulant and to alleviate hunger (the original formula of Coca-Cola included cocaine and an extract from kola nuts; the cocaine content was subsequently replaced).

History

The origins of these different brews go back into the mists of time. Legend says that a mythical emperor of China discovered tea after some fresh leaves from a tea plant accidentally blew into boiling water. Tea arrived in Japan in 805 CE. It may have been brought to Europe by Portuguese or Dutch traders in the late 16th or early 17th centuries. It was introduced to England by Catherine Braganza, wife of King Charles II, in 1658; Samuel Pepys records drinking his first cup in 1660. The invention of the Wardian case (a miniature greenhouse) by Nathaniel Bagshaw Ward, an apothecary and botanist, permitted the transport of 20,000 tea plants from Shanghai to Assam, starting the Indian tea industry (Wall, 1932).

Tea is now widely grown in the tropics and sub-tropics – there is even a flourishing tea plantation in Cornwall, UK, and several starting up in Scotland with seed sourced from high-altitude areas in Nepal and Georgia.

Coffee seems to have originated in the Horn of Africa. There is a legend that an Ethiopian shepherd (or in another version, a monk) noticed that the goats grazing on a coffee shrub became frisky and did not sleep, and experienced similar effects when he chewed some coffee beans. From Ethiopia, coffee was exported to Yemen then into Arabia.

Prospero Alpini, a Venetian physician, is credited with being the first to describe coffee in a printed

Coffea arabica in *Medical Botany* by Stephenson and Churchill, 1836

DE PLANTIS AEGYPTI
B O N.

Coffea arabica called 'Bon' in Prosper Alpini's *De Plantis Aegypti*, 1592

CACAHVATL vulgo CACAO.

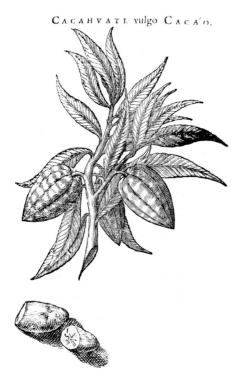

Theobroma cacao – woodcut from Piso, *Mantissa aromatica*, 1658

publication, *De Plantis Aegypti* [Of the Plants of Egypt] (Alpini, 1592). He saw it growing in a *viridario* – a pleasure garden or wooded glade – while living in Egypt in 1580-1583 as physician to the Venetian consul in Cairo. He describes it as a tree called Bon or Ban and that the Arabs made a drink called Caova from the seed. He says it originated from Arabia Felix (the fertile part of Arabia). In Egypt, at that time, it was used to strengthen the stomach and induce menstruation, and to benefit the liver and spleen.

The first coffee houses originated in Mecca and Medina in 1470: by 1554 coffee had reached Istanbul, then Cairo and (later) Venice. The first English coffee houses were established in Oxford (1650) and London (1652); coffee arrived in the North American colonies about the same time. The Dutch began to grow coffee in Indonesia and Ceylon [Sri Lanka] (after failing in India), the French initially in Martinique then throughout the Caribbean, and the Portuguese in Brazil. The British succeeded in establishing coffee plantations in India, and subsequently in Kenya, Uganda and Jamaica ('Blue Mountain' coffee).

Cocoa beans, from the *Theobroma cacao* tree (so named by Linnaeus, 'Theobroma' meaning the 'food of the Gods'), were used in Central America some 4000 years ago to make a drink called Xocolatl (from which the word 'chocolate' is derived); this was a bitter, unsweetened drink which also contained other ingredients such as vanilla and chilli peppers. In 1528, the Spanish conquistador Cortés brought chocolate to the Spanish Court, where it was kept secret until 1615, when Anne, the daughter of King Philip III of Spain, wed the French King Louis XIII. Chocolate was introduced into England c. 1650, allegedly by Sir Hans Sloane, President of the Royal College of Physicians and of the Royal Society (though this has been disputed). He is also reputed to have invented milk chocolate. 'Sir Hans Sloane's milk chocolate' was sold under his name, with the claim that it cured consumption (tuberculosis), among other disorders; it was probably made up by a local apothecary without Sloane's permission (Moore 2010). As demand for chocolate grew, plantations were established in the West Indies, South America, Philippines, Asia and West Africa.

Pharmacology of caffeine and its analogues

When coffee, tea and chocolate entered the pharmacopoeias of Britain at the beginning of the 18th century, they were still regarded as food rather than medicine. Quincy (1718), however, noted that coffee was used as a hangover cure and to help people stay awake but that it could cause tremor.

The benefits of strong coffee in cases of asthma were noted by Buchan in his popular book *Domestic Medicine* (Buchan, 1814) which quoted Sir John Pringle (1707-1782) who regarded a strong infusion of roasted coffee as 'the best abater of the periodic asthma that he has ever seen'. William Withering (the discoverer of the benefit of foxglove in heart failure) stated that spasmodic asthma could be relieved by 'coffee made very strong' (Withering letter, 1786). Its use in asthma was also endorsed by Laennec, the inventor of the stethoscope, in his 1821 textbook on diseases of the chest. However, no rationale for its beneficial effects was forthcoming until Hyde Salter, an influential English physician, noted that coffee had a similar effect to a sudden shock, implying a stimulant action on the sympathetic nervous system (Salter, 1859). As we now know, sympathetic stimulation leads to relaxation of smooth (that is, non-skeletal) muscle, including that in the bronchial tubes, leading to bronchodilatation (just as stimulation of the parasympathetic nervous system leads to bronchoconstriction).

Tea and chocolate do not seem to have had the same reputation for treating asthma as did coffee; this may simply be because it is easier to prepare a really strong brew of coffee.

Detailed investigations of the mode of action of coffee had to await the identification of its active ingredients. In the same year that he isolated quinine from *Cinchona* bark (1819), Friedlieb Ferdinand Runge isolated caffeine from coffee beans at the suggestion of the great German literary figure, von Goethe. In 1827, an active substance initially named *théine* was isolated from tea but was subsequently shown to be caffeine, which was first prepared synthetically in 1861. Theobromine was first identified in 1841, in cacao beans; and theophylline was first extracted from tea leaves in 1888.

Varying, but small, amounts of theophylline and theobromine are found in tea, coffee and cocoa; however, in humans, caffeine is metabolised initially to theobromine and theophylline (as well as a major metabolite, paraxanthine) so the body is exposed to all three active agents (it is worth noting that many mammals, including dogs and cats, metabolise theobromine very slowly and can become very ill or even die after eating a few bars of chocolate).

The diuretic effects of caffeine and theobromine were studied in 1886 at the University of Strasburg; theophylline in 1902 after it had first been synthesised. Theobromine and theophylline were shown to be more potent than caffeine (Sneader, 1985); all three cause diuresis by reducing renal tubular reabsorption. Theophylline is poorly soluble and difficult to formulate, but in 1907 a more soluble compound of theophylline with ethylene diamine was made and named aminophylline. Theophylline was shown in 1926 to be effective clinically at reducing persistent oedema in patients already taking digoxin and also, in 1946, to have a direct cardiostimulant effect (Hollman, 1992). Theophylline and aminophylline were thus initially used clinically for their benefits in patients with heart problems. However, caffeine had been shown in 1912 to relax airway smooth muscle both in isolated preparations and in animal models (Persson, 1985), and theophylline was also shown to be effective in isolated bronchial muscle preparations in two studies published in 1921 and 1922; in the latter, four asthmatics were also given a combination of theophylline and theobromine, with benefit (Schultze-Werninghaus, 1982). Despite these findings, and the long history of use of coffee in asthma, this indication appears to have been ignored until about 1936 when theophylline suddenly became a fashionable asthma treatment.

Theophylline/aminophylline remained a mainstay of asthma therapy for the next 40 years or so, as an infusion or as tablets (sometimes combined with other constituents such as ephedrine), until the advent of inhaled beta2-receptor agonists and corticosteroids (theophylline and some of its analogues have also been shown to cause bronchodilatation by the inhaled route, but this effect is less than that of a standard dose of beta-agonist). Derivatives with similar effects (such as diprophylline and doxofylline) were synthesised but made little impact. Theophylline and aminophylline preparations are still widely available, some as slow-release formulations.

Caffeine itself has been shown to have bronchodilator properties (Welsh *et al*, 2010) though not marketed for this indication. It is, however, approved for the treatment of apnoea (a lack of respiratory effort) in premature infants (Schoen, 2014), acting presumably through a stimulant effect on the breathing centre in the brain. Caffeine is also a component of some over-the-counter compound analgesic formulations.

Theobromine has also been shown to be an effective bronchodilator, albeit less potent than theophylline (Simons, 1985), and to have potential in treating chronic cough (Usmani, 2005), for which it is approved for use in South Korea.

Chemistry

Mode of action

The mechanisms of action of the methylxanthines are still not fully understood. They include inhibition of the phosphodiesterase (PDE) enzyme system which breaks down the cell messengers cyclic AMP and cyclic GMP; the resulting increase in cyclic AMP concentrations leads to relaxation of bronchial smooth muscle (although this is seen only at high doses of the drug) and a broad anti-inflammatory effect. The methylxanthines also increase the activity of the enzyme histone deacetylase-2 (HDAC-2) which has the effect of reducing the expression of pro-inflammatory genes, and also reversing the reduction of the anti-inflammatory effects of corticosteroids seen in smokers. Methylxanthines also bind to adenosine receptors, thus blocking adenosine-mediated bronchoconstriction.

Adverse events and toxicology

Caffeine, theophylline and some other PDE inhibitors have been shown to cause testicular toxicity in rats; fortunately, this is a species-specific finding and is not seen in man. Caffeine does not have adverse effects on the foetus, but high intake may slightly increase the risk of miscarriage.

In normal doses, caffeine can cause a mild diuresis and insomnia. Larger doses can raise the heart rate and blood pressure, and cause agitation. Overdoses can lead to mania, cardiac arrhythmias, seizures and, rarely, cardiac arrest.

Theophylline has a narrow therapeutic window; that means the effective dose is close to the one which causes adverse effects. These are similar to those seen with caffeine, including the risk of death. Interactions can also occur between theophylline and several other medications, which may increase the plasma concentration of theophylline and the risk of adverse reactions.

Some animal studies have shown a variety of adverse effects on the foetus, but human case studies have not confirmed these findings. Theophylline is excreted in breast milk and may cause mild irritability in nursing infants.

An interesting consequence of an overdose of sildenafil (and there are several) is cyanopsia – objects appear to be tinted blue (*q.v.* xanthopsia with *Digitalis* overdose).

Chemistry

Development of methylxanthine derivatives as modern medicines

The PDE system referred to above includes a number of different subtypes, or isoenzymes.

Pentoxifylline is a non-specific PDE inhibitor which is used to reduce the pain and cramping in the limbs caused by intermittent claudication, a type of muscle pain resulting from peripheral artery disease. In addition to causing peripheral vasodilatation, it improves red blood cell deformability (allowing them to squeeze through the smaller blood vessels) and reduces platelet aggregation and blood clot formation. Several inhibitors of specific PDE isoenzymes have been synthesized in the search for novel medicines. PDE3 inhibitors include milrinone, which is a vasodilator and increases the contractility of the heart and was marketed for the treatment of congestive heart failure, and cilostazol which is a vasodilator and inhibits platelet aggregation, and, like pentoxifylline, is used in intermittent claudication. PDE4 inhibitors include roflumilast which has anti-inflammatory properties and is approved for use in severe chronic obstructive pulmonary disease. The PDE5 inhibitor sildenafil was being studied as a potential anti-angina medication when an interesting side effect was noted in the volunteers in clinical trials, as a result of which it and similar analogues were developed instead as treatments for erectile dysfunction. In addition to affecting the penile blood flow, it also dilates the blood vessels in the lungs and has been used as a treatment for pulmonary arterial hypertension.

Potential health benefits of coffee, tea and chocolate

In addition to the research on active ingredients of coffee, tea and chocolate discussed above, many studies have looked at potential beneficial effects of whole brews from these plants. The problem with any positive findings is to identify which active component (or combination of components) is responsible for the observed effect. In addition to the methylxanthines (which have been the main topic of this review), coffee, tea and chocolate contain numerous potentially interesting constituents, particularly a group of natural antioxidants named flavonols. One of these (epigallocatechin gallate, found in green tea) inhibits the activity of VEGF-R which is a survival factor for cancer cells and which promotes tumour angioneogenesis. Consumption of green tea has been linked with a reduction in the risk of death from cancer. Likewise, phenolic acids in coffee reduce the risk of breast cancer substantially, while other compounds called diterpenes have shown beneficial effects on prostate cancer in the laboratory and in animal models.

Coffee consumption is associated with reduced all-cause mortality in the general population; higher coffee intake has also been found to be associated with a reduced risk of heart failure, liver fibrosis and cirrhosis, and liver cancer. Coffee (including decaffeinated coffee) contains phenylindanes which prevent the clumping of proteins in the brains of patients with Parkinson's and Alzheimer's diseases. Green tea consumption can be beneficial in improving the prognosis for survivors of stroke and heart attack. Tea drinking has also been associated with a reduced risk of developing tuberculosis. Flavonols in cocoa appear to be protective against vascular disease. There is evidence that consumption of chocolate (particularly dark chocolate) may be associated with a reduced risk of clinically relevant depression; interestingly chocolate contains a compound, anandamide, which binds to the same receptors in the brain as cannabis. Cocoa flavonols also improve cognitive function (an entertaining study found a close correlation between national chocolate consumption and the number of Nobel Prize winners produced in each country) (Messerli, 2012).

References

Buchan W. *Domestic Medicine or New Practical Family Physician*. London, Thomas Johnson and Peter Edwards, 1816

Hollman A. *Plants in cardiology*. London, *British Medical Journal* P13, 1992

Messerli F. Chocolate consumption, cognitive function, and Nobel laureates. *New Engl J Med* 2012;367:1562-4

Moore W. Hans Sloane's bitter taste of success. *BMJ* 2010;340:c3210

Pepys, S, *The Diary of Samuel Pepys*, trans and ed. Robert Latham and William Matthews, 11 vols (London: Bell & Hyman, 1970-1983), (for Tuesday 25 September, 1660)

Persson C. On the medical history of xanthines and other remedies for asthma: a tribute to H H Salter. *Thorax* 1985;40:881-886

Quincy J. *Pharmacopoeia Universalis Extemporanea or a Complete English Dispensatory*. London, A. Bell, 1718

Salter H. On some points in the treatment and clinical history of asthma. *Edinburgh Medical Journal* 1859;4:1109

Schoen K, Yu T, Stockmann C, *et al.* Use of Methylxanthine Therapies for the Treatment and Prevention of Apnea of Prematurity. *Paediatr Drugs* 2014;16(2):169-177. doi: 10.1007/s40272-013-0063-z

Schultze-Werninghaus G, Meier-Sydow J. The clinical and pharmacological history of theophylline. *Clinical Allergy* 1982;12:211-215

Simons F, Becker A, Simons K, Gillespie C. The bronchodilator effect and pharmacokinetics of theobromine in young patients with asthma. *J Allergy Clin Immunol* 1985;76:703-7

Sneader W. *Drug discovery: the evolution of modern medicines*. Chichester, John Wiley & Sons, pp153-4, 1985

Usmani O, Belvisi M, Patel H, *et al.* Theobromine inhibits sensory nerve activation and cough. *Faseb J* 2005;19:231-3

Wall C. *The London Apothecaries*. London, The Apothecaries' Hall, p26, 1932, reprinted 1955

Welsh EJ, Bara A, Barley E, *et al.* Caffeine for asthma. *Cochrane Database of Systematic Reviews* 2010;Issue 1. Art. No.: CD001112. doi: 10.1002/14651858.CD001112.pub2

Withering W. Unpublished letter from 1786 reproduced in Mann R. *Modern drug use: an enquiry on medical principles*. Lancaster, MTP Press, pp355-58, 1984

Further reading

Kaszkin M, Beck K-F, Eberhardt W, *et al*. Unravelling green tea's mechanism of action: more than meets the eye? *Molecular pharmacology* 2004;65:15-17

Latosińska M, Latosińska J (eds). *The Question of Caffeine*. London, IntechOpen Ltd., 2017. Available online at http://www.intechopen.com/books/the-question-of-caffeine

Coffea arabica, coffee beans

Theobroma cacao, fruit and flowers

Camptotheca acuminata
The source of the anti-cancer drugs, camptothecin, topotecan and irinotecan

The anti-cancer medicine camptothecin was originally found in the bark and stems of an Asian tree, *Camptotheca acuminata*, in the Nyssaceae family.

Introduction

Camptothecin is a complex quinoline alkaloid that has been an important anti-cancer medicine since the late 1990s. Topotecan and irinotecan are synthetic derivatives of camptothecin with a similar mode of action.

It was discovered in a screen of plant extracts for potential medical value in the 1950s (Wall & Wani, 1995; Wall *et al*, 1996; Yang, Cragg & Newman, 1996; Martino *et al*, 2017). This was part of a more general search for plant products of economic value over several decades by the US Department of Agriculture and, later, more medically focused work by branches of the US National Institutes of Health, which explored leaves, stems and branches, fruits and roots from trees and other plants from many countries for any potential therapeutic value (Demain and Vaishnav, 2011). The screen led to camptothecin from the *Camptotheca acuminata* tree, paclitaxel from the yew (*Taxus baccata*, *q.v.*), harringtonine from *Cephalotaxus harringtonia* (*q.v.*) and podophyllotoxin and etoposide from *Podophyllum peltatum* (*q.v.*).

Plant profile

Camptotheca acuminata is a deciduous tree up to 20m tall found in areas of southern China and Tibet with fertile, moist soil and a moderate temperature range. It is a member of the Nyssaceae family. It is also known as the happy tree, tree of joy, cancer tree or tree of life, as well as water tung and water chestnut.

The species was first defined in 1873 by Joseph Decaisne, Director of the Jardin des Plantes in Paris, on the basis of specimens collected in China by Father Armand David in Lushan, Jiangxi Province, China during his 1868-1870 plant collecting expedition. The name *campto* comes from the Greek word for bent and *theca* means a case. They refer to the distinctive inward bend of the anthers (stamens).

The initial screen of simple extracts from the bark and stems showed notable activity against lines of tumour cells *in vitro*, including human lung and intestinal cancers. Many years of research by academic and industrial chemists, pharmacologists and clinicians were required to isolate the novel compound responsible, which was named camptothecin. It proved difficult to work with because of its relative insolubility, the low concentration found in *Camptotheca* stems, bark and leaves, and its chemical instability in some of the solvents commonly used in chemical isolation and analysis. Many years of work revealed the complete structure of the molecule, its mode of action down to the molecular level, and the best solution for delivering it to patients. Then followed several more years of clinical research into suitable treatment regimes for the several types of cancers that it was most effective in treating.

Chemistry

Camptothecin is a topoisomerase type 1 inhibitor. Topoisomerases are enzymes which ensure that the helical DNA in our chromosomes winds together correctly when cells are dividing. By blocking this action, the cell does not divide properly and dies.

There were two other major problems to be solved before it could be made available as an approved medicine. The worst was trying to find a suitable source for manufacture of camptothecin as the highest concentration in extracts of the bark or stem wood of the tree was only about 0.3%. Collection in the wild usually killed the trees because the bark was removed for processing; over-collection has now made them so rare that the species is on the IUCN Red List of endangered species.

Industrial cultivation of the trees was not feasible because they took up to 20 years to reach a useful size. Chemists and biologists collaborated in trying to cultivate cells *in vitro* but only quite low levels of camptothecin were produced. Chemical syntheses were difficult but were eventually developed into a useful but expensive process that has been used by some manufacturers. Camptothecin itself is produced by natural infection of tree cells with endophytic fungi and there has been some practical success with in vitro cultivation of several species of unusual fungi either on their own or used to infect cultured cells from the plant.

Alternative species were sought that contained a sufficiently high concentration of camptothecin to permit direct extraction, and which grew at a rate that would permit commercial production from wild or better-cultivated plants. The chemical was found in a potentially useful concentration in several other species from southern China and Tibet, notably several species of

Nothapodytes and other genera, although none appeared to be a practical source as they were also rare and slow growing (Ramesha *et al*, 2008; Swami *et al*, 2021). The supply of camptothecin now depends both on partial chemical synthesis and extraction from cultures.

Efficacy and toxicity in the clinic, as well as the tumour types most responsive to treatment, were worked out partly in the clinic. Greater efficacy and tolerability were obtained from use of synthetic chemical analogues that were more stable in the body, thus permitting better control of lower doses, and greater separation between effective and toxic doses; the most important are topotecan and irinotecan (Stehlin, 1996; Ulukan & Swaan, 2002). In certain cases, combinations with other anti-cancer drugs and radiotherapy have proved particularly valuable. The types of cancer most often treated with these compounds include colorectal, lung and stomach cancers, and certain lymphomas (Swami *et al*, 2021).

The lengthy and tortuous pathway followed in the development of these drugs is not uncommon for complex natural products.

Camptothecin is unusual in that folk use appeared to be restricted to small groups of people in remote areas and then for general, rather than specific illnesses (Li and Zhang, 2014). The popular name, Happy Tree, comes from the belief that an extract of the tree could relieve chronic phlegm and catarrh, often associated with repeated headaches and coughing, resulting in a Happy Patient, but this was not recorded until 1843 although there was probably folk usage long before then.

There has been recent interest in the broad antiviral activity of camptothecin and its analogues due to the same mechanism of action – inhibition of topoisomerase I and consequential disruption of protein synthesis (Kacprzak, 2913; Pepin *et al*, 2017). It has yet to be shown that that activity can be controlled in such a way that the antiviral action can be exploited in the clinic without causing excessive toxicity.

References

Demain AL, Vaishnav P. Natural products for cancer chemotherapy. *Microbial Biotechnology* 2011;4(6):687-99

Li S, Zhang W. Ethnobotany of *Camptotheca* Decaisne: New Discoveries of Old Medicinal Uses. *Pharmaceutical Crops* 2014;5:(Suppl 2:M7)140-5

Kacprzak KM. Chemistry and Biology of Camptothecin and its derivatives. In: Ramawat KG, Me´rillon JM (Eds.) *Natural Products*. Berlin Heidelberg, Springer-Verlag, 2013

Martino E, *et al*. The long story of camptothecin: from traditional medicine to drugs. *Bioorganic & Medicinal Chemistry Letters* 2017;27(4):701-7. doi: 10.1016/j.bmcl.2016.12.085

Pépin G, Nejad C, Ferrand J, *et al.* Topoisomerase 1 Inhibition Promotes Cyclic GMP-AMP Synthase-Dependent Antiviral Responses. *American Society for Microbiology* 2017;8(5)

Ramesha BT, *et al.* Prospecting for Camptothecines from Nothapodytes nimmoniana in the Western Ghats, South India: Identification of High-Yielding Sources of Camptothecin and New Families of Camptothecines. *Journal of Chromatographic Science* 2008;46:362-368

Stehlin J (ed). The Camptothecins: from Discovery to the Patient. *Ann NY Acad Sci* 1996;803, Issue 1: ix-x, 1-328

Sawami MK, *et al.* Biotechnology of camptothecin production in Nothapodytes nimmoniana, Ophiorrhiza sp. and Camptotheca acuminata. *Applied Microbiology and Biotechnology* 2021;105:9089-102

Ulukan H, Swaan PW. Camptothecins. *Drugs* 2002;62:2039-57. doi: 10.2165/00003495-200262140-00004

Wall M, Wani M. Camptothecin and Taxol: Discovery to Clinic. Research Triangle Institute, Progress Report No. 21, June 26, 1967. *Cancer Research* 1995;55:753-760

Yang SS, Cragg GM, Newman DJ. The Camptothecin experience: from Chinese medicinal plants to potent anti-cancer drugs. In: Lin Y *Drug Discovery and Traditional Chinese Medicine Science, Regulation, and Globalization.* Springer Science, 1943

The popular name, Happy Tree, comes from the belief that an extract of the tree could relieve chronic phlegm and catarrh, often associated with repeated headaches and coughing, resulting in a Happy Patient, but this was not recorded until 1843 although there was probably folk usage long before then.

A rare success story of finding new medicines from plants

In the 1950-1990s the US Department of Agriculture and the US Department of Health, examined tens of thousands of plant-sourced chemicals for medicinal value.

Four stand out for their special value – irinotecan from camptothecin from *C. acuminata*; paclitaxel from *Taxus baccata*, harringtonine from *Cephalotaxus harringtonia* and etoposide from podophyllotxin in *Podophyllum peltatum*.

The thousands of hours of labour and huge costs that are spent by 'Big Pharma' in finding effective medicines for us, should never be underestimated.

CHAPTER 7 – SUSAN BURGE

Capsicum annuum
The source of capsaicin

Introduction

Capsicum annuum, the chilli pepper, is a tender evergreen sub-shrub that produces the pungent fruit (the chilli pepper pod) that makes the mouth feel hot and provides the taste that we associate with foods from countries such as India, Thailand and Mexico. The intensity of heat depends on the variety as well as the ripeness of the pod (green – relatively mild; red – strongly hot) and is measured in Scoville units. A standard chilli pepper used in England would be around 5000 Scovilles. The cause of the hot 'kick' associated with chilli peppers is a group of related chemicals (alkaloids) called capsaicinoids, of which capsaicin is the main active component. *Capsicum annuum* is a member of the Solanaceae family, and unrelated to other culinary peppers such as black pepper, *Piper nigrum*, from the Piperaceae family.

Capsaicin is licensed for use topically to relieve some types of chronic pain, particularly neuropathic pain (pain caused by damage to or abnormal activity of small sensory nerves). Locally applied capsaicin may have a role in other conditions including non-allergic rhinitis (unexplained sneezing with a blocked or runny nose) and bladder overactivity.

Capsicum annuum cultivars include the sweet bell peppers, which contain no capsaicin and are an excellent source of vitamin C (*q.v.*).

Plant profile

Capsicum annuum originated in the tropical climate of Southern and Central America and in these warm zones it is naturally a short-lived evergreen perennial. In temperate zones, however, it needs to be treated and grown as a frost-tender annual. The plant has branched stems reaching about 1m high and the single, off-white flowers typical of the Solanaceae family are produced from July to September in Europe. Today there are many cultivated varieties grown for their diverse shapes (cherry, bell or cone) and brightly coloured fruits that can be yellow, orange, red, green or even black when ripe.

The seeds should be sown under glass in spring in free-draining compost with good light and humidity, and temperatures above 15°C. Although *C. annuum* is better grown under glass it can be planted outside from early summer after all danger of frost has passed. It will require a warm, sunny, sheltered position in fertile, well-drained but moisture-retentive soil. After the first fruits have set the plants should be fed once a week with a high-potassium liquid fertilizer.

Commercial cultivation

There are about 35 species of *Capsicum*, but most commercially-cultivated pepper species are in the *C. annuum* group which includes *C. chinense*, *C. frutescens* and, in South America, *C. baccatum* and *C. pubescens*. In temperate regions both the hot chilli peppers and the sweet bell peppers are grown as annuals or cultivated as perennials in greenhouses. The mature fruits of the former ('chilli' pods) are the source of capsaicin which is produced as a defence mechanism against being eaten by herbivores. The sensory nervous system of birds is not affected by capsaicin so they can eat the fruits without discomfort and disperse the seeds. Chemicals and other agents (elicitors) that stress growing crops or plant cells in culture may maximise the yield of capsaicin (Chapa-Oliver and Mejía-Teniente, 2016).

Capsicum evolved away from *Solanum* in South America about 19 million years ago in Colombia, Ecuador and Peru, spread clockwise round Brazil extending to Argentina and later north again. New species began evolving around 10-12 million years ago, moving into Mexico 6 million years ago, where *C. annuum* arose, and to the Galapagos. New species are still evolving. The major factor in the success of *Capsicum* is its importance in food, which can be dated back six or seven millennia in South America. Cultivation, breeding and selection since then has resulted in over 50,000 cultivars, with the sweet bell peppers at one end of the culinary market and the hottest chilli pepper, the cultivar 'Trinidad Moruga Scorpion', at the other.

History

Capsicum was first described by Hieronymus Bock (1539) and first illustrated with three forms by Leonard Fuchs (1542) who called it *Capsicum* and *Siliquastrum*, the latter (from the Latin *siliqua*, a seed pod) being the name of a plant known to Pliny in the 1st century CE who would have had no knowledge of New World plants. Fuchs had no idea where it came from except that it was commonly grown in pots in Germany. It was known to Henry Lyte (1578), the English herbalist, as *Capsaicum*, Indian or Calechut pepper, in his translation of Dodoens' *Crüÿdeboeck* (1554) . Lyte and Dodoens write that the leaves looked like those of *Atropa belladonna*, that the flowers were white with green centres; the fruit was green, turning red and purple, and the seeds were of a pale yellow colour, hot and of a biting taste, like pepper. It was cooked with meat to impart its flavour and to heat the stomach, and medicinally used for scrofula (tuberculous lymph nodes) and facial spots. Lyte said it killed dogs which were given it to eat. Fifty years later many varieties were in cultivation in England, the seeds being germinated in pots in a bed of hot horse dung. The fruits could be purchased at Billingsgate vegetable market where it was known as Indian or Ginnie pepper (Gerard, 1633). While the Portuguese word *ginné* is Guinea in West Africa, Ginnie pepper may refer to South America where guinea pigs came from. 'Indian pepper' probably refers to where it was grown (India) rather than the West Indies/South America where it originated. It was also known as Calicut pepper, now Kozhikode in Kerala State, India, a centre of the spice trade where *Capsicum* was grown commercially.

Portuguese merchants sailing home from Brazil and Spanish treasure ships returning from Peru in the 16th century, after crossing the Pacific stopped to pick up spices in the East Indies and traded *Capsicum* with the merchants there. It was quickly adopted into Asian cuisine and its use and cultivation spread there and around the world with great speed.

The Spanish physician, Nicholas Monardes (1575a), writing on the plants of the 'West Indies' [Latin America] waxes lyrically about *Capsicum*, calling it Indian pepper (*Pepe del Indie*), but says it is very expensive from there and cheaper from Calicut and Maluco (Maluku Islands of the Maluccas). He adds that if you grow it yourself from seed it costs nothing. Called *Secamone*, it was known to be grown in Egypt from a lovely woodcut by Prospero Alpini (1592). In 1606, the Capuchin friar Gregorio da Reggio, in charge of the pharmacy, infirmary and garden in the monastery of Monte Calvario, Bologna, wrote to Carolus Clusius that he was delighted with the 25 varieties of *Piper Americanum* (American pepper) he had been growing for some years, and doubted that anyone else had so many (Clusius, 1611). John Parkinson (1640) lists 20 sorts of *Ginney Pepper* or *Capsicum* as coming from 'the West Indies called America' especially Brazil and 'our Sommer Islands' [Bermuda] and notes that it was grown all over Europe. Parkinson was the first to mention its use as an analgesic, writing that taken orally for four to five days it relieved postpartum

Capsicum annuum – woodcut from Egenolph's *Plantarum Arborum*, 1562

uterine pain. His graphic description of its effects – even inhaling the vapour from them – is unrivalled, noting that it is 'hot and dry in the fourth degree and beyond if there be any beyond it'.

Nomenclature

Called *Capsicum* by Fuchs (1542), it had numerous names, most commonly variations on pepper, *Piper*, with a place name, *e.g. Indicum* [from India], attached. The name 'chilli' was first used by the Dutch physician, Jacob Bontius (de Bondt), for plants he saw growing in gardens in Batavia (now Jakarta). He described it as '… *fructum ricine Americani, quod lada Chili Malaii vocant*' ['the fruit of the American Ricinus, which shrub the Malays call Chili'] (Bontius, 1642) [*Ed*: Malays are an Austronesian ethnic group that live in the area of Malaysia to Indonesia]. Sixteen years later, Bontius describes it in more detail with a woodcut (Bontius, 1658). The plant received its current name *Capsicum annuum* from Linnaeus (1753) who had previously included it in his *Materia Medica* with the pharmaceutical name of *Piperis Indici* [Indian pepper], coming from Brazil, Mexico and Bermuda, and used to treat asthma, colds and anorexia (Linnaeus, 1749).

Capsicum annuum as Secamone in Prospero Alpini's *De Plantis Aegypti*,1592

Capsicum annuum from Garcia ab Horto's *Aromatum*, 1593

Medicinal use

Analgesia

When capsaicin is applied to the skin it stimulates small sensory nerve fibres whose purpose is to detect heat and tissue damage. The electrical activity in these stimulated nerve fibres is perceived as a burning pain by the brain. If the nerve fibres are exposed to capsaicin repeatedly, after a time the capsaicin receptor stops responding so that the nerves are desensitized and no longer capable of causing burning sensations. The same receptors are present in nerve cells in the mouth. Desensitisation explains why one can build up tolerance for hotter and hotter curries by eating chilli peppers regularly.

L A D A   C H I L I.

Capsicum annuum from Bontius' *Historia Naturalis et Medicae*, 1658

Chemistry

Capsaicin – mode of action

Capsaicin stimulates nerve fibres by binding to a specific receptor, the transient receptor potential cation channel subfamily V member 1 (TRPV1), also known as the capsaicin receptor. The capsaicin receptor is widely distributed in the body and appears to play a role in the regulation of normal body functions in the digestive, cardiovascular and respiratory systems. It has been hypothesised that activation of TRPV1 by capsaicin might help conditions ranging from gastric ulcers and irritable bowel syndrome to hypertension and diabetes. Research continues – truly a hot topic! (Hui *et al*, 2019; Wang *et al*, 2020; Du Q *et al*, 2019; Patowary *et al*, 2017).

The ability of capsaicin to desensitise nerves has proved helpful in the treatment of some patients with chronic neuropathic pain. Neuropathic pain may be caused by damage to the same small sensory nerve fibres or by abnormal, spontaneously-generated activity in these nerve fibres and typically causes tingling, burning or increased sensitivity of the skin. Topical capsaicin may relieve the distressing neuropathic pain experienced after an attack of herpes zoster (shingles) or caused by nerve damage in diabetes (diabetic neuropathy) and other peripheral neuropathies by desensitising the nerves. Topical capsaicin is also licensed for the symptomatic relief of osteoarthritis, prescribed as a cream or self-adhesive patch. Patients must use capsaicin regularly to maintain the desensitisation (Patowary *et al*, 2017).

Capsaicin is a potent irritant. At first, applications of cream to normal skin produce a burning sensation but this sensation lessens after several days of regular use because the sensory nerves no longer respond. Capsaicin should not be applied to broken or inflamed skin or applied near the eyes (see below).

Experimental evidence from both animal and cell culture studies show that oral capsaicin may reduce inflammation in some forms of autoimmune neuropathies, conditions in which the immune system damages normal nerve tissues. This potential of capsaicin as a treatment for autoimmune neuritis requires more exploration (Grüter *et al*, 2020).

Idiopathic rhinitis

Inhaled pepper spray makes the nose stream and burn so it seems somewhat surprising that capsaicin delivered in a nasal spray could be useful medicinally. Rhinitis or inflammation of the nasal passages is associated with sneezing, a streaming or blocked nose and postnasal drip. Allergy to pollen (hay fever) is a common cause of rhinitis, but some people experience rhinitis without any obvious cause (idiopathic non-allergic rhinitis). Capsaicin may be an option in the treatment of this form of rhinitis. Capsaicin receptors in the sensory nerve endings in the lining of the nose (nasal mucosa) are over-activated in non-allergic rhinitis, leading to swelling and inflammation in the nasal mucosa. Capsaicin nasal spray appears to desensitise the sensory nerves in nasal mucosa, as it does in the skin, so the inflammation, swelling and nasal drip subside. The nasal spray is available for purchase over the counter in the UK. Capsaicin spray is not effective in allergic rhinitis. (Gevorgyan *et al*, 2015; Fokkens *et al*, 2016)

Bladder hypersensitivity

Patients with damaged nerves caused by spinal cord injuries or multiple sclerosis may suffer from intractable incontinence caused by an overactive bladder. Bladder contraction and emptying is regulated by sensory nerves which have capsaicin receptors. Small studies performed some years ago indicated that capsaicin instilled into the bladder might reduce the desire to pass urine, but more clinical studies are required to prove that treatment is effective (Rackley & Shenot, 2018).

Toxicity

Oral capsaicin, from chilli peppers in curry, causes a burning sensation in the mouth and irritates the lining of the stomach. As capsaicin is not water-soluble, the discomfort is not alleviated by drinking water, but a yoghurt-based drink is able to remove residual capsaicin and soothe the mouth and stomach. Oral capsaicin is absorbed into the circulation and excreted in urine. After a particularly hot curry the burning sensation can be experienced in the lungs, bladder and urethra. Hot curries are also a common cause of diarrhoea.

Pepper sprays containing synthetic capsaicinoids are used by law enforcement officers for riot control but are illegal for use by the public. Exposure of the eyes to pepper spray or to topical capsaicin may cause very painful burning irritation with redness, profuse watering and even temporary blindness. Inhalation of pepper spray causes temporary shortness of breath that is associated with a burning discomfort in the lungs.

Other plant sources of capsaicin-like substances

The milky latex of the resin spurge, *Euphorbia resinifera*, which is native to Morocco, contains an extremely irritant compound, resiniferatoxin, which has structural similarities to capsaicin. Capsaicin and resiniferatoxin are collectively termed vanilloids. Resiniferatoxin is an even more potent activator of the capsaicin receptor and, like capsaicin, desensitises the sensory nerves. It is being investigated for use in chronic neuropathic pain as well as in bladder disorders.

References

Chapa-Oliver AM, Mejía-Teniente L. Capsaicin: From Plants to a Cancer-Suppressing Agent. *Molecules* 2016;21:931

Du Q, Liao Q, Chen C, *et al.* The Role of Transient Receptor Potential Vanilloid 1 in Common Diseases of the Digestive Tract and the Cardiovascular and Respiratory System. *Front Physiol* 2019;10:1064

Fokkens W, Hellings P, Segboer C. Capsaicin for Rhinitis. *Curr Allergy Asthma Rep* 2016;16:60

Gevorgyan A, Segboer C, Gorissen R, *et al.* Capsaicin for non-allergic rhinitis. *Cochrane Database Syst Rev* 2015;7:CD010591

Grüter T, Blusch A, Motte J, *et al.* Immunomodulatory and anti-oxidative effect of the direct TRPV1 receptor agonist capsaicin on Schwann cells. *J Neuroinflammation* 2020;17:145

Hui S, Liu Y, Chen M, *et al.* Capsaicin Improves Glucose Tolerance and Insulin Sensitivity Through Modulation of the Gut Microbiota-Bile Acid-FXR Axis in Type 2 Diabetic db/db Mice. *Mol Nutr Food Res* 2019;63:e1900608

Patowary P, Pathak MP, Zaman K, *et al.* Research progress of capsaicin responses to various pharmacological challenges. *Biomed Pharmacother* 2017;96:1501-12

Rackley RR, Shenot PJ. *Pharmacologic neuromodulation.* In: Krames ES, Peckham PH, Rezai AR (eds) *Neuromodulation: Comprehensive Textbook of Principles, Technologies, and Therapies*, 2nd edn. Academic Press, pp257-65, 2018

Wang Y, Tang C, Tang Y, *et al.* Capsaicin has an anti-obesity effect through alterations in gut microbiota populations and short-chain fatty acid concentrations. *Food & Nutrition Research* 2020;64:3525 doi: 10.29219/fnr.v64.3525

Zhang S, Wang D, Huang J, *et al.* Application of capsaicin as a potential new therapeutic drug in human cancers. *J Clin Pharm Ther* 2020;45:16-28

CHAPTER 8 – TIMOTHY CUTLER

Catharanthus roseus
The source of vincristine and vinblastine

Introduction

Catharanthus roseus, in the family Apocynaceae, is commonly known as the Madagascar periwinkle and sometimes is referred to as the Cayenne jasmine, old maid and rosy periwinkle. It has provided the world with a most important range of anti-cancer drugs, collectively known as the vinca alkaloids, which are still the mainstay of treatment in a variety of malignant diseases, in particular leukaemias, sixty years after their discovery.

There are eight species of *Catharanthus*, all endemic to Madagascar except *C. pusillus* from India. Numerous cultivars and hybrids have been raised, with a range of colours from purple to white, making it a very popular garden plant worldwide.

Plant profile

Catharanthus roseus is an attractive evergreen sub-shrub with glossy leathery-green leaves. The flowers comprise five petals with a long tubular throat and vary from dark pink to white. Growing on disturbed ground and in dry places, the plant has become naturalised in many sub-tropical and tropical countries although it is native to the island of Madagascar. Despite its success elsewhere, in Madagascar itself the plant is an endangered species mainly due to habitat loss from increasing agriculture. It can be grown from seed or cuttings and in temperate climates it is best treated as a tender biennial and planted out in the second year. Once planted it resents disturbance.

History and nomenclature

Catharanthus roseus has escaped from Madagascar to be grown worldwide as a tropical ornamental with multiple uses in herbal medicine. It reached Europe as seeds sent by the governor of Madagascar, Étienne de Flacourt (1607-1660), to the garden of King Louis XV at Versailles where it was grown by Antoine Richard, the king's gardener. The first evidence of this plant in the UK was the arrival of some seeds at the Chelsea Physic Garden in 1757 sent by Richard to Philip Miller, the renowned horticulturalist who was Curator at Chelsea for 50 years. He grew the plant, had it illustrated in an important folio of 300 copper plates, and named it descriptively as *Vinca foliis oblongo-ovatis integerrimis tubis floris longissimo, caule ramoso fruticoso* (Miller, 1755-1760).

Fortunately, it was not long until Carl Linnaeus simplified botanical nomenclature with his binomial system and the plant gained the simpler name of *Vinca rosea* (Linnaeus, 1759) because of its superficial resemblance to other well-known periwinkles. Once it was fully re-evaluated it was assigned to a completely different genus, *Catharanthus*, by the Scottish botanist George Don in 1837 (from the Greek *Katharos* meaning pure and *Anthos* for flower) (Don, 1837).

Linnaeus makes no mention of any medicinal use in his *Materia Medica* (Linnaeus, 1782), but Forbes Royle, a Scottish surgeon with the East India Company in Bengal, observed that *Vinca parviflora* (now *Catharanthus pusillus*) was used as a skin rub in India for lumbago (Forbes Royle, 1839). Source literature on early uses of *C. roseus* as a herbal remedy is poorly documented, and its absence from the encyclopaedic *Pharmacographia* of drugs of vegetable origin in Britain and India (Flückiger & Hanbury, 1874) indicates, at least, that its use at that time was not widespread.

In the 1920s-1940s, popular accounts from New Zealand, Australia, the Philippines, South Africa and the Caribbean about the beneficial effects of *C. roseus* in diabetes and the control of raised blood pressure were reported, although experimental studies showed no effect on blood glucose levels (Watt & Breyer-Brandwijk,1932). Although a herbal preparation of its leaves, called Vinculin, was marketed in the UK for diabetes in the 1930s, the American Medical Association listed Vinculin among a series of ineffective drugs for diabetes in 1936.

In the 1950s, extracts from the leaves of this plant were studied by two independent research groups, one headed by Robert Noble and Charles Beer at the University of Western Ontario, Canada, and the other by Gordon Svoboda at the headquarters of Eli Lilly and Co. in Indianapolis, USA. Both were researching substitutes for insulin which had first been isolated by Banting and Best at the University of Toronto in 1921 and mass-produced by Eli Lilly since 1922. Robert Noble had been sent leaves from his brother Clark Noble which he had received from one of his patients, a Miss Farquharson, who had been visiting Jamaica where they were used to make a tea for treating diabetes.

Neither research team could substantiate any significant effect on blood sugar levels by extracts from the plant leaves, but the Canadian team revealed dramatic reductions in white blood cell counts and bone marrow activity in laboratory rats, and Charles Beer isolated an active constituent which was named vincaleukoblastine, now known as vinblastine, in 1958.

The team at Eli Lilly, having given up on their attempts to find a useful anti-diabetic agent from the periwinkle, submitted an extract into a routine screening programme, which included tests on mice against the P-1534 leukaemia cell line, and they showed a dramatic reduction of the number of leukaemia cells. Their work led to the commercial production of both vinblastine and subsequently a related alkaloid which they named leurocristine, now known as vincristine, which came into use in 1963 (Johnson, 1963). Thus, quite independently, two distinct approaches, one following a specific experimental observation and the other stemming from a broad screening programme, led to similar conclusions and at last to the production and use of two drugs of plant origin which were to transform the treatment of several malignancies.

Catharanthus roseus contains several terpene indole alkaloids of great biochemical value, including ajmalicine, catharanthine and vindoline as well as vinblastine and vincristine (Barrales-Cureño, 2021). These have probably evolved as protective chemicals to deter herbivores, as they are extremely bitter to taste.

Chemistry

The chemical structures of vinblastine and vincristine are complex and the mechanisms by which the plant produces these chemicals is not fully understood. As such, they are still only available through extraction from the plant, although catharanthine and vindoline can be used as precursors in a semi-synthetic process, which has led to the introduction of three more drugs, vindesine, vinorelbine and, most recently, vinflunine. Once the whole pathway has been elucidated, it will be possible to genetically engineer other organisms to make these alkaloids in order to increase world supply and reduce the price of these vital life-saving drugs. At present it takes 530kg of dried leaves of *C. roseus* to produce 1kg of vincristine and half a ton to produce 1gm of vinblastine, the bulk of it being grown in Texas, and there is a world shortage of it (Barrales-Cureño HJ, 2019).

All the vinca alkaloids bind to intracellular tubulin at a specific site and prevent the formation of the mitotic spindle, so cell division is disrupted and leads to cell death. There are similarities here with other anti-cancer agents such as the taxanes (*q.v. Taxus*) and podophyllotoxin (*q.v. Podophyllum*) and also colchicine (*q.v. Colchicum*), but the binding sites differ. Another significant effect of the vinca alkaloids is the inhibition of angioneogenesis (the formation of new blood vessels needed by rapidly growing malignant tumours), so enhancing their anti-tumour effectiveness.

Medical uses

Vincristine has been a transformative drug in the treatment of childhood acute lymphoblastic leukaemia. Since the 1960s, it has turned a uniformly fatal disease into one with a survival rate of higher than 90%. It is also used in combination therapy against neuroblastoma (a brain cancer), myeloma (a type of leukaemia), rhabdomyosarcoma (a muscle cancer), Wilms' tumour (childhood kidney cancer), thyroid cancer, Hodgkin lymphoma and other lymphomas (Coufal & Farnaes, 2011).

Chemistry

Toxicity

The side effects of vincristine include peripheral neuropathy (damage to sensory and motor nerves), bone marrow suppression, constipation (caused by damage to autonomic nerves), nervous system toxicity, nausea and vomiting. Nerve damage is much more common from vincristine than all the other vinca alkaloids and the only remedy is dose reduction or cessation. All the alkaloids are routinely given by the intravenous route where they can cause great irritation to the vein walls. They are excreted via the hepatobiliary system, so care needs to be taken with the dosage, especially in cases of liver dysfunction or hepatic metastatic disease.

Vinblastine has been used as an integral part of regimens in the treatment of testicular cancer, breast cancer, Kaposi's sarcoma, germ-cell cancers, Hodgkin and non-Hodgkin lymphomas. Side effects involve much more bone marrow suppression than with vincristine and also the more common side effects of many anti-tumour agents including nausea, vomiting, constipation, shortness of breath and chest or tumour pain.

Vindesine was the first of the semi-synthetic vinca alkaloids. It has similar effects and side effects to vinblastine and has been used in acute lymphocytic leukaemia, the blast crisis of chronic myeloid leukaemia, malignant melanoma, paediatric solid tumours and metastatic renal, breast, oesophageal and colorectal cancers.

Vinorelbine was developed in France in the 1980s and licensed in the USA in 1994. It has significant anti-tumour activity in breast cancer and also bone cancer. In the USA, it is approved for use in advanced lung cancer. Its side effects are similar to those of vinblastine. It can be given intravenously, but in Europe it is also licensed for oral use, the only one of the vinca alkaloids to be available in this form.

Vindoline. Recent research on vindoline, one of the originally described extracts of *C. roseus*, has shown it lowers blood sugar levels *in vivo* in laboratory-induced diabetes in both mice and rats (Yao *et al*, 2013). This might explain the views of folklore herbal medicine of the benefits of the Madagascar periwinkle in diabetes. Further work is ongoing to find more applications in the improvement of beta-cell dysfunction (the cells in the pancreas that produce insulin) and thus potential treatment for type 2 diabetes.

Thus, the wheel of fortune of medical research has turned full circle from the initial search for an anti-diabetic agent based on folklore, which was discarded, to the serendipitous discovery of the most important class of anti-cancer drugs called the vinca alkaloids and now to the latest revelation confirming the blood sugar lowering effects of one of the original extracts!

Footnote/Epilogue

The author experienced the emotional trauma of working on a paediatric ward in a major London teaching hospital in the early 1970s when childhood leukaemia was universally fatal. Because of this he pursued a career specialty (dermatology), where there were plenty of children as patients but not the dire outcome seen then in leukaemia. Had the vinca alkaloids come into use a few years earlier to transform the outcome of this disease and relieve the emotional trauma on the wards of the death of so many young children, he would almost certainly have pursued a career in paediatrics.

References

Barrales-Cureño HJ, *et al. Alkaloids of pharmacological importance in Catharanthus roseus*. In: Kurek J (ed) *Alkaloids: Their importance in nature and human life*. IntechOpen, 2019

Barrales-Cureño HJ, *et al. Metabolomics and fluxomics studies in the medicinal plant Catharanthus roseus*. In: Aftab T, Hakeem KR (eds) *Medicinal and Aromatic Plants*. Academic Press, Elsevier, pp61-86, 2021

Coufal N, Farnaes L. *The Vinca Alkaloids*. In: Minev BR (ed) *Cancer Management in man: Chemotherapy, Biological therapy, Hyperthermia and Supporting Measures, Cancer Growth and Progression 13*. Springer Science + Business Media B.V, pp25-36, 2011

Don G. *Catharanthus roseus* (L.) G. Don Gen. Hist. 1837;4(1):95

Flückiger F, Hanbury D. *Pharmacographia. A History of the Principal Drugs of vegetable origin met with in Great Britain and British India*. London, Macmillan and Co., 1874

Forbes Royle J. *Illustrations of the Botany and other branches of the Natural History of the Himalayan Mountains and of the Flora of Cashmere*. Messrs Allen & Co., 1839

Johnson IS, Armstrong JG, Gorman M, *et al*. The Vinca alkaloids. A new class of oncolytic agents. *Cancer Res* 1963;23:1390-1427

Miller P. *Figures of the most beautiful, useful, and uncommon plants described in The Gardeners Dictionary*. London, for the Author vol 2, fig 186, p124, 1755-1760

Moudi M, *et al*. Vinca Alkaloids. *Int J Prev Med* 2013;4(11):1231-5

Watt JM, Breyer-Brandwijk MG. *The Medicinal and Poisonous Plants of Southern Africa*. Edinburgh, E & S Livingstone, 1932

Yao XG, *et al*. Natural product vindoline stimulates insulin secretion and efficiently ameliorates glucose homeostasis in diabetic murine models. *Journal of Ethnopharmacology* 2013;150(1):285-97

CHAPTER 9 – ANTHONY DAYAN

Cephalotaxus harringtonia
The source of harringtonine, homoharringtonine (omacetaxine) and cephalotaxine

Cephalotaxus harringtonia, known as the Japanese or Chinese plum yew or cowtail pine, in the Taxaceae family, is the source of the anti-cancer alkaloids, harringtonine, homoharringtonine and cephalotaxine. *Cephalotaxus fortunei* and other species are also sources of homoharringtonine (Pérard-Viret et al, 2017).

History
Cephalotaxus harringtonia was introduced to Belgium in 1829 from Japan by the German physician and plant collector, Philipp Franz von Siebold (1796-1866). It was named for the 4th Earl of Harrington who had planted avenues of it at his home, Elvaston Castle, Derbyshire.

Plant profile
Cephalotaxus harringtonia is a medium-sized conifer up to 3-4m in height with half the spread. It occurs naturally in moist, deciduous, broadleaf forests and thickets on limestone slopes at 600-1000m in China, Japan, Korea and Taiwan. It is cold hardy through zone 6 (Tripp, 1995). It is regarded as an endangered species because of overcollection for drug manufacture.

The potential of extracts of *C. harringtonia* as a cancer chemotherapy was first recognised in a simple test for activity against experimental leukaemia cells as part of a broad survey of plant extracts by a division of the US National Cancer Institute. A major alkaloid, cephalotaxine, was isolated in 1963 from *C. drupacea*, following which numerous other active substances of a similar nature, including harringtonine, were identified from *Cephalotaxus* species. Subsequent research was focused on identification of these substances, their structures, mode of action and how to manufacture them sustainably. Dozens of cephalotaxines – chemicals with a basic core skeleton but different side chains – were isolated and many more were synthesised by chemical modification of the natural ones, all of which are only found in the genus *Cephalotaxus* (Perdue *et al*, 1970; Powell *et al*, 1972; Pérard-Viret *et al*. 2017).

Sufficient supplies of harringtonine became available, by extraction from *C. harringtonia*, in the late 1960s to early 1970s, and clinical studies on patients were done to explore it, and others, for anti cancer chemotherapy. Initially harringtonine appeared promising, but it proved to be very toxic; further laboratory and clinical research showed that certain esters of it, especially homoharringtonine, were more active and better tolerated. The best clinical results were obtained in treating chronic myeloid leukaemia, but interest waned in the 1990s when new effective drugs, such as tyrosine kinase inhibitors and imatinib mesylate came into use. However, resistance, either acquired or innate, to these drugs reawakened interest in the cephalotaxines, and homoharringtonine was approved as a medicine in the EU in 2009 and in the USA in 2012 to treat chronic myeloid leukaemia and other leukaemias for which other drugs were, or had become, ineffective. Combination therapy of cephalotaxines with other drugs is now practiced in order to overcome the problem of drug resistance (Kantarjian *et al*, 2001; Pérard-Viret *et al*, 2017; Hao *et al*, 2021).

Chemistry

The novel mechanisms of action of homoharringtonine were revealed once suitable molecular, biochemical techniques became available in the 1970s-1980s. Harringtonine and its esters inhibit normal protein synthesis in cancer cells by targeting their ribosomes, resulting in inhibition of polypeptide synthesis initiation, blocking the progression of the cell cycle from G1 to S phase, and G2 into M phase, causing the cell death called apoptosis. Related chemicals in this group with a methylene side chain, such as harringtonine, homoharringtonine, isoharringtonine and deoxy-harringtonine, are most active while those lacking the methylene group are inactive. (Huang, 1975; Pérard-Viret *et al*, 2017; Srivasta & Raghuwwanshi, 2021).

A suitable and stable manufacturing process was difficult to devise. Collection of trees from the wild was not sustainable because of their relative rarity, and cultivation was not feasible because of the slow growth of all plum yew species. Harvesting the bark for manufacture of the drugs killed the trees. Over the years several different approaches to manufacture have been adopted since use of bark from wild and cultivated trees was not practicable; extraction from cultured cells has been tried but semisynthetic procedures have been more successful (Pérard-Viret *et al*, 2017).

Toxicity of these drugs has been a problem, but the derivative, homoharringtonine (aka omacetaxine), a semisynthetic cephalotaxine that acts as a protein translation inhibitor, is used to treat chronic myeloid leukaemia that is resistant to tyrosine kinase receptor antagonists.

The exact mechanism of action of homoharringtonine has not been fully elucidated. Upon administration, homoharringtonine targets and binds to the 80S ribosome in eukaryotic cells and inhibits protein synthesis by interfering with chain elongation. This reduces levels of certain oncoproteins and anti-apoptotic proteins (Pub.Med., 2022).

The history of the development of harringtonine and its derivatives is complicated and much work was duplicated because independent paths were followed by American and Chinese researchers. The former employed conventional western scientific and clinical strategies without reference to the Chinese work and the latter proceeded independently but largely concurrently because of the geo-political climate in the 1960s-1980s; compare the accounts of the western, mostly American research reviewed in Pérard-Viret *et al* (2017) and the Chinese work in Hao *et al* (2021).

Extracts of *C. lanceolata*, *C. fortunei* var. *alpina*, *C. griffithii*, and *C. hainanensis* have been employed in traditional Chinese medicine as remedies for a very wide range of disorders including swellings (possibly tumours), cough, cancers, internal bleeding, traumatic injuries, rheumatism and parasitic diseases, depending on the area of China being considered and the practices of local traditional healers (Hao *et al*, 2021).

There is current interest in harringtonine derivatives (and in camptothecin *q.v.*) as potential antiviral agents provided their toxicity can be overcome (Oliveira *et al*, 2017).

References

Hao D-C, Hou X-D, Gu X-J, *et al*. Ethnopharmacology, chemodiversity, and bioactivity of Cephalotaxus medicinal plants. *Chinese Journal of Natural Medicines* 2021;19(5):321-38

Kantarjian H, Talpaz M, Santini V. Homoharringtonine History, Current Research, and Future Directions. *Cancer* 2001;92(6):1591-1605

Oliveira AF, de Silveira C, Róbso RT, *et al*. Potential Antivirals: Natural Products Targeting Replication Enzymes of Dengue and Chikungunya Viruses. *Molecules* 2017;22:505

Pérard-Viret J, Quteishat L, Alsalim R, *et al*. Cephalotaxus alkaloids *The Alkaloids: Chemistry and Biology*. 2017;78:205-352

Powell RG, Togovin SP, Smith CR. Isolation of antitumor alkaloids from *Cephalotaxus harringtonia Ind Eng Chem, Prod Res Develop* 1974;13:129-32

Pub. Med. Cephalotaxine. https://pubchem.ncbi.nlm.nih.gov/compound/Cephalotaxine accessed 3 April 2022

Purdue RE, Spetzman LA, Powell, RG. Cephalotaxus – Source of Harringtonine, a promising new-anticancer alkaloid. *The American Horticultural Magazine* 1970;49(1):19-22

Srivastava A, Raghuwanshi R, *et al*. (Eds.) *Evolutionary Diversity as a Source for Anticancer Molecules*. Academic Press, 2020

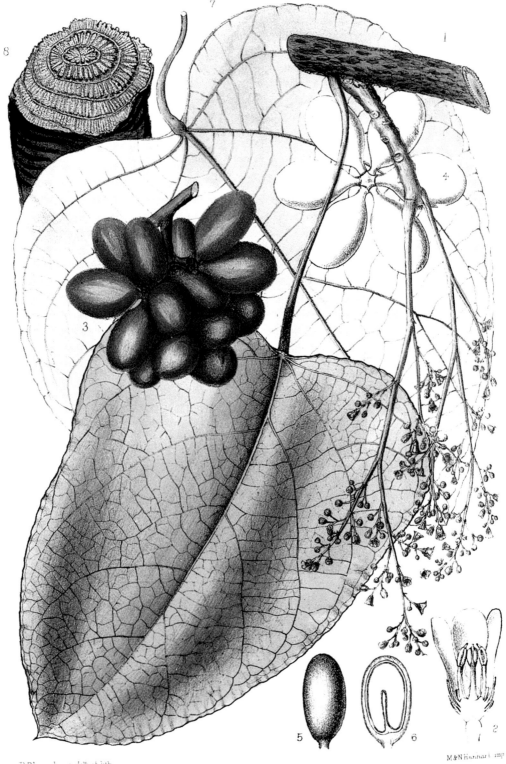

D.Blair ad nca del. et lith.

M&N Hanhart imp.

CHONDRODENDRON TOMENTOSUM, *Ruiz & Pav.*

CHAPTER 10 – ARJUN DEVANESAN

Chondrodendron tomentosum
The source of tubocurarine

Chondrodendron tomentosum, the source of curare from which tubocurarine was extracted, is a tropical plant from the Menispermaceae family, not found in the College Garden nor in its herbarium. However, tubocurarine is of such fundamental importance in anaesthesia that it is important to describe it and its plant source in this book about plants from which medicines are derived. *Strychnos toxifera*, another tropical plant which also contains curare, is represented in the College herbarium (see below).

Curare, the sap of *Chondrodendron tomentosum*, a tropical vine, was used as a dart poison in South America, and found to work by paralysing all voluntary muscle activity, most importantly respiration. The discovery that its action in blocking the nerve supply to muscles could be reversed by another poison, physostigmine from the Calabar bean, *Physostigma venenosum* (*q.v.*), led to the widespread use of curare in anaesthesia. Studies of the chemical structure of curare also led to an understanding of the way nerves control muscle contraction and subsequent development of synthetic muscle relaxants still in use today.

Plant profile
Chondrodendron tomentosum is a vigorous woody liana with large heart-shaped leaves covered with fine silky hairs. It grows up forest trees in the tropical rainforest of Central and South America, to a height of 30 metres. The branched scapes of small yellow flowers are succeeded by clusters of small dark fruits.

History and nomenclature
It was first formally described botanically by Hipólito Ruiz and Joseph Pavon (1798) who found it in tropical forests in Pillao and Chacahuassi, Peru, in 1787. They noted its bark as having an extremely bitter taste but were ignorant of its use as an arrow poison. It is known currently in Colombia as *nascata*, and because of the tomentose (velvety) leaf as velvet leaf. Curare vine is a more useful epithet. Up until the late 18th century, there were a number of names for the extract from it which came to be known as curare. One of Sir Walter Raleigh's lieutenants called the

Chondrodendron tomentosum, Credit: R Bentley & H Trimen, Medicinal Plants (1880). Wellcome Collection. Public Domain Mark

87

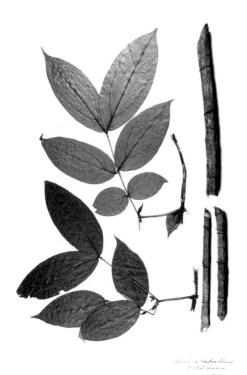

Strychnos toxifera from the Pharmaceutical Society's herbarium at the Royal College of Physicians, collected in Guyana, 1902

poison *ourari*, possibly from the Venezuelan native word *uria*, meaning bird and *eor*, meaning kill (Raleigh, 2010). Others, like the Ticuna of Guiana, called it *woorara*. In Demerara, it was called *wourali*. Curare is a later European group name for the different plant compounds which cause muscle paralysis and were used as arrow poisons. Curare from *C. tomentosum* was transported in bamboo tubes, so was called tube curare, hence the chemical name of the active substance isolated from the crude sap became tubocurarine (or D-tubocurarine because of its dextro-rotatory crystalline structure).

Curare from *S. toxifera*, which is transported in calabashes (hollowed-out plant gourds), was called calabash curare and contains a different chemical, toxiferine. There was also pot curare, stored in pots, whose source and active chemical is not specified. These different curares contained different but related compounds of which tubocurarine is the most important to the history of modern anaesthesia.

The history of the discovery of curare from the South American vine *Chondrodendron tomentosum*, and its morphosis from an arrow poison to a vital muscle relaxant that revolutionised anaesthesia, has been well described by Professor MR Lee, a Fellow of this College and the Royal College of Physicians of Edinburgh (Lee, 2005). We acknowledge his detailed account. The earliest reports of this paralysing plant extract were received at the beginning of the 16th century from travellers returning from South America. At the end of the 16th century, it was mentioned by Sir Walter Raleigh and his contemporary, Admiral Sir John Hawkins. Charles-Marie de la Condamine, the French explorer and scientist, during his descent of the Amazon from Peru to the Pacific coast in 1743, saw Amerindians using blowguns with darts dipped in a black resin for hunting, which paralysed the prey. He brought some of this back to France. Condamine thought that sugar by mouth was the antidote, but the English physician Richard Brocklesby to whom he later gave samples, found that two chickens injected with this black resin died within minutes and sugar by mouth was ineffective. Two cats to whom he fed the dead chickens experienced no ill effects – a phenomenon noted by Condamine on the Amazon – *i.e.* the poison was harmless in food (Brocklesby, 1747).

The discovery that curare caused paralysis, and so stopped breathing which led to death, was demonstrated in 1811 by the British surgeon Sir Benjamin Brodie who kept a cat alive with artificial respiration via a tracheostomy until the curare wore off. A donkey was similarly kept alive with artificial respiration in an experiment by the naturalist Charles Waterton in 1814. While Waterton (1825), who had travelled extensively in South America, regarded the plant source as *Chondrodendron tomentosum*, the German explorer Alexander von Humboldt (1807) identified it as a species related to *Strychnos nux-vomica* (the source of strychnine) which Robert Schomburgk named as *Strychnos toxifera* (Schomburgk, 1841). Both were correct and there are several other plants in diverse genera which contain curare (although with different active chemicals).

Chemistry

Curare – mode of action

Through the 19th century the action of curare was much investigated, but it was not until the 1930s that it was found that acetylcholine, released from nerve endings at the neuromuscular junction, was shown to be responsible for muscular contraction and that this action of acetylcholine was blocked by curare. It had been found previously that physostigmine, the poison extracted from the Calabar bean (*Physostigma venenosum*), an alkaloid ordeal poison in Africa, inhibited the destruction of acetylcholine by the body's naturally occurring enzymes, the acetylcholinesterases. Hence, local acetylcholine levels at the nerve-muscle interface increased to a level where there was sufficient to overcome the acetylcholine block by curare. It then became possible to use curare in anaesthesia and stop the paralysis when required, rather than waiting for hours for the effect of curare to wear off naturally. Other anticholinesterase inhibitors were also trialled, including galanthamine from snowdrops – *Galanthus* (*q.v.*) – and found effective but had unpleasant neurological side effects, so are not used.

Tubocurarine from curare

The analytical chemist Harold King extracted D-tubocurarine, the active chemical, from specimens of *Chondrodendron* roots and identified its crystalline structure (King, 1940). In 1942 the company E.R. Squibb and Sons developed a method of preparing a commercial extract from *C. tomentosum*. Marketed as Intercostrin, it was initially used to modify the severe contractions that occurred in electroconvulsive treatment in psychiatric patients who, at that time, had to be ventilated artificially until the curare wore off.

Curare in anaesthesia

The introduction of curare to anaesthesia is attributed to Harold Griffith, a Canadian anaesthetist who mastered the use of endotracheal intubation for artificial respiration in patients unable to breath sufficiently, which was commonly needed when using the inhaled anaesthetic cyclopropane. Endotracheal intubation is the procedure whereby a flexible tube is inserted into the trachea via the mouth (as opposed to a tracheostomy incision in the neck) and so is relatively atraumatic and reversible. Griffith was confident that the same procedure would be useful in overcoming the major drawback of curare, viz. cessation of respiration, and in 1942, with Enid Johnson, demonstrated the use of curare on a young patient undergoing appendicectomy (Griffith, 1942). At that time, abdominal surgery needed deep levels of anaesthesia to ensure the abdominal wall muscles were sufficiently relaxed to allow surgical access. The usual anaesthetics in use had many side effects, particularly at high doses, and this made deep anaesthesia extremely dangerous. The use of curare allowed for a lighter anaesthesia to still achieve the required degree of muscle relaxation. However, cessation of breathing was an unavoidable side effect and manual respiration with a face mask was extremely labour intensive. Endotracheal intubation, therefore, made curare a viable anaesthetic adjunct.

In the early days of curare use in anaesthesia it was not known whether curare causes anaesthesia (loss of consciousness) as well as paralysis. After all, animal subjects were unable to report back, and curare was always used with other inhaled anaesthetics which caused unconsciousness themselves. In a case series reported in the 1940s, curare was used as a sole anaesthetic in young children undergoing surgery. Later, when they were older, they reported that they were aware and able to feel everything. Upon hearing this, in 1946 Frederick Prescott (Research Director at the Wellcome Research Institute) decided to try curare on himself to observe the effect. In his usual understated style, Prescott reported in *The Lancet* that 'to be conscious yet paralysed and unable to breathe is a very unpleasant experience' (Prescott, 1946).

The Second World War halted a lot of research into curare in Europe and the UK, but American doctors stationed in England brought word of the use of curare as a muscle relaxant for anaesthesia. John Halton, an anaesthetist in Liverpool, convinced an American colleague to bring back Intercostrin and trialled it on a series of patients which he reported on in 1946. His technique came to be known as the Liverpool technique: a triad of anaesthesia, analgesia and muscle relaxation which is now the standard model of general anaesthesia.

Tubocurarine is no longer used as a muscle relaxant, but its chemical structure formed the basis for most modern muscle relaxants in use, and greatly expanded our understanding of the physiology of the neuromuscular junction. Its contribution to the development of safe and effective anaesthesia cannot be overstated. It provided us with knowledge of the essential structure of molecules that block acetylcholine's action at the neuromuscular junction. From study of curare, chemists were able to produce other compounds with additional features which made them degradable to inert metabolites in the body and to muscle relaxants with these clinically valuable properties, such as atracurium and vecuronium. These are still in use today.

Summary

The role from an African ordeal poison – physostigmine – to an Andean arrow poison – tubocurarine – in revolutionising anaesthesia and surgery is a good example of how it is possible to exploit the poisonous effects of plants for medical treatments.

References

Brocklesby R. A letter from Richard Brocklesby MD and FRS to the President [Martin Folkes] concerning the Indian poison sent over from M. de la Condamine. *Phil Trans* 1747;44:408-12

Griffith HR, Johnson GE. The use of curare in general anesthesia. *Anesthesiology* 1942;3:418-20

King H. Alkaloids of some Chondrodendron Species and the Origin of Radix Pareira Bravae. *J Chem Soc* 1940:737

Lee MR. Curare: the South American arrow poison. *Journal of the Royal College of Physicians of Edinburgh* 2005;35:83-92

Prescott F, Organe G, Rowbottom S. Tubocurarine Chloride As An Adjunct To Anesthesia. *The Lancet* 1946;2(6412):80-84

Raleigh W. (2010) *The discovery of the large, rich, and beautiful empire of Guiana … 1595*. Cambridge University Press, 2010. (Facsimile of Schomburgk R (ed). London, Hakluyt Society p48, 1848)

Ruiz H, Pavón J. *Systema Vegetabilium Florae Peruvianae et Chilensis*. Gabrielis de Sancha, 1798

Schomburgk R. On the Urari, the arrow poison of the Indians of Guiana. *Ann Mag Nat Hist* 1841;7:407

Von Humboldt A, Bonland A. *Voyage aux régions équinoxiales du nouveau continent*. Paris, 1807

Waterton C. *Wanderings in South America, the North West of the United States and the Antilles in 1812, 1816, 1820 and 1824*. London, J Mawman & Co, 1825

CHAPTER 11 – MICHAEL DE SWIET

Cinchona
The source of quinine and quinidine

The bark of *Cinchona* trees is the source of quinine, at one time vital for the treatment of malaria, and quinidine, which was used to treat irregularities of heart rhythm.

Cinchona is a genus in the Rubiaceae family. It is a tropical tree originally from the forests of Ecuador and Peru. It has been cultivated in many other tropical countries as a source of quinine. Quinine is a chemical in the quinoline group of medicines.

Plant profile

Handsome trees growing 5-10m in height, they prefer damp, sandy, humus-rich, well-drained soil with an optimum pH of 4.5-6.5. They are found at 1000-2000m altitude in areas of high humidity, average temperatures of 20°C, frost free. Typical cloud forest trees on mountain slopes, they are cultivated in plantations, principally in Indonesia, India and Zaire as well as Latin America. The bark is still the source of quinine.

There are 65 species of *Cinchona*, but only four have been successfully cultivated. Many hybrids have arisen in the plantations, both by natural cross pollination and by selective breeding to raise varieties with higher quinine content.

History

Much of the early history of *Cinchona* relates to Jesuit missionaries in South America who brought the bark back to Europe. Agostino Salumbrino (1561-1642), a Jesuit brother and trained apothecary who lived in Loja in Ecuador and Lima in Peru, was an early observer of the Quechua people using the quinine-containing bark of the *Cinchona* tree to cure malaria. Bernabé Cobo (1582-1657) is credited with being the first to bring *Cinchona* bark to Europe in 1632. Its presence in Europe was first documented by Heyden (1643).

However, Nicoló Monardes (1493-1588), the Spanish physician living in Seville who made a study of plants coming from the Americas, wrote of the bark of the 'tree for treating diarrhoea' from the New World of which he had a piece – at least a century before Cobo (Monardes, 1565;

Cinchona officinalis in the Andes of Peru, with bark removed for treating fevers

Engraving of the flowers of *Cinchona condaminea* – now *C. officinalis* – in *Medical Botany* by Stephenson & Churchill, 1836

Monardes, 1575b). The description of the plant fits that of *Cinchona*. He writes that on the advice of the indigenous people, a bean-sized portion of powdered bark taken in red wine or water cured fevers and diarrhoea, and he had experimented with it with 'miraculous success', his diarrhoea ceasing quickly.

Linnaeus published the name *Cinchona* in 1749, based on the herbarium specimen and description of the tree found by the French naturalist Charles-Marie de la Condamine (1701-1774), travelling with the French botanist Antoine Laurent de Jussieu (1748-1836), in Cajanuma, near Loxa (now Loja), in Southern Ecuador. Condamine published *Sur l'arbre du quinquina* [On the *quinquina* tree] deriving the name from the local Quetchua word, *quina-quina* with two engraved plates (Condamine, 1738). He noted that the indigenous Americans had long hidden the source of this valuable medicine from the Spanish due to their not unnatural antipathy to their conquerors. He relates that they attributed the discovery of the febrifuge activity of the bark to the observation that their 'lions' (presumably pumas) ate the bark when afflicted by intermittent fevers – according to an ancient legend '*dont je ne garantis la verité*' [whose truth I cannot guarantee]! Its effectiveness in treating fevers was well known in Loja but had been forgotten by the rest of the country (and the world) until, he writes, the Countess of Chinchon in Lima, suffering from a long febrile illness, was given the bark – specially sent to her from Loja – and recovered. Thereafter it became very popular; she distributed it to the citizens of Lima and brought it back to Spain. This story originated from a manuscript by Bernabé Cobo, written in Lima in 1653, and published by Sebastiano Bado (1663). Linnaeus, having read Condamine's account, named it for the Countess.

The truth of the cure of the Countess of Chinchon and the discovery of the use of *Cinchona* bark to treat malaria has been heavily contested and many variants are in circulation. However, A W Haggis (1941) showed that the first Countess of Chinchon, Ana de Osorio, died in 1625 before her husband left Spain for Peru (in 1628) and the second Countess, Francisca Henriquez de Ribera, arrived in Lima, Peru, in 1629. Thereafter, the daily diary of the Count's secretary,

Dr Antonio Suardo, records in detail the life, illnesses and treatments of the Count and Countess for the next 11 years. While the Count was ill frequently, his wife only had two minor illnesses, no fevers, never received *quina-quina*, never distributed it to the populace, and could not have taken *Cinchona* bark back to Spain as she died in Colombia in 1641 on her way home.

Linnaeus included *Cinchona* in his *Materia Medica* (1749), noting that its pharmaceutical (trade) name in Europe was the tincture or powder of 'Chinae cortex' [China bark] and that it was extremely bitter, a tonic and a treatment for severe fevers, anorexia, renal stones, oedema, gangrene and 'hysteria' (it should be noted that the 'Chinae radix' [China root] of 17th century pharmacists was a completely different plant). He gave it its full name of *Cinchona officinalis* in 1753 and added its use to treat haemorrhage, tuberculosis, persistent cough and smallpox in his *Materia Medica* of 1782.

Up to 300 different *Cinchona* species have been described but it is now realised that many are hybrids. A recent search of The Plant List online revealed only 23 different species amongst 288 synonyms. When quinine first became available, malaria was also prevalent in temperate climates including much of Europe. Condamine writes that the bark was taken to Rome by the Jesuits and called Jesuit's powder – the name making it fiercely unacceptable to Protestant England and given as a reason why Oliver Cromwell refused it and died of malaria. King Louis XIV of France and the Dauphin were successfully treated with it for fever.

Robert Talbor (1642-1681), an English apothecary's apprentice, pioneered the use of *Cinchona* in malaria treatment in England in the 17th century (Talbor, 1672). He first used his secret remedy in the Fens and Essex, where malaria was rife. His contribution was to use smaller doses more frequently thus avoiding the toxic and potentially lethal effects of the single large doses used previously.

Talbor treated and cured King Charles II (Siegel *et al*, 1962). He received a knighthood and was made Royal Physician and offered membership of the Royal College of Physicians. He also treated the French and Spanish royal families.

Commercial *Cinchona* plantations.

Because of the huge demand for *Cinchona* bark, and to maintain their monopoly, Peru and surrounding countries started outlawing the export of *Cinchona* seeds and saplings, beginning in the early 19th century (Jaramillo-Arango, 1949). However, in 1860, a British expedition to South America brought back smuggled *Cinchona* seeds and plants. These were introduced in Darjeeling, India, and in Sri Lanka. By 1883, about 64,000 acres (260km²) were in cultivation in Sri Lanka, with exports reaching a peak of 15 million pounds (6.6 million kg) weight in 1886.

'*Cinchona ovata*' collected in Guanai-Tipuani, Bolivia in 1892. In the Pharmaceutical Society's herbarium at the Royal College of Physicians

Plantations were also established in Mexico to supply North America using seeds purchased from England. These seeds were the first to be introduced into Mexico. The Dutch government also smuggled seeds, and by the 1930s Dutch plantations in Java were producing 22 million pounds (10 million kg) of *Cinchona* bark annually, or 97% of the world's quinine production (Goss, 2014). These plantations grew *Cinchona ledgeriana*, the seeds of which were obtained from Bolivia by Charles Ledger in 1865. Its bark had a higher yield (8-13%) than the species being grown in British India and so enabled the Dutch monopoly.

Without quinine, Europeans would not have been able to colonise, for better or for worse, the tropics from South America to Africa, India and the Far East.

The problem of malaria

Malaria was a major problem in the Second World War. Allied powers were cut off from their supply of quinine when the Germans conquered the Netherlands where there were major stockpiles. The Japanese controlled the other sources of *Cinchona* bark in the Philippines and Indonesia. By December 1942, more than 8500 US soldiers were hospitalized with malaria. In one hospital, as many as eight out of every ten soldiers had malaria, not war-related injuries. The only other possible effective agent at that time was Atabrine but it was very toxic and, because it was reputed to cause impotence, the troops would not take it.

Tens of thousands of US troops in Africa and the South Pacific died due to the lack of quinine for the treatment of malaria. The Japanese also did not make effective use of quinine, and every Japanese soldier in Burma had at least one attack of malaria. However, before the Japanese takeover of the Philippines, the United States did manage to obtain four million *Cinchona* seeds from the Philippines and planted *Cinchona* in Costa Rica.

The alternative to growing *Cinchona* again in the Far East was to return to South America to obtain *Cinchona* bark where it originally grew. From 1942 onwards the US government sent teams of botanists and foresters to South America to find strains of *Cinchona* yielding high concentrations of quinine. This was difficult, not least because the various species of *Cinchona* interbred and it was not obvious which were high-yielding hybrids. There were few natives any longer skilled in stripping the bark, and the teams suffered terrible privation as they had to go further afield once the more accessible trees had been cut down. Nevertheless, by 1944 12.5 million pounds (5.7 million kg) of *Cinchona* bark had been shipped back to the USA. This method of collection was unsustainable, and it is only from the massive plantations, before and after the Second World War, that quinine production came from a sustainable resource.

Quinine

Friedlieb Ferdinand Runge, a student at the University of Jena in Germany, isolated quinine from *Cinchona* bark in 1819, followed a year later by the French pharmacists, Joseph Pelletier and Joseph Caventou.

Quinine is an alkaloid commonly marketed as the sulphate. It was made from *Cinchona* bark mixed with lime. The bark and lime mixture was extracted with hot paraffin oil, filtered, and shaken with sulphuric acid and neutralised with sodium carbonate. Quinine sulphate crystallizes out from the cooling mix. It was synthesized from precursors in 1944 by the American chemists RB Woodward and WE Doering. Since then, other ways of synthesising quinine have been developed but none have been commercially viable compared with extracting quinine from *Cinchona* bark. For example, quinine has been produced by culturing plant cells derived from *Cinchona*. The cells can be manipulated to release quinine, which is absorbed by a resin and then extracted. This technique is difficult to perform, expensive and, like other methods of quinine biosynthesis, has not proven to be commercially viable.

Therefore, extraction of quinine from *Cinchona* bark remains the only practical way of manufacturing this drug. But the use of quinine is declining because other less toxic agents are now available and the production of quinine is also declining, causing the disappearance of the previous huge plantations in countries throughout the tropics that were devoted to *Cinchona* cultivation. One plantation in the Democratic Republic of Congo is the site of one of the last remaining five quinine extraction factories in the world. It produces about 100 tonnes of quinine per year, about one third of the global demand. About half of the 100 tonnes is sent overseas, of which half is turned into medicine and half makes tonic water. The remaining half is used to make quinine tablets for local consumption. But the factory in the Congo is threatened by Indian merchants who buy the bark and send it to India where it is cheaper to extract the quinine.

An early bottle for quinine sulphate, for treating malaria

Quinine for the treatment of malaria

Quinine remained the antimalarial drug of choice until the 1940s when other drugs such as chloroquine that have fewer unpleasant side effects replaced it. In turn, the malaria parasite has become resistant to most of the standard antimalarial drugs so that artemisinin derivatives, which also have fewer side effects than quinine, have become the antimalarial drugs of choice (*q.v. Artemisia annua*). Quinine is still recommended by the WHO for the treatment of malaria but only when artemisinin is not available.

Chemistry

Quinine and chloroquine – how they work

Malaria is caused by the *Plasmodium* parasite which is transmitted in the saliva of the female *Anopheles* mosquito when it bites humans to obtain a meal of blood. The *Plasmodium* has a complicated lifecycle whereby it multiplies in human liver and red blood cells. It is thought that quinine acts by killing the developing *Plasmodium* at the schizont (multiplication) stage in red blood cells by inhibiting glycolysis (the mechanism whereby cells produce energy to function) in the parasite. Chloroquine is another quinoline drug developed for the treatment of malaria and this agent inhibits the haem detoxification pathway. Haem is produced by haemoglobin in red blood cells being broken down by the parasite. Haem is cytotoxic and when detoxification is inhibited it accumulates, killing the *Plasmodium*.

Cinchonism is the syndrome caused by quinine toxicity. The symptoms include nausea and vomiting, sweating, headache, tinnitus (ringing in the ears) and deafness, visual disturbances, irregular heartbeat, low platelet count and renal failure associated with haemoglobinuria (blood-stained urine). Some of the symptoms (called blackwater fever) may have been due to malaria itself. When there was uncertainty about quinine dosing, one practice was to increase the dose until the patient was aware of tinnitus.

Quinine and more recently the synthetic forms chloroquine and hydroxychloroquine have also been used for the treatment of certain autoimmune diseases such as rheumatoid arthritis and systemic lupus erythematosus.

Quinine is a neuromuscular blocking agent. It can block transmission of nerve impulses to the muscles. It may therefore prolong paralysis in patients given curare during anaesthesia. Quinine has also been used for the treatment of night leg cramps and restless leg syndrome. This is not recommended because of quinine toxicity in the high doses that have been used. But if the quinine is taken in very small quantities as in tonic water it is unlikely to do any harm.

Quinine is a very bitter substance, so much so that patients objected to swallowing the tablets. Possibly the very bitter taste of quinine protected the bark of the *Cinchona* tree from wood-boring insects and birds. One way to make quinine more palatable was to take it with gin. So, gin with a little quinine in soda water (Indian tonic water) with the addition of a slice of lemon became a much-favoured drink.

Quinine and quinidine for the treatment of cardiac rhythm disorders

In 1749, Jean Baptiste de Sénac noted that *Cinchona* bark could eliminate cardiac palpitations and under his influence quinine was used in France to augment digitalis therapy. Subsequently in 1912 Karel Frederik Wenckebach saw a patient who found that a dose of 1g of quinine would abort an attack of atrial fibrillation within 25 minutes whereas with no treatment the attack lasted for 2-14 days (Wenckebach, 1914). Unfortunately, this success was not repeated with other patients (Davies and Hollman, 2002; Wenckebach, 1923). Wenckebach was a leading authority on cardiac rhythm disorders. One form of rhythm disorder is still named after him.

Quinidine is an optical isomer of quinine and is also present in *Cinchona* bark. In 1918 Walter von Frey of Berlin reported in a leading Viennese medical journal that quinidine was the most effective of the four principal *Cinchona* alkaloids in controlling some cardiac disorders such as atrial arrhythmias.

Chemistry

Quinidine – mode of action

Quinidine is a class I anti-arrhythmic agent. It increases the cardiac action potential duration and prolongs the QT interval.

It has been used to restore and maintain sinus rhythm in patients with atrial fibrillation and to prevent ventricular dysrhythmia. However, because of its toxicity, which is similar to that of quinine but also includes specific cardiac malignant rhythm disorders such as Torsades de Pointes, it is rarely used nowadays. There are better drugs, and radio-ablative surgery to obliterate irritable foci in the myocardium is increasingly used instead.

Quinic acid

Quinic acid is another chemical that can be extracted from *Cinchona* bark and has been used as a starting point in the synthesis of oseltamivir phosphate (Tamiflu) that was produced for the treatment of avian flu (see also *Illicium verum*). The multistep process of producing oseltamivir has one stage where an explosive chemical is produced and this part of the synthesis from quinic acid was being conducted in Arizona where it blew the roof off the factory. The place of Tamiflu in the management of influenza has been questioned (Ebell, 2017).

References

Ebell MH. WHO downgrades status of oseltamivir. *BMJ* 2017:358. doi: 10.1136/bmj.j3266

de la Condamine, C-M. Sur L'Arbre du Quinquina. In: *Histoire de L'Academie Royale des Sciences*. Paris, L'Imprimerie Royale, pp 233-44, 1738

Davies MK, Hollman A. Quinine. *Heart* 2002;88:118

Goss A. Building the world's supply of quinine: Dutch colonialism and the origins of a global pharmaceutical industry. *Endeavour* 2014;38:8-18

Haggis AW. Fundamental Errors in the Early History of Cinchona. *Bulletin of the History of Medicine* 1941;10:417-459

Jaramillo-Arango J. A critical review of the basic facts in the history of Cinchona. *Journal of the Linnean Society of London, Botany* 1949;53:272-311

de Sénac JB. *Traité de la structure du coeur, de son action, et de ses maladies*. Paris, Jacques Vincent, 1749

Siegel RE, Poynter FNL. Robert Talbor, Charles II and Cinchona a contemporary document. *Medical History* 1962;6:82-5

Stephenson J, Churchill JM. *Medical Botany*. London, John Churchill, 3 vols. 1834-1836

The Economist. Eastern Congo has the world's largest quinine plantations. *The Economist Group Limited* Jun 8, 2019 Eastern Congo has the world's largest quinine plantations | The Economist accessed 16 July 2022

The Plant List – www.theplantlist.org accessed 16 July 2022

Wenckebach KF. *Die unregelmässige Hertztätigkeit und ihre klinische Bedeutung* [irregular cardiac activity and its clinical importance]. Leipsig, Berlin W. Engelmann, 1914

Wenckebach KF. Cinchona derivatives in the treatment of heart disorders. *J Am Medical Ass* 1923;81(6):472-4

World Health Organisation. Compendium of WHO malaria guidance – prevention, diagnosis, treatment, surveillance 2019 https://apps.who.int/iris/bitstream/handle/10665/312082/WHO-CDS-GMP-2019.03-eng.pdf accessed 26 May 2021

Further reading

Gänger S. *A Singular Remedy. Cinchona across the Atlantic World, 1751-1820*. Cambridge University Press, 2021

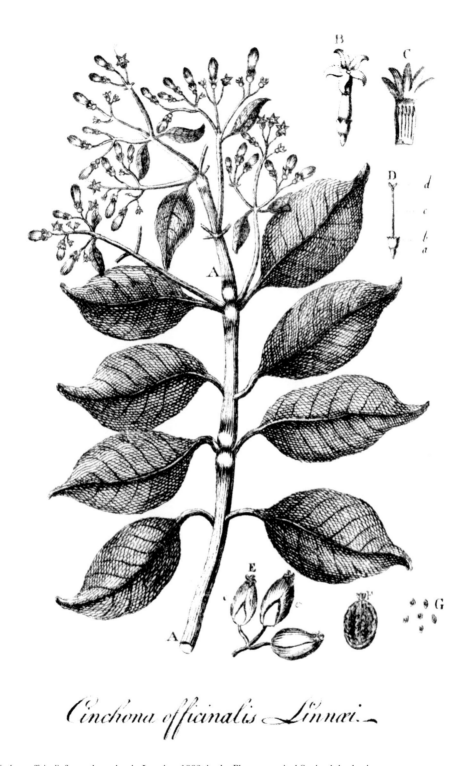

Cinchona officinalis Linnæi.

Cinchona officinalis from plantation in Jamaica, 1882, in the Pharmaceutical Society's herbarium at the Royal College of Physicians

CHAPTER 12 – HENRY OAKELEY

Citrus x *limon*
A source of vitamin C

Citrus x *limon* and vitamin C

Vitamin C, otherwise known as ascorbic acid, is a vital vitamin which we obtain from our diet, principally fruit and vegetables, and to a lesser degree from meat. The illness due to vitamin C deficiency is called scurvy. The lemon, *Citrus* x *limon*, is selected here because of its importance in the discovery of this vitamin and in the prevention and treatment of scurvy. Various *Citrus* bushes grow in in the College Garden as does *Drimys winteri*, Winter's bark, whose berries were once used as a source of vitamin C.

Almost all animals make their own vitamin C, in the liver for mammals, in the kidneys in birds and reptiles. Our inability to synthesise vitamin C occurred 63 million years ago when a genetic defect arose in one of our early ancestors, the apes and the monkeys, but not lemurs. In short, we all carry with us an inherited genetic defect that can cause a terrible illness and death in a matter of months if we do not eat fruit and vegetables. We are lucky with this 'disease'; no other genetic disorder is so easily treated.

Editor's note: For a decider in a tied family or pub quiz competition, the question as to which organ of the body in dinosaurs was used to manufacture vitamin C can be answered confidently as 'the kidney'; birds and reptiles being their evolutionary descendants. In other animals it is the liver.

Editor's second note: Fruit bats (whose diet is purely fruit) and guinea pigs also carry the defective form of the gene that should make vitamin C.

Plant profile

Citrus x *limon* is a natural hybrid between the citron from Central Asia, *Citrus medica*, and the bitter orange, *Citrus* x *aurantium*. It is a broadleaved evergreen shrub in the Rutaceae family that can grow up to 3m high. It has glossy aromatic leaves, small spines in the leaf axils, and beautifully fragrant flowers with white petals tinted purple on the undersides. The fruit has a thick, yellow rind, dotted with the oil glands which produce the familiar aroma of *Citrus*.

In Britain, the plant is best grown in a pot that can be placed outdoors in a sunny, sheltered position after the last frosts. A suitable compost is crucial to prevent yellowing of the leaves: a slightly acid mix of loam and leaf mould plus a little charcoal will work. Regular applications of specialist *Citrus* feed (summer and winter) is recommended to maintain healthy plants and promote fruiting. The plants need protection from frost and winter waterlogging so should be overwintered undercover in a cool, well-lit greenhouse. They can be pruned to shape in late winter.

Citrus x *limon* probably originated in Northern India and reached Southern Italy in the 2nd century CE, gradually spreading through the Near East and North Africa around 700 CE during the Muslim conquests, reaching Genoa in the 15th century and travelling with Christopher Columbus to America. Its pollen parent, the citron – *Citrus medica* – is mentioned by Theophrastus in his *Enquiry into Plants* (c. 350 BCE) as the 'Median' (*i.e.* from the land of the Medes) or 'Persian' apple growing in Persia. From there it came to Israel, as pollen grains have been found in plaster work in a Royal Persian Garden in Ramat Rahel near Jerusalem dated to 400-500 BCE.

Citrus x *limon*, hand-coloured woodcut from Egenolph's *Plantarum Arborum*, 1562

History

The earliest author to mention the lemon (meaning *C.* x *limon*) is the Egyptian Ibn al-Bayṭār (d. 1248) who reports that it is good for digestion, improves the appetite, is an antidote to poisons, and is useful for fevers, rashes, palpitations of the heart and vomiting. The 'limonis' (*limunium*) of Avicenna (1522) was a fruit with styptic properties, which was good for healing a weak stomach and bloody diarrhoea, and is regarded as being another plant (Adams, 1847). Sylvaticus's *Pandectarum* (1524), however, describes Avicenna's 'Limon' as a fruit with a beautiful fragrance that is probably *C. medica*, well known to Dioscorides, Pliny and Galen. Sylvaticus has 'Limoncelion' as a syrup that was good against pestilent fevers. As the modern drink Limoncello is made from Sorrento lemons it is likely that this is one of the earliest European references to lemons.

Chemistry

Vitamin C – our genetic defect

Vitamin C is manufactured in birds and animals that do not have the genetic defect, starting with glucose, to make a chemical called l-gulono-g-lactone. This is converted by an enzyme called gulonolactone oxidase, in association with flavin adenine dinucleotide, into ascorbic acid. While we have the gene (human gene 8p21) for making that enzyme, it has been corrupted by several mutations and has become a pseudogene that is non-functional so we cannot make vitamin C.

Vitamin C is needed as a co-substrate by an enzyme called peptidyl-prolyl hydroxylase which converts proline to hydroxyproline which is then made into collagen. Without the ability to make and renew collagen, new collagen is not formed. Old collagen breaks down, so blood vessels, sinews, bones and all tissues become abnormal – mainly fragile.

Evolutionary advantage

One of the functions (side effects?) of gulonolactone oxidase is to produce hydrogen peroxide (H_2O_2) in body tissues, a free radical which endangers cells. So, the absence of gulonolactone oxidase and consequently of this source of toxic H_2O_2 is advantageous. H_2O_2, along with other free radicals from other metabolic sources, is inactivated by vitamin C so we still need an external dietary source of vitamin C.

Vitamin C necessary to produce energy

Vitamin C is also a catalyst for carnitine synthesis, responsible for fatty acid transport to mitochondria and thence for the production of adenosine triphosphate. Adenosine triphosphate is the source of energy for all our body's cells. Without it, simply, they stop working.

Evolutionary compensations – recycling

Animals that produce their own vitamin C, *e.g.* goats, produce 200mg/kg body weight per day. Humans only need less than 1mg/kg per day because they, with the other animals that do not produce their own, have evolved a gene that allows their red cells to recycle vitamin C that has already been used, converting oxidised vitamin C, L-dehydroascorbic acid (DHA), back to the active unoxidized form and transporting it around the body. The primary transporter of glucose is a protein called Glut1 found in the membranes of body cells, and this is massively increased in the red cells of all animals that do not make vitamin C. In their red blood cells, Glut1 transports DHA preferentially to glucose in association with another protein called stomatin. Animals that produce their own vitamin C have a different protein, Glut4, which only transports glucose in their red blood cells and has no ability to transport or remake vitamin C (Cell Press, 2008).

Vitamin C deficiency – scurvy

Without vitamin C we cannot make or repair collagen, the fibrous stuff which holds us together. Blood vessels, cartilage, skin, our intestines, muscles, and bones all need collagen to be continually replaced; without it, scurvy develops. The principal features are rotting gums, teeth falling out, bleeding into the skin with ulcers which cannot heal and muscles wasting away. The heart and lungs and other organs fail, the body swells, bones break, and old wounds break down; even ones which have been healed for decades come apart and bleed.

Depression, exhaustion, coma and an appalling death ensue due to the inability of the body to heal itself and to produce energy.

Vitamin C stores and requirement

We can store 2000mg (2g) of vitamin C (that is about the weight of a sugar lump) and we need a minimum of 10mg a day to stay healthy, so everyone will get scurvy within 200 days – about six and a half months, 28 weeks – if they stop taking fresh vegetables and fruit. People who have a poor diet, usually the poor, prisoners, people in warzones or under siege, may have much lower stores and can develop scurvy in weeks. In the 16th and 17th century, sailing across the Atlantic to North America could take a few weeks, but contrary winds might double that time. The longer voyages to South America, and to Asia round the Cape of Good Hope, took months, and the diet of the sailors contained no fruit, vegetables or fresh meat, so scurvy on ships was the norm.

When scurvy develops it progresses inexorably, but as soon as a vitamin C-containing diet is commenced – even as low as 10mg per day – enough becomes available to reverse the damage and recovery from scurvy begins within a week.

The recommended daily requirement for an adult is 40-70mg a day (NHS website). The reason for this minute requirement is explained in the scientific section 'Evolutionary compensations – recycling' and is a marvellous example of how evolution has produced compensatory genetic change to deal with an acquired genetic defect.

So little vitamin C is required that just gargling with lemon juice will reverse the rotting gums and falling teeth within hours. If it is applied to scorbutic ulcers (those due to scurvy), they will begin to heal (Lyte, 1578).

Vitamin C content of fruit and vegetables

There is huge variation in the vitamin C content of raw fruits and vegetables:

Acerola cherries – 1678mg/100g	Brussels sprouts – 62mg/100g (cooked)
Hot green chilli peppers – 242mg/100g	Oranges & lemons – 53mg/100g
Guava – 228mg/100g	Limes – 20mg/100g
Black currants – 181mg/100g	Potatoes – 13mg/100g (boiled)
Sweet red bell pepper – 128mg/100g	Apples – 4.6mg/100g (freshly picked)
Kiwifruit – 93mg/100g	Bread – 0
Broccoli – 65mg/100g (cooked)	Cooked meat – 0

So, while 'an apple a day' would not keep the scurvy away, a spoonful of potato would.

History of scurvy

Caesar Germanicus and scurvy

Historically, scurvy was well known in the navies of the world, when long sea voyages began in the 15th century, but it had been known on land for centuries. The army of Caesar Germanicus (15 BCE-19 CE) after two years in Germany developed scurvy due to a poor diet. They were cured by eating a seaside plant called 'Britannica' thought to be *Cochlearia officinalis* (other English names are spoonwort, or scurvy grass), a member of the cabbage family, Brassicaceae, which tastes like water cress (Pliny, 70 CE). *Cochlearia officinalis* was still sold for treating scurvy in British and Irish apothecary shops at the beginning of the 19th century (Duncan, 1819).

Vasco da Gama

Over the centuries scurvy was described in besieged towns, and in crusaders in the 12th century, but it was sailors who were worst affected due to their poor diet during months at sea. The sailors with the Portuguese Vasco da Gama on his two-year (1497-1499) voyage to India via Cape Horn developed scurvy after 21 weeks and were cured by oranges and lemons in Mombasa. They were not so lucky on their return and 117 of his crew of 170 died of scurvy on the return journey. The Portuguese subsequently planted oranges on St Helena to provide fruit for their sailors.

Jacques Cartier

The French settlers in Stadacona, Canada, under the leadership of the Breton Jacques Cartier were almost wiped out by scurvy after the sea crossing, and a Canadian winter in 1535 without vegetables. They were rescued by eating the leaves and bark of (probably) spruce on the advice of a Native American (Savage, 1853).

Captain John Winter

In 1579 Captain John Winter in the Elizabeth, sailed with Francis Drake to raid Spanish ports on the Pacific coast of South America. When they reached Tierra del Fuego, his men were suffering from scurvy. They ate the fruits of a tree, *Drimys winteri*, common to the area, and within a few days were cured. They also used the bark as a condiment and took this home to England as 'Winter's bark' – *Cortex winteri* – for treating scurvy (Parkinson, 1640).

James Lancaster and Sir Richard Hawkins

There were worse disasters. James Lancaster sailed in 1591 to find the route to Malaya via the Cape with three ships and 600 men. 200 were killed by the inhabitants of the Comoros Islands and all the rest, except for 25, died of scurvy. In 1601, when he repeated the voyage, he warded off scurvy with lemon juice, but only had small quantities and still lost 140 out of 577 men. Admiral Sir Richard Hawkins (1622) reported that in 20 years he lost 20,000 men on the British

Navy ships that he commanded, but the value of fruits was being recognised and he bought hundreds of oranges for his men when sailing to Brazil in 1590 (Hakluyt, 1926).

James Woodall

Acting on behalf of the merchant navy, John Woodall, Surgeon General to the East India Company advised in 1617 on plants to be carried to prevent scurvy but concluded: '*Lemmons, Limes, tamarinds, oranges and other choice of good helps in the Indies … do far exceed any that be carried thither from England*' (Woodall, 1617). This was repeated by Parkinson (1640). A century later the British Navy had still not learnt the lesson, and Lord Anson's voyage round the world (1740-1743) as part of the 'War of Jenkins' ear' against Spain was devastated by scurvy, over 1400 men dying of it out of his initial complement of 1854 (Williams, 1967).

Captain Cook and Joseph Banks

Captain Cook and Joseph Banks' three-year voyage to Australia and back (1768-1771) was unique, at the time, in that no-one died of scurvy and only two contracted it. This was due to carrying lemon juice for all crew and purchasing fresh vegetables at every port, which the crew were obliged to eat.

James Lind

James Lind is credited, completely erroneously, as having introduced lemons to the British Navy following the publication in 1753 of his *Treatise on the Scurvy* in which a trial of diets in 12 patients with scurvy resulted in the two given two oranges and one lime a day recovering in six days while the others did not. He notes the indispensability of fresh vegetables, especially oranges and lemons, no less than 117 times, but in the end merely advises green vegetables and purgatives (Lind, 1753).

Gilbert Blane (1749-1834), courtesy of the Royal College of Physicians

Gilbert Blane

Carl Linnaeus' Materia Medica (1782) recommends lemons and oranges for scurvy, but it is to Gilbert Blane, physician to the British Fleet in the Caribbean (1779-1783) that the end of scurvy as a pandemic disease can be attributed. In 1780 he noted that of the 12,000 sailors in the British fleet that year, 59 had been killed by enemy action and 1577 by scurvy. In 1781 seven warships arrived from England after three months at sea, with 1600 cases of scurvy on board. With only 200 hospital beds on land, he sent fresh vegetables to the ships, and within a month all were cured. His book, *Observations on the Diseases incident to Seamen* (1785), is among the greatest and most fascinating publications in the history of medicine. He was a personal friend of Lord Rodney, Admiral of the Fleet, which may be why he was listened to. In 1797 lemons became a standard item of diet in the Royal Navy, although they had already become widely used in the merchant navy (Woodville, 1793).

Mortality from scurvy

It has been estimated that the British Navy lost a million men to scurvy during the 17th to 18th century, and it is astonishing that observations on its prevention and cure were ignored by the admirals and the medical establishment during the three centuries that had elapsed since Vasco da Gama's crew were cured with oranges.

Lemons and limes

The Navy's daily allowance of lemon juice prescribed for sailors contained the equivalent of 8mg of vitamin C, just enough to supplement body stores, but in 1850 the Navy changed to limes as they were more easily available from British colonies in the Caribbean, rather than at inflated prices from Italy and Spain. These contain half the amount of vitamin C compared with lemons and, unsurprisingly, scurvy returned to the British Navy.

Disasters continued, with lemon juice being stored in lead-lined barrels, which caused lead poisoning; the *British Pharmacopoeia* (1864) recommending the distillation of lemon juice – which destroyed vitamin C – and the British medical establishment, in the person of Sir Almroth Wright, insisting that polar explorers' scurvy was caused by contaminated pemmican. Consequently, polar explorers from 1900 to 1913 died of scurvy as they took no vegetables or lime juice (Nicolaysen, 1980). Scurvy became rare in the Navy only because ships became able to traverse the oceans in weeks instead of months.

In 1906 it was discovered that guinea pigs developed scurvy if they were fed a diet deficient in vegetables and were cured with apples, unboiled cabbage, lemon juice and potatoes. At last, there was an 'animal model' which helped the understanding of the disease.

In 1912 Casimir Funk at the Lister Institute in London classified beriberi, scurvy and rickets as deficiency diseases due to deficiency of substances which he called vitamins. Scurvy was a deficiency of vitamin C – a mythical substance not yet isolated which only existed in name.

Discovery of vitamin C

In 1928 an amazing man, a Hungarian called Albert Szent Györgi, having decided that he was going to be interested in plants found indication of a substance that was a reducing agent – or colloquially an 'antioxidant' – in certain plants. He knew there was a reducing agent in adrenal glands and, while working at Cambridge, isolated it from oranges, lemons and cabbages and adrenal glands. Not knowing what it was, he called it ignose from *ingnosco* (Latin, I do not know) and -ose to indicate it was a sugar – like glucose and sucrose. This name was rejected by the editor of the *Biochemical Journal* where he wished to publish his findings, so he called it Godnose but ended up calling it, rather more prosaically, hexuronic acid. He then moved to America where he spent a year in the Mayo Clinic collecting adrenal glands from the St Paul slaughterhouse and isolated 25g of hexuronic acid. He returned to Hungary. A researcher called Swirbely showed that hexuronic acid was vitamin C.

Vegetables were not available in sufficient quantities for Szent Györgi to isolate hexuronic acid/ vitamin C from them but having been served a red pepper, *Capsicum annuum*, for supper – which he did not eat – he took it to the lab at midnight and found it had 2mg vitamin C per gram and within weeks he had kilograms of vitamin C which was distributed all over the world for scientists to work with. Analysis of its structure and synthesis occurred, and the beginning of the real understanding of how vitamin C is produced and works began.

Pure vitamin C could now be used in clinical trials

In the 20th century conscientious objectors were given 70mg of vitamin C for 6 weeks and then a vitamin C-deficient diet and developed scurvy in 6 months. Prisoners in an Iowa jail with no pre-loading developed scurvy in 6 weeks. Both recovered in 10 days on vitamin C supplements of 70mg and 10mg daily respectively.

It is striking that for an adult weighing 70kg (70,000g, 70,000,000mg) only 10mg of vitamin C per day is (just) enough to maintain health: 40mg a day is recommended. Once the body stores of 2g have been achieved, vitamin C is excreted by the kidneys. Taking vast amounts of vitamin C does not have any extra beneficial effects and causes stomach upsets including diarrhoea. Mega-doses of vitamin C have been promoted for several medical conditions – particularly for the treatment of the common cold by the distinguished scientist and peace activist, Linus Pauling (1901-1994) – but this has not been accepted by the majority of scientists.

Vilhjalmur Stefansson

The fact that the Inuit in the Arctic did not eat vegetables and did not have scurvy was explained in 1928 when Vilhjalmur Stefansson lived for a year on an Inuit diet of fresh meat eating it only slightly cooked and eating liver, a huge source of vitamin C, and he did not develop scurvy.

Vitamin C is the best-known vitamin of them all, as it is sourced from fruit and vegetables. A balanced diet easily contains sufficient vitamin C for all our needs, and supplements are never needed. However, in affluent countries, scurvy occurs in people whose diet consists of sandwiches and processed meat, so children still need to be educated to 'eat their greens'.

References

Adams, F. *The seven Books of Paulus Aegineta, translated from the Greek*. 3 vols. London, Sydenham Society (vol iii), 1847

Blane G. *Observations on the Diseases incident to Seamen*. London, J. Cooper, 2 vols, 1785

British Pharmacopoeia, London, Spottiswode & Co., 1864

Cell Press. How Humans make up for an 'Inborn' Vitamin C Deficiency. *ScienceDaily* 21 March 2008

Duncan A. *The Edinburgh New Dispensatory* Edinburgh, Bell & Bradfute, 1819

Foster W. *Voyages of Sir James Lancaster to Brazil and the East Indies (1591-1603)*, a new edition. London Hakluyt Society (original edition 1877), 1949

Funk C. The etiology of the deficiency diseases. Beri-beri, polyneuritis in birds, epidemic dropsy, scurvy, experimental scurvy in animals, infantile scurvy, ship beri-beri, pellagra. *Journal of State Medicine* 1912;20:341-68

Hakluyt R. *A Selection of the Principal Voyages, Traffiquess and Discoveries of the English Nation by Richard Hakluyt 1552-1616* with preface by Laurence Irving. London William Heinemann, 1926

Lind J. *A Treatise on the Scurvy*. Edinburgh, Sands, Murray & Cochran, 1753 (reprint, Edinburgh University Press, 1953)

Nicolaysen R. Arctic Nutrition. *Perspectives in Biology and Medicine* 1980;23(2-1):295-310

Winthrop J, Savage J. *The History of New England from 1630 to 1649 by John Winthrop*. Boston MA, Little Brown & Co., 1853. Reprint, Maryland, Genealogical Publishing Company. 2 vols, [see vol 1, page 54], 2008

Williams G. *Documents relating to Anson's Voyage round the World 1740-1744*. Navy Records Society, 1967

Woodville W. *Medical Botany* vol 1-3. London, James Phillips, 1790-3

CHAPTER 13 – TIMOTHY CUTLER

Colchicum autumnale
The source of colchicine

Introduction

Colchicum autumnale is a bulbous plant in the Colchicaceae family, with an ancient history. It is the source of a modern medicine called colchicine, a powerful anti-inflammatory drug used to treat acute attacks of gout, and a range of uncommon conditions.

Nomenclature

Its Latin name comes from the ancient land of Colchis at the Eastern end of the Black Sea. It has several common names, including naked ladies referring to the fact that the purple, tubular, crocus-like flowers appear out of the ground alone in the autumn long after the leaves with the seed heads have died down. However, the other common names include meadow saffron and autumn crocus which can lead to confusion with another bulbous plant in a completely different family called the saffron crocus, *Crocus sativus*. This is the source of the world's most expensive spice, saffron, and it also flowers in the autumn, but it is never naked: its leaves are always present with the flowers. As saffron is sold as the styles, not the corms, confusion is unlikely, but *Colchicum* has been eaten in confusion for wild garlic, *Allium ursinum*, with fatal results.

Plant profile

Colchicum autumnale is a cormous, herbaceous perennial native to Europe, where it grows in meadows and damp woodland clearings. In Britain, loss of habitat meant that by 1930 the plant had been lost from most of its natural sites. The crocus-like flowers appear first, in the autumn, and they are followed by broad strap-like leaves and seed heads in the spring. As the seeds appear to come before the flowers, it was called *Filius ante Patrem* – the Son before the Father.

The flower with its lavender-pink petals and orange anthers is particularly striking and unexpected, emerging as it does so late in the year.

This species is hardy, self-fertile, and good for attracting wildlife, especially the bees and flies which pollinate it. A popular plant in gardens, *C. autumnale* will naturalise easily when given the right conditions.

COLCHIDO FLORIDO.

Colchicum autumnale flowers – woodcut from
Mattioli's *Discorsi*, 1568

COLCHICO SENZA FIORI.

Colchicum autumnale in the spring, with leaves and
seeds – woodcut from Mattioli's *Discorsi*, 1568

Historical use

The famous Greek physician Pedanius
Dioscorides, writing in his great work *Materia
Medica* about 70 CE, a standard work for the
next 1500 years, describes *Colchicum* as being so
beautiful and alluring to the inexperienced that
they eat it. He includes it in his book to warn
people that it 'kills by choking, like mushrooms'
adding that cow's milk is an effective antidote –
this is not the case, there is no antidote. In the
same era, Pliny [70 CE] writing in his *Natural
History* says that poisonous spiders that touch it
die, and concurs that cow's milk is the remedy
for people poisoned by it. No medical use was
proposed for it until the 16th century. It is of
interest that in China, in the 21st century, juice
from *Colchicum* has been used as a pesticide for red
fire ants.

The popularisation of the idea that eating the
bulbs of *Colchicum* was a treatment for arthritis
or gout begins in Dodoens' Flemish herbal,
Crŭÿdeboeck (Dodoens, 1554) which appeared
in English by Henry Lyte (Lyte, 1578). These
sources say that a similar-looking plant called
Hermodactylus was a cure for gout, sciatica
and joint pain which also caused purging,
but the apothecaries were selling meadow
saffron (*Colchicum*) under that name and that
the latter could kill in a day. The great French
physician Jean Fernel (1497-1558), in his book
on therapeutics, recommended *Hermodactylus*
for arthritis and gout but made no mention of
Colchicum (Fernel, 1593). This *Hermodactylus* was
imported from overseas and had round bulbs
the size of a chestnut which means it was not
the plant we called *Hermodactylus tuberosa* (now

Iris tuberosa) today. This, and the lack of any illustration of the plant, makes the identification of *Hermodactylus* very difficult, but 16th century herbals (Gerard, 1597) suggested that it could be a *Colchicum* species from Syria or elsewhere.

Hermodactylus, as a purgative and treatment for arthritis, was mentioned by Alexander of Tralles as coming from Anatolia/Turkey, in the 5th century in his chapter on anodynes (analgesics) and by the Arabian physicians through the Golden Age of Islam. Where *Colchicum* is mentioned, it is just noted to be poisonous (Tralles, 1575).

John Gerard, the famous London surgeon, plantsman and author of a most famous herbal in 1597, followed Dodoens, describing various *Colchicum*, with woodcuts illustrating the bulb with leaves and seed head in the spring, and the flowers coming on their own in the autumn. He notes that *Colchicum*, 'Medowe Saffron', is sold by the apothecaries under the name of *Hermodactylus* and taken by mouth as a purgative and used topically for gout. He identified the *Hermodactylus* of the apothecaries as a *Colchicum* from Illyria (the Balkans). He quotes Dioscorides [70 CE], 'meadow saffrons are very hurtful to the stomach and being eaten they kill by choking' (Gerard, 1597).

Colchicum was listed as a medical ingredient in the College's *Pharmacopoea Londinensis* (1618), which Culpeper translated in his *Physical Directory* (1649) giving its use as follows: 'Colchici. Of Medow saffron. The Roots are held to be hurtful to the stomach, therefore I let them alone'.

For a while, the concept of *Colchicum* (under the name *Hermodactylus*) as a treatment for arthritis did not survive the 16th century confusion.

Colchicum (as *Hermodactylus*) was omitted from Quincy's *Dispensatory* (1718), the standard pharmacopoeia of the era, so when, in the late 17th century, the famous London physician Thomas Sydenham suffered from gout himself and wrote of the exquisite joint pain he suffered which could not even bear the weight of a bedsheet upon it, there was no effective treatment. Ironically, Sydenham's antagonism to *Colchicum* and other medicines for gout apart from laudanum (morphine) contributed to its disappearance and deprived him of effective help. Fortunately, his young assistant at the time, a Dr Hans Sloane, was able to go out and see Sydenham's patients for him when he was incapacitated by it and thus started his long and very successful career, becoming one of the foremost physicians of his time as well as President of both the College of Physicians and the Royal Society.

By the end of the 18th century, *Colchicum*'s poisonous ferocity, and the toxicity experienced by merely handling it, was well known, but many physicians, including Bergius, Tournefort, Hummelberg and Geofroy, were convinced that *Hermodactylus* was *Colchicum* (Woodville, 1793).

Early 20th century bottle of colchicine wine, for gout

At the beginning of the 19th century, John Ayrton Paris in his *Pharmacologia* reports that a specimen of *Hermodactylus* was procured from Constantinople and found to be a species of *Colchicum*, and it was from then that its reputation as a treatment for gout was restored (Paris, 1820). It was in the College's *Pharmacopoeia* (1824) as 'Vinegar of Meadow Saffron' for 'dropsy' (heart failure) because of its purgative and diuretic effects, and also for asthma and gout, but to be used 'with great caution'. Within another 50 years William Whitla (1887) was writing 'of its wonderful effects' in curing the pain of gout without diuresis or purgation within an hour or two but was unable to explain its mode of action. Extract of *Colchicum* was made by crushing the corms, letting the liquid settle, decanting it and boiling it, then straining it through a flannel, and evaporating it down to a paste. 45kg of corms produced 4kg of paste. The dose for gout was 15-60mg.

Colchicine the medicine

The active ingredient of *Colchicum*, the chemical called colchicine, was identified in Paris in 1820 by the chemists Pelletier (1788-1842) and Caventou (1795-1877) who also isolated and identified quinine the same year. By then, the use of *Colchicum* for gout had become popularised by physicians in Vienna, and once the active ingredient was known, colchicine became the standard treatment for acute gout until the non-steroidal anti-inflammatory drugs (NSAIDs) started to replace it for most patients from the early 1970s onwards (Levy, 1991).

Colchicine is also a useful medicine in a variety of diseases which require specific anti-inflammatory treatment (Sullivan, 1998). These include:

- Behçet's syndrome, a rare condition characterised by mouth and genital ulcers, eye problems and arthritis
- Cutaneous small vessel vasculitis, which causes damage to small blood vessels in the skin
- Scleroderma, which can affect both skin and internal organs, although this is no longer a current treatment

- Acute viral pericarditis, an inflammation of the external layer of the heart
- Familial Mediterranean fever. In this rare condition, inflammation attacks several parts of the body including the joints and kidneys, leading to a build-up of a complex harmful protein called amyloid, which colchicine helps to prevent.

Colchicine has a powerful anti-inflammatory action expressed by several mechanisms but especially by interfering with activation and cell division (mitosis) of white blood cells called neutrophils which play a large part in the inflammatory reaction to uric acid crystals in joints which is the cause of acute gout.

Chemistry

Colchicine – mode of action

Colchicine stops cells dividing properly by interfering with the growth of microtubules (crucial to cell division and multiplication) within dividing nerve cells, ciliated cells, neutrophils and sperm. It forms complexes with the protein tubulin, the precursor of microtubules, and inhibits their production. Microtubule assembly and elongation are thus disrupted, limiting the production and activity of neutrophils that cause the painful inflammation in gout and damage in other diseases. As well as several other anti-inflammatory effects, colchicine inhibits cell-mediated immune responses by inhibiting immunoglobulin secretion, interleukin-1 production, histamine release and HLA-DR expression.

Colchicine has been chemically synthesised, but this is not cost-effective for drug manufacture, so it is still extracted from the dried corms of *Colchicum autumnale* which are grown commercially in various countries in Asia. Another source is a closely-related member of the same plant family which also contains colchicine, *Gloriosa superba*, the glory lily.

As John Gerard noted (1597), all parts of *Colchicum autumnale* are poisonous. Gloves should be worn when handling the plant and self-administration should never be undertaken as even a small amount above the normal therapeutic dose can cause death by multi-organ failure. There is no treatment for overdosage, although Parkinson (1640) considered erroneously that sweet marjoram, *Origanum majorana*, was an antidote. In the treatment of gout, for which it has been most frequently used, the most common side effects are vomiting and diarrhoea, a sure sign that the dosage should be urgently reduced or stopped altogether. Overdosage and toxicity are far more likely from intravenous administration than the more usual oral route. Great care should also be taken to avoid interaction with some prescribed drugs, especially statins, antivirals and antifungals which can all increase the exposure to colchicine and thereby risk damage to kidneys and other organs, particularly if there is pre-existing liver or kidney disease.

Derivatives of colchicine are uncommon, but one called thiocolchicoside is prescribed in some European countries (Czech Republic, France, Greece, Hungary, Italy, Malta, Portugal and Spain), as a treatment for painful muscular contractions, secondary to spinal damage. It is available for use by mouth or by injection into the muscles. In some countries it is also available as a preparation to be applied to the skin. Side effects include nausea, sleepiness, allergy, vasovagal reaction (faints), epileptic seizures and liver inflammation.

Chemistry

Demecolcine – mode of action

Demecolcine (Colcemid), is another, less toxic derivative acting on microtubule function. It also arrests cell division in metaphase and has been used to improve the results of cancer radiotherapy by synchronising tumour cells at metaphase, the radiosensitive stage of the cell cycle.

The toxicity of colchicine in higher doses is a concern. Low doses of colchicine in treating gout compared with glucocorticoids (such as prednisone) or NSAIDs such as ibuprofen or an interleukin-1β monoclonal antibody (canakinumab) were found to be all equally effective, but there were more side effects with high-dose colchicine (Wechalekar *et al*, 2014).

A Canadian clinical trial of 6000 people with COVID-19 infection (the Colcorona Trial) began in March 2020 to test the potential efficacy of colchicine over a 30-day period to reduce symptoms from the inflammatory phase of the disease (the cytokine storm) which is often fatal. The results were published in January 2021 and revealed that of 4159 patients in whom the diagnosis of COVID-19 was proven by a PCR test, the use of colchicine was associated with a significant reduction in the risk of death or hospitalisation compared with placebo (Reyes *et al*, 2020). Colchicine reduced hospitalisation by 25%, the need for mechanical ventilation by 50% and deaths by 44% (Institut de Cardiologie de Montreal, January 2021).

However, there have been similar European trials in COVID-19 disease in both Greece and Italy, some of which have shown a reduction in mortality and a reduction of oxygen dependency after treatment with colchicine. The UK based RECOVERY trial, the largest so far, has also assessed the possible benefits of colchicine therapy when used early on in the infection. On 5th March 2021, the following statement was published: 'recruitment to the colchicine arm of the RECOVERY trial has now closed. The Data Monitoring Committee saw no convincing evidence that further recruitment would provide conclusive proof of worthwhile mortality benefit either overall or in any pre-specified subgroup'.

Summary

The corms of *Colchicum autumnale* had long been known to be extremely poisonous, but when an imported plant, known as *Hermodactylus* and used since the 5th century for arthritis, was being sold by apothecaries in Europe as *Colchicum* in the 16th century, it was discovered to be *Colchicum autumnale* and its medicinal use for gout was realised. With colchicine available as pure chemical, its mode of action and its use in other diseases were identified. In the management of acute gout, it has been superseded by other less toxic medicines.

References

Institut de Cardiologie de Montreal *Colchicine reduces the risk of COVID-19 related complications* 23 January 2021. https://www.icm-mhi.org/en/pressroom/news/colchicine-reduces-risk-covid-19-related-complications accessed 1 June 2022

Levy M. Colchicine: A state of the art review. *Pharmacotherapy* 1991;11(3):196-211

Malkinson FD. Colchicine. New uses of an old, old drug. *Arch Dermatol* 1982;118(7):453-7

Paris JA. *Pharmacologia or the History of Medicinal Substances*, 4th edn. London, W. Phillips, 1820

Pharmacopoea Londinensis, A Translation of the Pharmacopoeia Collegii Regalis Medicorum MDCCCXXIV. London, Simpkin and Marshall (translation 'by a Scotch Physician residing in London'), 1824

Quincy J. *Pharmacopoeia Universalis Extemporanea or a Complete English Dispensatory*. London, A. Bell, 1718

Reyes AZ, *et al*. Anti-inflammatory therapy for COVID-19 infection: the case for colchicine. *Ann Rheum Dis* 2020;80:550-7

Sullivan TP. Colchicine in dermatology. *J Am Acad Dermatol* 1998;39:993-9

Wechalekar MD, *et al*. The efficacy and safety of treatments for acute gout: results from a series of systematic literature reviews including Cochrane reviews. *Rheumatol Suppl* 2014;92:15-25. doi: 10.3899/jrheum.140458

Whitla, W. *Elements of Pharmacy, Materia Medica and Therapeutics*, 4th edn. London, Henry Renshaw, 1887

Woodville, W. *Medical Botany* vols 1-3. London, James Phillips, 1790-93

CHAPTER 14 – MICHAEL DE SWIET

Digitalis purpurea
The source of digitoxin

Digitalis lanata
The source of digoxin

Digitalis purpurea, purple foxglove, and *Digitalis lanata*, woolly or Grecian foxglove, are now members of the Plantaginaceae family. The family is famous for being the source of a large number of different cardiac glycosides with actions on the heart, in particular two medicines, digitoxin from *D. purpurea* and both digitoxin and digoxin from *D. lanata*. Historically, the dried leaves of *D. purpurea* were important in treating cardiac dysfunction; currently only digoxin is used in the UK.

Plant profiles

One of our best loved and familiar wildflowers, the foxglove *Digitalis purpurea* is a biennial plant native to North Africa and Europe. Often glimpsed in woodland clearings, it colonises areas of open woodland, scrub and roadside verges. During the summer months, its tall spires of purple bell-shaped flowers stand 1-2m above a rosette of large coarse velvety leaves. The flowers may also appear in pink or white. It is hardy and will grow in sun but prefers partial shade and acid soil. It will sow itself around in a garden if the seed heads are left. If more control is desired, seed

Digitalis lanata, woolly foxglove

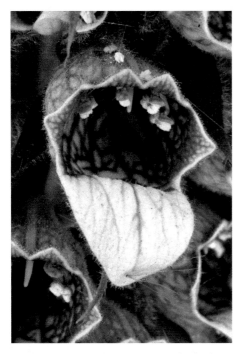

Digitalis lanata, woolly foxglove

Digitalis purpurea – woodcut from Parkinson's *Theatrum Botanicum*, 1640

can also be sown in late spring and plants grown on in a cold frame for planting out and flowering the following year. Pollinated by bees, it is a great food source for insects.

Digitalis lanata, the woolly or Grecian foxglove of Eastern Europe, is an evergreen biennial or short-lived perennial which is smaller, neater and altogether smarter looking than the common foxglove. Its lance-shaped leaves are narrower and darker and the flowers, held on a tall spike, are an elegant cream or pale yellow veined with brown. Less easy to grow than *D. purpurea*, it needs to be sown as it will not freely seed itself. It has similar cultivation requirements to *D. purpurea* but a greater desire for sun. It is drought-tolerant once established and does not like overly damp conditions.

History

Digitalis purpurea is such a poisonous plant that it never appeared in any herbal until the middle of the 16th century when Leonard Fuchs (1542) published an account and woodcuts of *Digitalis purpurea* and *Digitalis lutea* (now *D. grandiflora*). He noted that it was bitter to taste, and quoting Galen, said that this bitterness indicated it was a purgative.

Lyte (1578) and Gerard (1633) report that the leaves boiled in wine can be used to bring up phlegm and open the 'stoppings of the liver and spleen'. It is clear that neither Fuchs, Lyte nor Gerard had tried it. Parkinson (1640) reports its use, additionally, for healing wounds, scrofula and the treatment of epilepsy. L'Obel (1576) does mention vomiting and purgation, but it disappears from herbals and pharmacopoeias until the mid-18th century (James, 1752) when it is noted that 'the decoction of it purges very powerfully both upwards and downwards'.

Digitalis – modern uses

The use of the leaves of purple foxglove, *D. purpurea*, for treating heart failure was discovered by Dr William Withering. At that time heart failure was called 'dropsy' because swelling of the lower limbs was its most obvious feature.

Withering was born in Wellington, Shropshire and attended Edinburgh Medical School from 1762 to 1766. In 1767 he started as a consultant at Stafford Royal Infirmary. In 1775 he was appointed physician to Birmingham General Hospital at the suggestion of Erasmus Darwin, another physician. In 1785 he was elected a Fellow of the prestigious Royal Society and published his *Account of the Foxglove* where he reports on his work with *D. purpurea* in treating heart failure. He died on 6th October 1799.

Withering had been asked to comment on a herbal remedy that had been the property of an old woman and had decided that foxglove was the active ingredient. The story that it was the herbal mixture created by a 'Mother Hubbard' was an invention of the marketing department of the drug company, Parke-Davis, in 1928. Withering experimented with foxglove leaf on 156 of his charity patients over nine years (1776-1785). He gradually worked out that a small dose caused a diuresis and a large dose vomiting and that the toxic effects could accumulate with time if a non-toxic dose was given repeatedly. This is because the half-life of digoxin in the body is about 5 days and it is only slowly excreted. He found that the leaves varied in strength according to the time of year, and that dry leaf was easier to use than an infusion or decoction. It was effective for treating the oedema and breathing difficulties of heart failure. Withering recognised the other toxic side effects of foxglove, including xanthopsia (objects appear yellow), bradycardia (slow pulse), vomiting, diarrhoea and death (Withering, 1785).

Withering regarded *Digitalis* as a diuretic. He did not realise that the diuresis was a consequence of increased renal blood flow secondary to increased cardiac output. The concept of increased cardiac output would take another century to appear. However, many would consider Withering's critical approach to dosing and toxicology to be the foundation of modern drug research and treatment.

Erasmus Darwin (the grandfather of Charles Darwin) published an account in 1780 of nine patients treated by his recently deceased son (also Charles) who had been given *Digitalis* in the doses used by Withering, seven of whom had heart failure and three with atrial fibrillation. He did not mention Withering, the discoverer of this treatment, and this has been seen as a deliberate attempt to deny the fame to Withering. More critical commentators have accused Erasmus Darwin of 'intellectual dishonesty, plagiarism and rank opportunism' (Littler, 2019).

Digoxin

Digitalis purpurea (containing digitoxin) was used as *Dig. fol.* (*Digitalis folium*, tablets of dried *Digitalis* leaf) until digoxin was purified from the leaves of *D. lanata*, the woolly or Grecian foxglove of Eastern Europe. This was first achieved by Dr Sydney Smith at Burroughs Wellcome in Britain in 1930. This has the advantages of purity, and of not varying in strength according to the time of year.

Chemistry

Glycoside assay

For the dried leaves, pharmaceutical assays had to be done to assess glycoside (including digoxin) content using rabbits, and later with mice of standardised sex and weight, the fatal dose being an index of their potency. In the early 1960s the radioactive rubidium assay was introduced as more convenient and faster. It relied on the action of the glycosides in producing a dose-(concentration-)related inhibition of uptake of radioactive rubidium by red cells in a carefully standardised incubation medium.

This became unnecessary once purified digoxin was used and the dose could be measured in milligrams. However, one advantage of the plant extracts was that they did not cause poisoning as often as the purified chemical, because overdoses induced nausea and vomiting within minutes of ingestion, preventing the patient from consuming more.

It is possible to synthesise digoxin, but this is an expensive process and digoxin is still obtained from the plant. This is done by anaerobic fermentation of the leaves of chopped up *D. lanata* to liberate the digoxin which is then extracted with 10% alcohol in water in a percolator and purified to remove other cardiac glycosides.

Chemistry

Digoxin (and digitoxin) – mode of action

Digoxin (and digitoxin) exert their effects by inhibiting the ATPase activity of a complex of transmembrane proteins that form the sodium/potassium ATPase pump which normally pumps sodium) out of cardiac muscle cells (Fisner and Smith, 1991). Inhibition of this causes a rise of intracellular sodium, and the body responds by increasing intracellular calcium which, in turn, results in an increase in the force of heart muscle contractions. Thus, digoxin makes the heart contract more strongly.

Digoxin is still used in modern times for the treatment of heart failure and atrial fibrillation (Kelly & Smith, 1993). Atrial fibrillation is an abnormal heart rhythm in which the heart beats rapidly and chaotically – the pulse becomes 'irregularly irregular' – due to loss of conduction of the cardiac impulse between the atria and the ventricles.

However, digitoxin and digoxin have steep dose-response curves; thus, minute increases in the dosage of these drugs can make the difference between an ineffective dose, a therapeutic dose and a fatal one. There is also concern about the toxic effects which remain (nausea, vomiting, malaise and visual disturbance and include disorders of heart rhythm) which may also be fatal. Furthermore, there are now alternative and/or better forms of treatment for heart failure and atrial fibrillation. For these reasons, digoxin, once a routine part of treatment for heart failure, is becoming little used in modern therapeutics.

Early 20th century bottle of tincture of *Digitalis folium*, foxglove leaf, for treating heart failure. The dose was 1-3 grains (65-200mg)

Editor's note: Around a small hamlet in the Andes of Peru, miles from civilisation, roads, towns or medical care, I came across a wood with *D. purpurea* Excelsior Group (these have flowers arranged all the way round the stem, not just to one side) in full flower. They must have come from a seed packet from Suttons Seeds as *D. purpurea* in any form is not native to South America. The villagers knew that foxglove was 'good for the heart' and were planning to make foxglove soup but abandoned the idea when told of its toxicity. A story was in circulation in the 1980s of a young man whose mother had heart failure, unresponsive to medication. As a dutiful son, he brought her from her village in Eastern Europe, and took her round the cardiologists in Harley Street to be told that medication would not improve her condition. He made her foxglove tea using several leaves, a toxic dose. Older physicians remember the 1960s when *Dig. Fol.* (*Digitalis folium* – dried foxglove leaf) at a dose of 100-200mg per day was a therapeutic dose, i.e. a very, very, small amount of leaf was required. His 'home-brew' well-exceeded the known fatal dose of 2.5g (Martindale, 1967). A warning of the dangers of self-medicating with 'herbal remedies'.

References

Eisner DA, Smith TW. *The Na-K pump and its effectors in cardiac muscle*. In: Fozzard HA, *et al* (eds) *The Heart and cardiovascular system*, 2nd edn. New York, Raven Press, pp863-902, 1991

James R. *Pharmacopoeia Universalis*, 2nd edn. London, J. Hodges. 1752

Kelly RA, Smith TW. Digoxin in heart failure: implications of recent trials. *J Am Coll Cardiol* 1993;22:107A-112A

Littler WA. Withering, Darwin and digitalis. *QJM: An International Journal of Medicine* 2019; Volume 112, Issue 12

Martindale W. *Extra Pharmacopoeia*, 25th edn. London & Bradford, Percy Lund, Humphries & Co. Ltd., 1967

Thomson AT. *The London Dispensatory*. London, Longman, Hurst, Rees, Orme & Brown, 1811

Withering W. *An Account of the Foxglove*. Birmingham, M Swinney for GGJ and J Robinson, London, 1785

A therapeutic dose of foxglove leaf is about 2 sq.cm of a leaf

Portrait of Dr William Withering (1741-1799) holding a plant of foxglove, *Digitalis purpurea*

CHAPTER 15 – ANTHONY DAYAN

Dioscorea polystachya
Glycine max
The sources of diosgenin from which steroids were synthesised

Introduction

In the College Garden we grow the Chinese yam, *Dioscorea polystachya* (previously *D. batatas*), in the Dioscoreaceae family. *Dioscorea mexicana* and *D. composita* are more important as medicinal plants. We also grow soya, *Glycine max*, and they have all been starting points for the synthesis of steroids. Sisal agave, *Agave sisalana*, from the Americas and the leaves of jute, *Corchorus olitorius*, from south Asia and Africa are two further major sources of steroid production but are not grown at the College. They all contain chemicals called sterols which have molecules resembling steroid molecules. *Dioscorea* contains diosgenin, *Agave* and *Corchorus* contain hecogenin, and *Glycine max* contains stigmasterol. These sterols are extracted from the plant and converted into steroids. *Glycine max* is also a source of vitamin K (*q.v.* in appendix on vitamins).

Plant profiles

The *Dioscorea* that grow at the Royal College of Physicians are *D. polystachya* (previously grown as *D. batatas*) and *D. communis* (previously *Tamus communis*) but the latter is not used as a source of steroids.

Dioscorea polystachya, commonly known as Chinese yam, is a herbaceous perennial vine native to Central and South China, the Kuril Islands and Taiwan. It likes to grow on rocky slopes in the sunshine. It is an attractive, fast-growing plant which will rapidly climb a 3m frame and produce a large taproot. It has very glossy, dark green

Glycine max, soya bean

heart or fiddle-shaped leaves with beautiful parallel veining and in September it produces small, white, cinnamon-scented flowers. Although it may not survive in very exposed conditions, the plant is hardy in the UK where it will grow best in full sun. *D. polystachya* can be propagated from the small brownish bulbils which develop in the leaf axils but, as the plant is dioecious, both male and female plants need to be grown to produce seed.

The sisal, *Agave sisalana*, is native to Mexico. It is an evergreen plant in the Asparagaceae family with a rosette of sword-shaped leaves about 1.5-2m long. Young leaves have toothed margins. It is not currently grown at the Royal College of Physicians.

Edible soya, *Glycine max*, an annual in the Fabaceae family, grows in the Royal College of Physicians' Garden. Its natural habitat is the open woodland, grassland, scrub and verges of East Asia. The leaves are trifoliate and the flowers similar in appearance to those of sweet peas: they are white to violet or pink and in flower from July. This plant will tolerate a wide variety of growing conditions but does best in full sun and moist soil. In common with other members of the family, this species has a symbiotic relationship with certain soil bacteria which form nodules on the roots and fix atmospheric nitrogen.

Artist's impression of Dioscorides of Anazarbus (40-90 CE) from Elizabeth Blackwell's *A Curious Herbal* (1739)

History

Dioscorea commemorates Pedanius Dioscorides (70 CE), the Greek physician whose *Materia Medica* provided the framework for therapeutics for over 1500 years. A hugely successful genus, over 800 species are known, and widely cultivated. The tubers of various species have fed Africa and the Pacific regions for millennia, and continue to do so, with around 80 million tonnes produced annually, significantly contributing to the improvement in the nutrition of peasant farmers, and indirectly also to the world's population explosion.

They are used in traditional Chinese medicine to stimulate the stomach, spleen, kidneys and lungs. The sap is applied topically for snake bites and scorpion stings. The sap of species of wild yams known as 'bitter yams', that contain toxic alkaloids and oxalates, have been used as arrow poisons.

Glycine max, the soya of commercial agriculture, was developed by domestication of the wild soya, *Glycine soja*. Ancestors of *Glycine* have been cultivated in China since 5000 BCE. In the past 500 years their cultivation spread across South and Southeast Asia, and then into North America 250 years ago. In the last 30 years, most cultivated soya has been genetically engineered for resistance to the weedkiller Glyphos ['Roundup']. Some countries, mainly in the EU and Central Europe, forbid the sale of genetically engineered soya as a human food even though most permit it as an animal feed. Soya beans are poisonous if eaten raw (but not to cows) and must be cooked by boiling to provide a high-quality protein food for human consumption. They are widely used to make a multitude of products from soaps to plastics, and with its use as animal feed annual production is some 334 million tonnes. It is now being made directly into synthetic meat instead of feeding it to cows for meat production. Additionally, it is the primary source of biodiesel in the USA. Health benefits of a soya-rich diet are mainly inconclusive.

Steroid hormones – an overview

There are two groups of steroids that govern aspects of our bodies: corticosteroids which include glucocorticoids like cortisol and mineralocorticoids like aldosterone; and the 'sex hormones': oestrogen, progestogens and androgens such as testosterone. Cortisol is important for the control of blood glucose, the functioning of the immune system, coping with stress and for dealing with inflammation; it has some mineralocorticoid activity; aldosterone's function is in regulating the levels of salt and water in our bodies and consequently blood pressure. Both are made in the adrenal cortex. Testosterone, the principal hormone in the development of the male reproductive system and appearance, is produced in the testes. Oestrogen is the female equivalent; progesterone's role is in maintaining pregnancy and in some aspects of the menstrual cycle; both are produced in the ovaries.

It has taken many centuries to understand what steroids are and how they work. But this knowledge has advanced rapidly in the past century as a consequence of advances in physiology and chemistry. At the same time these hormones, and novel analogues with related properties, have become very important medicines for their normal, physiological actions and because higher doses may provide hoped for or sometimes unexpected therapeutic benefits.

Medicinal uses of steroids depend on the availability of plentiful supplies of the pure substances made into appropriate and convenient dosage forms. Early on, cortisone, a very close analogue of cortisol, was found to have strong anti-inflammatory and immunosuppressant actions, and more recent analogues, such as betamethasone, have exploited these in the local or systemic treatment of a wide range of diseases. As knowledge of the normal actions of the sex steroids advanced in the 1920s-1940s it was soon realised that female fertility could be changed and potentially controlled by appropriate treatment with these hormones. In time that led to the invention of the female oral contraceptive pill based on stable and potent oestrogenic and progestogenic compounds, which enables safe, temporary and reversible reduction in female fertility. This has caused fundamental, ongoing changes in the position and roles of women in the developing and developed worlds.

All these developments have depended on the availability of different steroids following discoveries of novel pathways of drug synthesis and the manufacture of medicines. Plants have been central in these academic and industrial discoveries because most steroid molecules are too complex to be made by conventional synthesis from simple starting materials.

The development of therapeutic steroids

The general nature of the cyclical control of ovulation in the female by oestrogen and progesterone, and of spermatogenesis in the male by the chemically related testosterone was realised by the late 1930s, and that pointed to the possibility of their control by administration of these hormones or related substances. The problem was that total synthesis of the natural hormones or reliance on their extraction from animal sources was not practicable. In the late 1930s-1940s adrenal corticosteroids and sex steroids were prepared from ovaries, testes or bile from animals being processed in industrial abattoirs, and involved a number of chemical steps to produce the desired steroids. It was a very expensive process, and steroids so produced were only available in limited quantities.

The first commercially successful production of progesterone, and later of testosterone and oestrogen, was based on new chemistry applied to diosgenin from an unusual species of Mexican yam, *D. mexicana*, popularly called *cabeza de negro* after its black, woody caudex (the expanded base of the plant) covered with striking polygonal plates. This is a relatively uncommon plant in parts of Mexico and Central America, where the fleshy tubers of wild specimens are sometimes eaten by locals. Although it has a relatively high concentration of diosgenin it was not an acceptable industrial source of diosgenin because it grew slowly, and large-scale cultivation proved unreliable. It was replaced by another yam that grew in Mexico – *D. composita*, locally called *barbasco* – that could be farmed industrially and contained a higher concentration of diosgenin.

This was done by Russell Marker, an unusual US organic chemist, who devised a theoretical route for the cheap, large-scale synthesis of progesterone in 1940-41. Several major US pharmaceutical companies refused to support his work on sex steroids because they did not wish to be involved with contraception, so he obtained local funding for his own small laboratory in Mexico City. In 1943, after successfully and cheaply preparing kilograms of progesterone, which had previously only been available in grams and at high cost, he, Carl Djerassi and two Mexican colleagues founded a new company, Syntex ('Synt' from synthesis and 'ex' from Mexico), to continue to develop and manufacture several steroid hormones. The chemical and commercial exploitation of *barbasco* was accomplished there, followed by other companies in Britain, Germany and the USA. Also, once the potent anti-inflammatory effect of cortisone became known in 1949 there was great interest in using it in much greater quantities than could be prepared in conventional ways, and in modifying its pharmacology to enhance desirable actions and suppress or minimise harmful effects. Brilliant work in the 1940s-1950s, much at Syntex, showed how to make large quantities of the natural steroids and chemical analogues with better properties, from plants.

These plant derivatives included converting stigmasterol, a waste product from soybean processing, which was converted into cortisone; and diosgenin from the Mexican yam and hecogenin from sisal and jute waste which were converted into progesterone and other sex steroids. Recycling these waste products into a myriad of pharmaceutical compounds has been a triumph for the pharmaceutical industry. Derivatives with better therapeutic properties were developed subsequently, such as the fluorocorticoids, and the contraceptive steroids mestranol and norethisterone. They have come into very wide medical and social use.

Plant-derived materials are still the basis of steroid industries in many countries. However, the modern trend for using genetically-engineered lower organisms for industrial production of chemicals is displacing both collection from the wild and the harvesting of cultivated plants.

Anabolic steroids and aldosterone are not derived directly from plants, and some of the former were made entirely synthetically, particularly in 'sport' laboratories where the aim was to make chemicals which would help athletes win competitions rather than look primarily for a therapeutic benefit. This type of doping, which carries serious risks to the health of the athlete, is illegal but remains uncomfortably common.

Summary

The yam, *Dioscorea*, followed by soya, *Glycine*, provided the food for the population explosion in Africa and the Pacific region, and then (ironically) the contraceptive pill and other steroids, and in so doing they changed society and medicine for ever.

Reading list

The history of the development of knowledge about steroids, their roles in health and diseases and their exploitation as medicines is complex and quite technical in places. For those reasons instead of references to publications by individuals and groups of discoverers, papers are noted here that review broader aspects of the history of this group of compounds from an industrial and pharmaceutical viewpoint.

American Chemical Society 1999. The 'Marker Degradation' and Creation of the Mexican Steroid Hormone Industry 1938-1945. https://www.acs.org/content/dam/acsorg/education/whatischemistry/landmarks/progesteronesynthesis/marker-degradation-creation-of-the-mexican-steroid-industry-by-russell-marker-commemorative-booklet.pdf accessed 10 March 2021

Blackwell E. *A Curious Herbal*. London, John Nourse, 1739

Burns CM. The History of Cortisone Discovery and Development. *Rheumatol Clinics N America* 2016;42:1-14

Colton FB. Steroids and Research at Searle: Early Research at Searle. *Steroids* 1992;57:624-630

Davis KS. The Story of the Pill. *American Heritage Magazine* 1978;29:1-6

Dhont M. History of Oral Contraception. *Eur J Contraception Reprod Health Care* 2010;15(S2):S12-S18

Djerassi C. Steroid research at Syntex: 'The Pill' and cortisone. *Steroids* 1997;57:631-41

Freeman ER, Bloom DA, McGuire EJ. A Brief History of Testosterone. *Amer J Urol* 2001;165:371-3

Herráiz I. *Chemical Pathways of Corticosteroids, Industrial Synthesis from Sapogenins*. In: Barredo J-L, Herráiz I (eds.) *Microbial Steroids: Methods and Protocols, Methods in Molecular Biology*. New York, Springer Science, pp15-27, 2017

Herzog H, Oliveto EP. A history of significant steroid discoveries and developments originating at the Schering Corporation (USA) since 1948. *Steroids* 1992;57:617-23

Hirschmann R. The cortisone era: aspects of its impact. Some contributions of the Merck Laboratories. *Steroids* 1992;57:579-92

Hogg JA. Steroids, the steroid community, and Upjohn in perspective: a profile of innovation. *Steroids* 1992;57:593-616

Islam MM. Biochemistry, Medicinal and Food values of Jute (Corchorus capsularis L. and C. olitorius L.) leaf: A Review. *Int J Enhanced Res Science, Technology & Engineering* 2013;2:35-44

Miller WL. A brief history of adrenal research. Steroidogenesis – The soul of the adrenal. *Molec Cellular Endocrinol* 2013;371:5-14

Newerla GJ. The History of the Discovery and Isolation of the Female Sex Hormones. *New Engl J Med* 1944;230:595-604

Nieschlag B, Nieschlag S. The history of discovery, synthesis and development of testosterone for clinical use. *Eur J Endocrinol* 2019;180:R201-R212

Patrick G. *History of Cortisone and Related Compounds*. In: eLS. Chichester, John Wiley & Sons, Ltd., 2013. doi: 10.1002/9780470015902.a0003627.pub2

Quirke V. Making British Cortisone: Glaxo and the development of Corticosteroids in Britain in the 1950s–1960s. *Stud Hist Philosoph Biol Biomed Sci* 2005;36:645-74

Rasmussen N. Steroids in Arms: Science, Government, Industry, and the Hormones of the Adrenal Cortex in the United States, 1930-1950. *Medical Hist* 2002;46:299-324

Slater LB. Industry and Academy: The Synthesis of Steroids. *Historical Studies in the Physical and Biological Sciences* 2000;30:443-80

Zaffaroni A. From paper chromatography to drug discovery: Zaffaroni. *Steroids* 1992;57(12):642-8

The first commercially successful production of progesterone and later of testosterone and oestrogen was based on new chemistry applied to diosgenin from an unusual species of Mexican yam, *D. mexicana*, popularly called *cabeza de negro* after its black, woody caudex (the expanded base of the plant) covered with striking polygonal plates. This is the caudex of *Dioscorea elephantipes*.

CHAPTER 16 – NOEL SNELL

Ephedra sinica
The source of ephedrine, pseudoephedrine and amphetamines

Introduction

Ephedra sinica, a Chinese shrub, was used at least 5000 years ago for the treatment of cough and to induce sweating. In more recent times one of the active principles, ephedrine, has been used in the treatment of asthma, and another, pseudoephedrine, as a nasal decongestant. Ephedrine has also been used to enhance alertness and as an aid to weight loss. Active derivatives of ephedrine include the amphetamines (including Ecstasy).

Plant profile

It is in the family Ephedraceae and evolved around 123 million years ago, with fossils found in the Early Cretaceous period, and extant *Ephedra* arising 8-32 million years ago. It forms an evolutionary link between conifers and flowering plants.

Ephedra gerardiana flower

Members of the genus *Ephedra* form a tangle of thin-jointed branches that appear to be leafless but in fact the tiny reddish-brown sheaths around the stems at each node are leaves. The flowers which appear in early summer are also very small; female plants have small green, globular flowers and fleshy red fruits – unlike *Hippuris* and *Equisetum*.

The species is dioecious (single-sexed) so both male and female plants must be grown if seed is required. A more efficient way to propagate this plant is by division as the mature plant sends out

Ephedra gerardiana fruit, red in colour as described by Pliny, 70 CE

deep underground rhizomes which develop new stems and can be separated to multiply stocks. An adaptation to dry environments, this habit can cause it to be problematic in a border where it can spread unnoticed and emerge where it is not welcome.

Ephedra sinica is a small evergreen perennial shrub, a native of South Siberia through to North and Northeast China where it is found growing in desert, waste land and sandy places on plains and on mountain slopes. Other species such as *E. gerardiana* named for James Gilbert Gerard (1795-1835), an army surgeon in India, who collected plants in the Himalayas, and *E. distachya*, are also grown in the Garden of the Royal College of Physicians.

Nomenclature and botany

Ephedra comes from the Greek word for mares' (or horses') tails, *ĕphĕdra*. *Sinica* means Chinese. *Hippuris* and *Equisetum* are two unrelated genera, meaning mares' tails in Greek and Latin respectively, that are unrelated to *Ephedra*. Because of their similarity of form, in particular their jointed, leafless stems (the leaves being reduced to vestigial scales) they were confused by our ancestors – knowing this is fundamental in understanding *Ephedra*'s medicinal use in the past two millennia.

History

Ephedra sinica (known as *Ma-huang*) was included in a catalogue of herbs by the Emperor Shen Nung, c. 2700 BCE, and over 4000 years later was described in the dispensatory of Li Shih-Chen, the *Pen-ts'ao Kang Mu*; it was used as a circulatory stimulant and in the management of cough and fevers. The medicinal use of *Ephedra* plants is reported to be in the Egyptian Ebers papyrus in 1550 BCE, but the authors have not been able to substantiate this. In India *E. gerardiana* and *E. intermedia* were used as central nervous system stimulants during religious ceremonies.

The history of its use is poorly recorded in European literature. It appears in Dodoens (1583) in considerable detail, but not his earlier work (1551), as *Ephedra sive* [or] *Anabasi*, called *Caucon* by Pliny but distinguished by having fleshy red fruits located round the joints of the stems, so indicating it is *Ephedra* and not *Equisetum*, horsetail, which does not have fruits. He gave no illustration. Gerard's *Herbal* (1633) has put it back into *Equisetum* as a tenth variety of horsetail: Parkinson (1640) illustrates it, titled *Equisetum Montanum Creticum*, mountain horsetail of Candy [Crete] – now known as *Ephedra foeminea*. He writes that it is said to grow to the height of a plane tree but more commonly to that of 'shorter and lesser trees and shrubs'. His use of 'it is said to grow' suggests that he had not seen such a plant, for no European *Ephedra* grow that tall. No medicinal uses are given. It is, however, in Dioscorides [70 CE] and Pliny [70 CE] that the earliest European references to its current use as a styptic can be found.

Pliny, in Book 26, chapter 7, writes: 'The Herb *Caucon*, also called *Ephedra* (and here the translator has made a marginal note 'By these names he calleth also Horsetail'), and by some *Anabasis*, groweth ordinarily in open tracts exposed to the wind … leaves it hath none but is garnished by a number of hairs … [and is] full of joints and knots. Let this herb be beaten to powder and given in red wine … it is good for the cough, shortness of wind and the wrings of the belly'. In his chapter on horsetails (Book 26, chapter 13), he continues about *Ephedra* that the juice sniffed up the nose stops heavy nosebleeds; stops coughs and also shortness of breath in those who 'are forced to sit upright for to draw his breath' – usually a symptom of heart failure. Among other virtues he describes it as a cure for inguinal hernias and to stop bleeding including heavy periods and bloody diarrhoea.

Dioscorides, calling it *Ippouris*, *Anabasion* and *Ephudron*, says that 'being beaten small and sprinkled on, it closeth bleeding wounds'. While its habitat, according to Dioscorides, in 'moist places and ditches', have led translators to regard it as an *Equisetum*, its pharmacological properties of vasoconstriction and bronchial dilatation indicate it is *Ephedra*.

14. *Equiſetum montanum Creticum.* Mountaine Horſe taile of Candy.

Ephedra – woodcut from Parkinson's *Theatrum Botanicum*, 1640

Galen [200 CE] calling it *Hippuris* and *Cauda equina* (Latin for horse tail) gives the uses that Dioscorides attributes to *Ephedra*. It is not recorded in the European pharmaceutical literature from the 16th until the end of the 19th century.

Native Americans used local *Ephedra* species (*E. antisyphilitica*, *E. californica* and *E. nevadensis*), applied topically or taken as a tea, for the treatment of syphilis and other disorders. This misplaced belief in the plant's activity against syphilis persisted and, as 'whorehouse tea', it used to be served in brothels in the Western USA. Infusions of *Ephedra* were also consumed by Mormons in the USA as it was not forbidden to them by their religion, unlike drinks containing caffeine. *Ephedra antisyphilitica* and most other North and South American species are thought not to contain ephedrine or pseudoephedrine.

Pharmacology of ephedrine and its analogues

Ephedrine was first isolated in 1885, in an impure form, by a Japanese scientist, G Yamanashi, working in Osaka. Pure crystals of ephedrine were subsequently obtained by another Japanese worker, N Nagai, who elucidated its structure and was the first to synthesize it. A collaborator, K Muira, studied its pharmacology and found that it dilated the pupil (mydriasis) when applied topically. The racemic mixture of the synthetic product was patented by Nagai as methylmydriatin in 1918.

Ephedrine was used as a mydriatic agent which had the advantage over atropine (*q.v.*) of a shorter duration of action; it was marketed by Merck as Mydrin. However, its other potential uses were not pursued at the time, as it was considered too toxic to the circulatory system. This was despite the fact that two more Japanese researchers, H Amatsu and S Kubota, reported in 1913 that the effects of ephedrine were essentially similar to those of adrenaline, including a bronchodilator effect on the airways; as a result of their work, an ephedrine-containing medicine named Asthmatol was marketed in Manchuria for asthma, but was not commercially successful.

In 1923 two researchers at the Peking Union Medical College, Ku Kuei Chen and Carl Schmidt, decided to investigate the properties of *Ma-huang* (*Ephedra sinica*) which had been included in a catalogue of herbs by the Emperor Shen Nung, c. 2700 BCE. At the time they were unaware of the previous isolation of ephedrine and the investigations into its pharmacology, as the few published reports were in Japanese or German. They identified ephedrine and in a series of studies showed that it caused a rise in the heart rate and blood pressure, and constriction of the peripheral arteries, very like the effects of adrenaline (to which it is structurally similar) but more prolonged; ephedrine also had the advantage of being absorbed orally, as adrenaline has to be given by injection or inhalation (Chen & Schmidt, 1924). It is now known that in addition to its direct effects on the sympathetic nervous system, ephedrine also has indirect effects due to enhanced release of the transmitter norepinephrine (noradrenaline) from sympathetic neurones. Local vasoconstriction is the basis for its decongestant effects when applied topically to the nasal mucosa, and also for its use in prolonging the action of local anaesthetics, for example in dental surgery.

Ephedrine was then studied in asthma and allergic rhinitis (hay fever) by Leopold and Miller in Philadelphia; although the trials were not placebo-controlled they showed clinical benefit in the majority of cases (Leopold & Miller, 1927). The use of inhaled ephedrine in the UK was first reported by a London general practitioner, Percy Camps (1929). Ephedrine was approved for use in the USA and became very widely used for asthma, to the extent that exports of the plant from China to the USA increased substantially. In 1926 one pharmaceutical manufacturer attempted to corner the market in Chinese *Ephedra*, leading to temporary shortages which were overcome by increased exports of related species from India. Synthetic methods of producing ephedrine have since been described and instituted.

About this time, an allergy specialist in Los Angeles, George Piness, asked his laboratory team to see if a cheap substitute for ephedrine could be found. They synthesised several active compounds, eventually coming up with an orally active substance, amphetamine. It had previously been synthesised in 1897, and its pharmacology studied by Barger and Dale in 1910, but not further pursued. The compound was taken on by Smith, Kline & French who placed its volatile free base in a plastic inhaler and marketed it in 1932 as the 'Benzedrine Inhaler' for the relief of nasal congestion, which proved very profitable, probably because of its stimulant effects.

The dextrorotatory stereoisomer of amphetamine, dexamphetamine, had enhanced stimulant properties and was marketed as Dexedrine in 1935, for narcolepsy. During the Second World War dexamphetamine and a more potent analogue methylamphetamine (Methedrine) were issued to combatants to heighten alertness and avert fatigue. During the post-war period the amphetamines became widely abused and were gradually withdrawn from the market. Another derivative, 3,4-methylenedioxymethamphetamine, better known as Ecstasy, 'E', or 'Molly', was used in the 1970s as an adjunct to psychotherapy and became popular as a recreational drug in the 1980s; it has stimulant and psychedelic effects and is often handed round at rave parties. Its use is both illegal and dangerous, as hyperthermia and dehydration can develop after a single dose; cases of fatal hyponatraemia (low blood sodium) have also occurred. The active principle, cathinone, of another recreational psychostimulant, khat, (the chewed fresh leaves of the shrub or small tree, *Catha edulis*), is structurally very similar to ephedrine and amphetamine. It is widely grown and used in the Horn of Africa and Yemen.

Ephedrine itself remained a popular oral treatment for asthma into the 1940s and 1950s, often in combination with caffeine or theophylline (*q.v.*). Its use declined as more specific sympathomimetic agents such as salbutamol, administered from pressurised metered-dose inhalers, were developed.

Chemistry

Ephedrine – mode of action

Ahlquist, in 1948, classified the sympathetic nervous system receptors into alpha and beta types, and concluded that ephedrine acted at both, explaining its cardiovascular effects in addition to its action as a bronchodilator; salbutamol and its analogues act more specifically on the beta-2 receptor subtype and have fewer cardiac effects.

'Following administration, ephedrine activates post-synaptic noradrenergic receptors. Activation of alpha-adrenergic receptors in the vasculature induces vasoconstriction, and activation of beta-adrenergic receptors in the lungs leads to bronchodilation' (PubChem).

Ephedrine is still available as a nasal decongestant (as is pseudoephedrine, Sudafed), and has a place in the treatment of low blood pressure resulting from spinal and epidural anaesthesia. Herbal products containing *Ephedra* can still be found and have been widely used to increase energy levels and promote weight loss. However, there is a high risk of side effects; a study in 2003 in the USA found that products containing *Ephedra* represented only 0.82% of herbal product sales but accounted for 64% of all reported adverse reactions (Bent *et al*, 2003). Chronic use of both ephedrine and the amphetamines can lead to habituation and the need for higher doses to achieve the desired effects, with an increased risk of toxicity. The use of ephedrine and *Ephedra*-containing substances is generally banned by sports associations. Adverse effects of ephedrine and its analogues include tachycardia, hypertension, nausea, headache, anxiety, insomnia, psychosis, salivation, sweating, and difficulty passing urine; overdoses can lead to cardiac failure and fatal arrhythmias, stroke, hyperthermia, and convulsions. Ephedrine is secreted in breast milk and may cause irritability and sleep disturbance in infants.

Pseudoephedrine is produced by fermenting dextrose in the presence of benzaldehyde.

It is interesting that the use of *Ephedra* as a vasoconstrictor and bronchodilator was known 2000 years ago but disappeared from our knowledge only to be rediscovered in the early 20th century, by which time the identification of *Ephedra* from similar plants had been well established. It continued to be a useful and widely employed medicine for 70 years, until its toxicity was better appreciated, and its uses restricted.

References

Abourashed E, El-Alfy A, Khan I, Walker L. Ephedra in perspective – a current review. *Phytother Res* 2002;17:703-12

Bent S, Tiedt N, Odden M, Shlipak M. The relative safety of Ephedra compared with other herbal products. *Ann Intern Med* 2003;138:468-71

Chen K, Schmidt C. The action of ephedrine, the active principle of the Chinese drug Ma Huang. *J Pharmacol Exper Therapeut* 1924;24:339-57

Leopold S, Miller T. The use of ephedrine in bronchial asthma and hay-fever. *JAMA* 1927;88:1782-86

PubChem. Ephedrine https://pubchem.ncbi.nlm.nih.gov/compound/Ephedrine accessed 29 April 2021

Further reading

Lee M. The history of Ephedra (Ma-huang). *J R Coll Physicians Edinb* 2011;41:78–84.
doi: 10.4997/JRCPE.2011.116

'Mare's tails', *Equisetum*, have the same segmented stems as *Ephedra*, but are a much older family and not related. This is an 8 metre tall *Equisetum giganteum* in Peru.

CHAPTER 17 – ARJUN DEVANESAN

Erythroxylum coca
The source of cocaine

Introduction

Erythroxylum coca, the coca bush, is a vital part of Aymara Andean culture, and the whole plant is the source of the stimulant and local anaesthetic cocaine – the only naturally occurring local anaesthetic. It is the latter effect that led to cocaine's most significant contribution to modern surgery and anaesthesia, though it was its stimulant and addictive effects that eventually led to its decline in medicinal use.

It is illegal to grow *Erythroxylum coca* in the UK without a licence from the Home Office. It is in the Erythroxylaceae, a family closely related to the mangroves, Rhizophoraceae.

Plant profile

Erythroxylum coca is an evergreen shrub from the north of South America where it grows up to 3m tall in full sun on very acid soils. It is raised from seed, except for *E. coca* var. *ipadu* which does not set seed and, unlike var. *coca*, is widely and easily cultivated from cuttings.

History

The Aymara people of Peru tell the story of Kuka, a woman of such extraordinary beauty that none could resist her. She used her power to seduce innumerable men who, growing jealous, appealed to the Great Inca. Kuka was sacrificed, cut in half and buried, and from her grave grew a small, bright green shrub. When its leaves were chewed it imbued the chewer with miraculous vigour and health and so it was named Coca. Mama Kuka is now revered as a goddess of the Incas, associated with health and joy. Despite its notoriety in modern society, coca leaves are an integral part of Andean culture, still consumed today either by chewing or in tea, though illegal in most of South America. They are used in rituals, to relieve pain and altitude sickness, and as a general stimulant. Chalk, usually carried in a hollowed-out gourd, is added when chewing the leaves to increase their maceration and extraction of the cocaine.

Coca has been cultivated in South America for thousands of years. There are 259 species in the genus *Erythroxylum*, two of which, *E. coca* and *E. novogranatense*, are the principal sources of cocaine. At the height of the Inca empire, vast amounts of coca were farmed and used for everything from the highest religious ceremonies, divination and medicine to recreational consumption amongst the Inca elites. Towards the end of the Inca empire, restrictions on consumption were relaxed until the Spanish Conquest of 1532. Garcilaso de la Vega (1503-1536), the first great poet in the Golden Age of Spanish literature, recorded the following conversation between two Spaniards, a gentleman and a farmhand near Cusco, Peru. The first asked, 'Why do you eat coca, like the Indians do, when Spaniards find it so disgusting and detestable?' The other, who was carrying his two-year-old daughter on his back, replied 'In truth, sir, I detest it no less than anyone, but need forced me to imitate the Indians and chew it. Without it I would not be able to bear the burden. With it I have strength and vigor to be able to undertake my labors' (National Geographic, 2016).

Cristóbal de Molina, a Spanish priest who lived in Cusco shortly afterwards, and one of the first missionaries, watched and described Inca ceremonies. They burned leaves and blew coca fumes toward the sun, they drank coca tea and fell into trances, they gave it to their sick and buried it with their dead (Molina, 1572). Some rituals involved human sacrifice, and the ubiquitous coca was used there as well. In order to convert the Incas to Christianity, the priesthood decided that coca had to go. The colonists who exploited the Inca for their labour, however, had other ideas. For a start, they realised that coca increased productivity in their workers and reduced general malcontent. At first, this was put down to Inca superstition, and in 1653 Father Cobo wrote that the Inca 'say that [coca] gives them strength, and they feel neither thirst, hunger nor tiredness. I think it is mostly superstition, but one cannot deny that it gives them strength and breath, as they work twice as hard with it'. The Spanish soon turned coca into a lucrative crop which made its way to the West in the 19th century (National Geographic, 2016).

Coca in Europe

In 1859 in Vienna, Albert Niemann had isolated the active ingredient in coca leaves, which he named cocaine. Niemann was a student of the great chemist Wöchler and obtained some coca leaves which Dr Scherzer had collected on his Peruvian voyage on the Austrian frigate Novara. He and his student Wilhelm Lossen are credited with much of the early work on cocaine chemistry though a number of other German and Austrian scientists including Heinrich Wackenroder and Friedrich Gaedecke were the first to produce an active extract. Niemann noted: 'Its solutions have an alkaline reaction, a bitter taste, promote the flow of saliva and leave a peculiar numbness, followed by a sense of cold when applied to the tongue'. Being part of the Austrian scientific intelligentsia of the time, this knowledge made its way to Sigmund Freud who widely promoted cocaine as a medicinal panacea in 1884 in his scientific paper, *On Coca*, and other papers on its clinical use. He also suggested the use of cocaine on 'the diseased eye' to one of his students, though without a thought for its anaesthetic properties. Freud's admirers often attribute the discovery of cocaine's local anaesthetic properties to him, but it was Niemann who first noted this property, and Carl Koller (1884) who first suggested its use as a local anaesthetic for the eye (Seelig, 1941).

The consumption of coca in Europe remained largely benign up to the 1880s. A popular mode of consumption was coca wine, the most famous of which was Vin Mariani. While the formula was proprietary, the French government published a method for preparing coca wine so that any pharmacist could produce it. It was made from Bolivian leaf (*Erythroxylum novogranatense*) and contained on average 150-300mg of cocaine per litre of wine. Semi-refining cocaine was discovered in 1885 by a chemist working for Parke-Davis (Merck's fiercest competitor). Coca leaves lose their potency in transit, but after 1885 degradation in transit was greatly reduced, prices fell, and the availability of semi-refined cocaine increased markedly. Merck was Europe's main cocaine producer, and in 1883 its total output was about 1.65kg. This increased to over 70 tons by 1886. *Erythroxylum coca* contains at least fourteen distinct alkaloids, most of which are eliminated in the process of semi-refinement, leaving mainly the soon-to-be-discovered white crystalline cocaine.

Vin Mariani was imitated all over the world, and famous fans included Queen Victoria, HG Wells, Jules Verne, Ibsen, Zola, the singer Sarah Bernhardt, Thomas Edison and even two Popes (Pius X and Leo XIII). In America, John Pemberton experimented with cocaine to cure his morphine addiction (which began after his return from the Civil War). However, prohibition loomed large and so during 1886 Pemberton developed an alcohol-free preparation using coca leaves. After accidentally mixing the concoction with carbonated water and deeming it a success, Pemberton marketed Coca-Cola as an alternative to ginger ale and tonic for 'nervous trouble, dyspepsia, mental and physical exhaustion, all chronic and wasting diseases, gastric irritability, constipation, sick headache, neuralgia'. In 1904 only coca leaves from which cocaine had already been extracted were used in the manufacture of Coca-Cola, and this popular drink became free of cocaine.

Cocaine as a local anaesthetic

It was the Peruvian surgeon, Moréno y Maïz, who performed the first studies on animals using cocaine as a nerve-blocking anaesthetic, paving the way for its use as a regional anaesthetic for surgery. In 1880, Basil von Anrep published his studies on humans and in the conclusion recommended cocaine as a surgical anaesthetic. However, it was ultimately when Carl Koller demonstrated the use of cocaine as a topical anaesthetic for cataract surgery that it entered into widespread use. Partly because cataract surgery had been agonising before cocaine anaesthesia, partly because Koller was part of a circle of highly influential Viennese doctors and scientists, and finally because the newly invented telegraph could send medical reports around the world instantaneously, the world of medicine was interested in cocaine as a local anaesthetic within a few months.

Following Koller, William Stewart Halsted and Richard John Hall developed cocaine as an injected local anaesthetic for dental surgery. News of Koller's discovery, presented at the Heidelberg Congress of Ophthalmology, travelled to America by telegraph and was published in the New York Medical Records in 1884 to be picked up by Halsted, a successful surgeon in New York. That said, he also regularly visited Vienna and so may have heard about the local anaesthetic effects of cocaine directly from Koller. Conveniently, the hypodermic syringe had also been recently invented, allowing Halstead and Hall to inject cocaine into the brachial plexus and posterior tibial nerves for upper and lower limb surgeries. Halsted acquired a 4% solution of cocaine from Parke-Davis but quickly noticed undesirable side effects (these include cardiac irregularities, psychiatric disturbances and addiction) at this concentration and so diluted it to use in lower doses which also proved effective.

Addiction and side effects

The side effects had, by this time, become a regular feature in the *British Medical Journal* and *The Lancet*. Driven by the availability of semi-refined cocaine and the hypodermic syringe, an epidemic of cocaine addiction had already begun. Freud himself spent years recovering from cocaine addiction. Many of the surgeons who pioneered early clinical work on cocaine in surgery also became addicted. For Halsted this was to be the end of his career. He began to display erratic social and professional behaviour and was eventually admitted to hospital several times. William Welch, a friend and colleague of Halsted, managed to rehabilitate him and found him a place at Johns Hopkins Hospital in Baltimore working in Occupational Therapy, though this was not to last. In 1969, a sealed wax box was opened at Johns Hopkins Hospital containing a note written by William Osler detailing how Halsted eventually cured his addiction to cocaine by replacing it with morphine. It was this drug that eventually led to his death in 1922. In modern medicine, cocaine is still sometimes used in nasal surgery as a constituent of Moffett's solution. Its general use as a local anaesthetic, however, has largely disappeared and has been replaced by safer modern alternatives (*q.v. Hordeum jubatum* re lidocaine).

Most people these days will know of cocaine as a drug of abuse rather than as a medicine. This change in identity largely happened in the 1980s with the invention of crack cocaine – a free base form of cocaine which can be smoked. In the past, an average cup of coca tea would have contained about 5mg of cocaine whereas a single line of cocaine that is snorted recreationally is usually 20-30mg and the usual maximum safe dose for local anaesthetic use was 3mg/kg (210mg for a standard 70kg adult). 1g of cocaine is usually lethal if there is no prior history of cocaine use but may not be if tolerance has built up. Cocaine tolerance develops rapidly, especially with crack cocaine which reaches the brain more rapidly and at higher concentration. Acute cocaine toxicity is a common reason for hospitalisation and occurs at variable doses depending on prior use. In 2011, cocaine was the most common recreational drug to result in hospital admissions in the United States with 505,224 emergency department visits reported by Substance Abuse and Mental Health Services Administration (Richards & Le, 2020). It decreases cardiac function and increases excitability of the cardiac and central nervous system – often leading to fatal heart rhythm disturbances, agitation, hallucinations, seizures, strokes and heart attacks. Chronic cocaine toxicity also causes a number of long-term effects, most notably changing the muscular architecture of the heart making it more prone to fatal electrical disturbances, and changing the walls of blood vessels which makes people more prone to heart attacks and strokes.

Terminal care – the Brompton cocktail

Long after cocaine was replaced by procaine (synthesised in 1899) as the local anaesthetic of choice, it was still widely administered orally in combination with morphine for post-operative pain and pain management in terminally ill cancer patients. This mixture was known by a number of names – most famously as the Brompton cocktail after its origins at the Brompton Hospital – and achieved almost mythical status as a palliative remedy until its eventual demise following the work of Robert Twycross in the 1970s (Twycross, 1977). This was partly due to the rapid development of tolerance to cocaine which meant that after the first week or so it was only the morphine and diamorphine in the mixture that had any effect. In line with the general phasing out of cocaine in medical practice, the Brompton mixture is also no longer used.

References

Koller C. Vorläufige Mitteilung über locale Anästhesierung am Auge. *Klin. Mbl. Augenbeilk* 1884;22:Beilageheft,60-63

López-Valverde A, De Vicente J, Cutando A. The surgeons Halsted and Hall, cocaine and the discovery of dental anaesthesia by nerve blocking. *British Dental Journal* 2011;211:485-7

MacCallum WG. *Biographical memoir of William Stewart Halsted* 1852–1922. National Academy of Sciences of the United States of America Biographical Memoirs, pp151-70, Volume XVII – Seventh Memoir. Presented to the Academy at the Autumn Meeting, 1935

March SB. Cocaine: history, use, abuse. *J R Soc Med* 1999;92:393-7

National Geographic. Coca: a Blessing and a Curse. https://www.nationalgeographic.com/history/magazine/2016/11-12/daily-life-coca-inca-andes-south-america/ accessed 16 July 2022

National Institute on Drug Abuse. What is the scope of cocaine use in the United States? 2016. https://nida.nih.gov/publications/research-reports/cocaine/what-scope-cocaine-use-in-united-states accessed 16 July 2022

Richards JR, Le JK. *Cocaine Toxicity*. StatPearls Publishing, 2020

Seelig MG. History of cocaine as a local anesthetic. *JAMA* 1941;117(15):1284

Twycross RG. Value of cocaine in opiate-containing elixirs. *British Medical Journal* 1977a;2:1348

Euphorbia peplus
The source of ingenol mebutate

Introduction

Euphorbia peplus, commonly known as common or petty spurge, milkweed, radium weed and cancer weed, has an irritant, toxic, white milky sap which is the source of ingenol mebutate, a topical agent used until recently for the treatment of actinic keratosis, an early superficial precursor of a skin cancer. It was withdrawn from use because of serious side effects. It is one of the few herbal medicines whose use for treating skin tumours continued for two millennia and which eventually made a modern pharmaceutical with the same actions. It is in the Euphorbiaceae family.

Plant profile

Euphorbia peplus is a familiar and opportunistic weed in English gardens. It is a very small, glabrous, hardy annual, widely distributed throughout Europe, the Mediterranean and temperate Asia where it can be found anywhere on cultivated arable land, in gardens and on other disturbed soil.

Smooth, pale stems support the lime-green flowerheads which are arranged in three rayed umbels above an axil of whorled leaves. The inflorescence is known botanically as a cyathium, where a cup-like group of modified leaves encloses a single female flower and several male flowers. There are no petals or sepals in the flower which does, however, contain conspicuous nectar-producing glands.

Nomenclature and historical medicinal uses

The common name, spurge, is derived from the Old French *espurge*, due to the former use of the plant's very irritating and toxic milky sap as a purgative. This sap has been used for two millennia to burn off small warty tumours on the skin. The generic name *Euphorbia* is derived from Euphorbus (c. 10 BCE-20 CE), the Greek physician to the Berber King Juba II of Numidia and Mauritania (50 BCE-23 CE). Euphorbus was the brother of Caesar Augustus's physician, Antonio Musa, who is commemorated in the genus and family in which bananas belong – *Musa* and Musaceae (Stearn, 1994).

PEPLVS.

Euphorbia peplus from Mattioli's *Discorsi*, 1568

King Juba II was interested in plants and, according to Pliny [70 CE], gave *Euphorbia* its name in 12 BCE. In the 16th century it was commonly known as *Tithymalus*, but Linnaeus restored the name *Euphorbia* in 1753. One story is that Euphorbus cured King Juba from a stomach condition with the sap of *Euphorbia*. The other is that *Euphorbia regis-jubae* was discovered by the king in the Atlas Mountains of Morocco and that, because Euphorbos was a fat man and the plant was fleshy, the king thought it an apt name as *euphorbus* is derived from the Greek 'eu' meaning good and 'phorbe' meaning fodder.

Dioscorides [70 CE] lists several different spurges, with their corrosive, burning, milky sap and warns of the dangers and precautions to take to avoid contact with the skin while processing them. All are purgative and emetic, and pills made of the sap, leaves, fruit or roots were coated with wax or oil to prevent them burning the throat when swallowed, or the juice was included in figs – so being one of the earliest medicines to have an enteric coating. The sap was used to burn off warts, cancers and boils, and to treat fistulas. Despite his warning to avoid touching one's eyes after handling it, Dioscorides recommends the sap for treating the membranes that grow over the eyes – not something even to consider trying.

Galen [200 CE] also described the properties of the plant as burning, caustic and thinning, and that the sap was similar. Apuleius [6th century] used it in combination with other medications for toothache, stomach-ache, warts, fistulas and skin lesions. Mesue [1100] describes it as hot and dry in the fourth degree (a strength rarely equalled and never exceeded) and he too lists a multiplicity of uses for nervous complaints, epilepsy, arthritis and more, repeating Dioscorides' use for eyes. Jacques Dubois, in his translation of Mesue (1553), comments that mixed with tragacanth and mastic it was a treatment for the Italian disease (syphilis). These were still recorded by Pemel (1652) with warnings that it must be given in combination with other plants to reduce its ferocity, never used on eyes and only given to strong healthy men, so anticipating the fate of ingenol mebutate which very early on was noted by dermatologists to cause incredibly painful reactions.

Gerard (1597) had a bad personal experience with the sap of *E. peplus*, not heeding the warnings of his forebears: 'walking along the seacoast at Lee in Essex, I took but one drop of it into my mouth, which nevertheless did so inflame and swell in my throat that I hardly escaped with my life'. Only on reaching a nearby farmhouse was he able to obtain some milk 'to quench the extremitie of our heate, which then ceased'. He had not read Pliny's warning 1500 years before: 'Tast it never so little at the tongues end, it setteth all the mouth on a fire, and so it continueth a long time hot … until at the end it parcheth and drieth the chaws [jaws] and throat also far in'.

Editor's note: An eight-year-old grandson of one of the authors was advised in his woodcraft class that gently tasting dog's mercury, *Mercurialis perennis* (Euphorbiaceae), for bitterness would indicate if it was edible. He experienced the same fate as Gerard.

Gerard recounts the diverse uses of spurge other than as a purgative, 'the juice or milke is good to stop hollow teeth, being put into them warily so that you touch neither the gums nor any of the other teeth in the mouth'. The toxic latex probably destroyed the nerve endings in the roots of the teeth and so helped resolve toothache. He also advised 'the same cureth all roughness of the skin, manginesse, leprie, scurfe and running scabs, and the white scurfe of the head. It taketh away all warts, knobs and hard callousness of fistulas, hot swellings and carbuncles'. He concluded by advising, 'these herbes by mine advice would not be received into the bodie'.

Culpeper (1649) described several species of spurge and commented on the purgative qualities in all of them. Describing petty spurge, he wrote, 'The whole plant is full of a caustic milk, burning and inflaming the mouth and jaws for a great while together. It is a strong cathartic, working violently by vomit and stool, but is very offensive to the stomach and bowels, by reason of its sharp corrosive quality, and therefore ought to be used with caution'.

The medicinal uses of spurge continued as an emetic, purgative, and treatment for warts, boils, fistulas and ulcers into the 18th century, and as a cutaneous blistering agent (when this was thought to be a useful treatment). As a folk medicine it was used to remove warts if the scabs that formed were continuously removed (Quincy, 1718). In the mid-20th century one of the authors (HFO) found this ineffective, although his brother was successful with the sap of dandelions (*Taraxacum officinale*).

The *Euphorbia* of the pharmacists in the 19th century was imported into Britain and its identity unknown until 1863 when it was discovered to be *E. resinifera* grown in Morocco. But by then its medicinal use had become obsolete and it was only used as a paint for preservation of the bottom of ships (Flückiger & Hanbury, 1874).

Contemporary medicinal use

Euphorbia's move back into pharmacy began with 20th century observations in Australia, where sun damage to the skin is very prevalent. Home remedies for both actinic keratosis, one of the earliest visible manifestations of chronic sun damage, and basal cell carcinoma were well recognised. One of the most popular of these was the sap of *E. peplus*, which was widely available, being an abundant weed. In keeping with most of the 2000 species in this genus, its irritant milky sap contains a diverse mixture of toxic terpenoids which have almost certainly evolved as a defensive mechanism against insect predators and fungal pathogens (Ogbourne & Parsons, 2014).

The possibility that the latex sap contained an important ingredient for the treatment of skin cancer was brought to the attention of Prof Peter Parsons of the Queensland Institute of Medical Research, Brisbane, in 1996 by Dr James Aylward, who claimed his family had used it successfully for self-treatment of skin cancer since the mid-1900s. They were convinced of its efficacy with good cosmetic results. Despite initial scepticism, Prof Parsons' interest was provoked, and he proceeded to investigate the action of the sap on a number of skin tumour cell lines *in vitro*. Growth inhibition was observed in all the skin tumour cell lines but not in normal fibroblasts. A phase I/II clinical trial of the sap applied topically on three different skin cancers had favourable results (Ramsay *et al*, 2011). This encouraged Parsons to scale up the investigation which was passed onto a new biotechnology company called Peplin which was created for the purpose in 1998. By 2000, sufficient funds had been raised for full-scale isolation, purification, development and production of the active ingredient which came to be known, in 2004, as ingenol mebutate. Dr Aylward's unwavering confidence that the research programme would lead to the isolation of an active ingredient led him to establish cultivated crops of *E. peplus* in good time so that sufficient raw material would be available for the production process once the research was complete.

Laboratory research showed that the active ingredient extracted from *E. peplus* had equal anti-cancer bioactivity whether obtained from fresh or dried material. It was shown in early clinical use to induce necrosis (cell death) in skin cancer cells within hours of application, by interfering with mitochondrial function.

Chemistry

Mode of action

Probably more important in the longer term for its clinical effect on skin cancer, was the massive acute inflammatory response where applied, leading to a neutrophil-mediated, antibody-dependent cellular cytotoxicity, which caused further cell death of residual deep-seated skin cancer cells which had not been killed by the initial application. Experimental mice depleted of circulating neutrophils failed to respond to application of ingenol mebutate gel to xenograft tumours.

Its speed of effect when applied topically allowed it to be used once daily for only three days in clinical use and this showed 100% effectiveness in clearing actinic keratoses in the earliest clinical trials. This is in sharp contrast to existing topical therapies which are applied for at least four weeks.

Commercial production

The international dermatology company Leo Pharma acquired Peplin in 2009 and obtained approval for ingenol mebutate gel from the US Food and Drug Administration (FDA) in 2012 under the trade name Picato as a new treatment for actinic keratosis on the face, scalp, trunk and extremities. This was followed shortly by approval in Europe and the UK. Actinic keratosis tends to be seen much more commonly on elderly facial skin which has undergone age-related thinning as well as sun damage. This makes it especially vulnerable to the inflammatory effects of ingenol mebutate gel, and reactions in some cases were severe. Even so, it had a high success rate in clearing actinic keratoses from the skin, and the inflammation subsided spontaneously within a few weeks. Great care was needed to avoid the skin close to the eyes when applying the gel, as contact can lead to severe ocular inflammation.

Serious side effects and withdrawal

By 2019 reports were appearing of skin tumours developing in areas of the skin treated previously with ingenol mebutate gel, including varieties of non-melanoma skin cancer such as keratoacanthoma, basal cell carcinoma and squamous cell carcinoma, neuroendocrine tumours and atypical fibroxanthoma. This led to a safety review by the European Medicines Agency which decided that the risks of ingenol mebutate gel outweighed its benefits and the medication was withdrawn in February 2020 in the UK and Europe. It was subsequently withdrawn in Canada and by October 2020 the manufacturers, Leo Pharmaceuticals, announced that they were permanently withdrawing the product worldwide.

Several ingenol esters are known to be potent tumour promoters in experimental rodents, and ongoing surveillance of patients who have been treated with ingenol mebutate gel will be required for many years.

Summary

For two millennia, since Dioscorides, the sap of *E. peplus* had been used to treat warts and cancers, and this belief has persisted into the folklore of modern times and been confirmed in clinical trials. That it could be very painful was not a new observation, but follow-up when its active chemical was extracted and used revealed that the treatment was worse than the disease it treated as it induced other skin cancers.

References

Flückiger F, Hanbury D. *Pharmacographia. A History of the Principal Drugs of vegetable origin met with in Great Britain and British India*. London, Macmillan and Co., 1874

Ogbourne SM, Parsons PG. The value of nature's natural product library for the discovery of New Chemical Entities: The discovery of Ingenol Mebutate. *Fitoterapia* 2014;98:36-44

Quincy J. *Pharmacopoeia Universalis Extemporanea or a Complete English Dispensatory*. London, A. Bell, 1718

Ramsay JR, Suhrbier A, Aylward JH, *et al*. The sap from Euphorbia peplus is effective against human nonmelanoma skin cancers. *Br J Dermatol* 2011;164:633-6

Stearn WT. *Dictionary of Plant Names for Gardeners*. London, Cassell Publishers Limited, 1994

The cut leaf of *Euphorbia peplus* weeps corrosive and toxic sap

Gerard (1597) had a bad personal experience with the sap of *E. peplus*. He wrote that:

'walking along the seacoast at Lee in Essex, I took but one drop of it into my mouth, which nevertheless did so inflame and swell in my throat that I hardly escaped with my life'. Only on reaching a nearby farmhouse was he able to obtain some milk 'to quench the extremitie of our heate, which then ceased'.

Pliny (c. 70CE) wrote:

'Tast it never so little at the tongues end, it setteth all the mouth on a fire, and so it continueth a long time hot ... until at the end it parcheth and drieth the chaws [jaws] and throat also far in.'

CHAPTER 19 – SUSAN BURGE

Galanthus nivalis
The source of galanthamine/galantamine

Introduction

Galanthus nivalis, snowdrops, are familiar winter-flowering bulbous perennials with small white nodding flowers. In common with several other members of the amaryllis family, Amaryllidaceae, *G. nivalis* contains galanthamine, a chemical that increases the level of acetylcholine in the brain. Acetylcholine is a chemical messenger (neurotransmitter) that enables the brain to function. In the UK, galantamine is the name for the oral, 'Prescription Only' medicine licensed for use in mild to moderate Alzheimer's disease, a common form of dementia in which reduced levels of acetylcholine in the brain may contribute to difficulties such as loss of memory.

Daffodils (*Narcissus*) and snowflake (*Leucojum*) are other plant sources of galanthamine/galantamine.

Nomenclature

Galanthamine is the name of the chemical recognised by the International Union of Pure and Applied Chemistry. It comes directly from the genus name. Galantamine is the official pharmacopoeial name recognised by regulatory agencies worldwide.

Plant profile

Galanthus nivalis is a dwarf, bulbous, clump-forming perennial found in the moist woodlands and riverbanks of Central and Southern Europe through to Western Asia and the Caucasus. It is not native to the UK but it has been so long cultivated that it has become naturalised here.

It is one of the most popular of all cultivated bulbous plants, no doubt because the flowers emerge in winter: droplet-shaped, white tipped with green, often nodding above the snow and with a delicate honey scent. Numerous selected forms and named cultivated varieties are now grown by devotees of the genus who even have their own name: galanthophiles.

It needs partial shade and humus-rich, moist but well-drained soil that does not dry out in summer. A top dressing of leaf mould is recommended. *Galanthus* species hybridize readily in gardens and seed may not come true. To increase, lift and divide clumps of bulbs as soon as the leaves begin to die down after flowering.

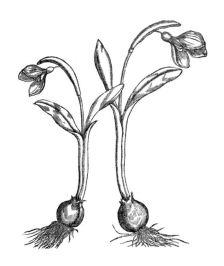

Woodcut of *Galanthus nivalis* from Mattioli's *Discorsi*, 1568

Historical names and uses

The history of the medicinal use of *Galanthus* is difficult to trace as it was classified as *Narcissus* which, prior to the 16th century, contained tulips, daffodils, snowdrops and snowflakes (now *Tulipa*, *Narcissus*, *Galanthus* and *Leucojum*, respectively). Mattioli (1569) illustrates them with woodcuts and refers the reader to Pliny [70 CE] who has two 'daffodils' called *Narcissus*, one of which is purple (presumably tulips), the other 'grass green' which causes vomiting and stupefaction. Snowdrops are white and green and Turner (1568), with a woodcut of a *Leucojum*, believes this is what Pliny is referring to, and that its common English name was *Laus tibi* (Latin for 'Glory to you'). In the same chapter Pliny refers to the 'flower of the white violet to wit the bulbous stock Gillofre … [and] yellow wallflowers' that induced menstruation and were diuretic. Neither stock gillyflowers (*Matthiola*) nor the similar-looking wallflowers (*Erysimum*) are bulbous, and this appears to be continuing a name, *Viola alba*, possibly meaning snowdrops, used by Theophrastus [c. 350 BCE].

Lyte (1578), following Dodoens (1554), calls snowdrops *Leucoion bulbosum triphyllum* (*triphyllum* meaning three petals) which flowers in February and *Leucojum* as *Leucoion bulbosum hexaphyllum* (six petals) which flowers in April, and says they have no medicinal value. He says they are Narcissus violets, the *Viola alba* of Theophrastus, and should be called, in English, white bulbous violets. The names bulbous violet and *Leucoion bulbosum praecox minus Byzantinus* continue into the 17th century (Parkinson, 1629) and even later, as *Narcisso-Leucoium* (Brookes, 1772). It gained a local, English name of summer fools in Gerard's *Herbal* of 1597 – a translation from the Dutch name – and the name 'snow drops' (sic) appears for the first time in Thomas Johnson's edition (Gerard, 1633).

No medicinal value was attributed to it until the mid-20th century, but its confusion with other Amaryllidaceae had led it to be given the properties of daffodil bulbs. It never entered the pharmacopoeias in any form and is not mentioned by Culpeper (1649).

It was left to Carl Linnaeus, the Swedish botanist, in 1753, to finally separate it from *Leucojum* and name the snowdrop as *Galanthus nivalis*, 'snowy milk flower', from the Greek – *gala* meaning milk, and *anthos* meaning flower, and the Latin – *nivalis* meaning snowy. He does not include it in his

Materia Medica (1749, 1782). Snowdrops were being cultivated in gardens in Britain in the 1500s and wild snowdrops were first recorded here in the 1770s, but the drifts of snowdrops in our woodlands must have originated from garden escapes. Bulbs may have been brought to Britain in the 15th century by Italian monks and introduced into the gardens of monasteries.

Legends and superstition

Snowdrops, the fair maids of February, Candlemas bells, feature in many myths, legends and superstitions. Snowdrops may have played a role in protecting Odysseus from the magic of Circe – see below after the section on acetylcholine in Alzheimer's disease. The white flowers are associated with purity, innocence and renewal. Large colonies of snowdrops are found in monastic sites across Britain and, in the Catholic Church, snowdrops are regarded as a symbol of Candlemas (2nd February), the Feast of the Purification of the Virgin Mary. Victorians believed snowdrops must never be brought into the house for that would bring ill-fortune or even death. But in the language of flowers, the snowdrop symbolises hope: spring is round the corner.

Drifts of *Galanthus nivalis* in Welford Park, a monastic grange until 1536

Discovery of galantamine

Galantamine was first discovered in *G. nivalis* by the Russian scientists Proskurnina and Areshknina in 1947, who also established its chemical structure in 1952 using *G. woronowii*, a species of snowdrop native to the Caucasus. According to unconfirmed reports (Heinrich, 2010), the potential medicinal value of these snowdrops was noted in the 1950s by a Bulgarian pharmacologist, who observed villagers from the Ural Mountains rubbing ground-up snowdrop flowers into their skin to relieve pain, especially headaches.

The structure of galantamine was established from *Galanthus woronowii* in 1952 (photo by Susan Burge)

The Russian pharmacologist Mashkovsky investigated galantamine isolated from *G. woronowii* and discovered its mode of action (see scientific paragraph later). By the late 1950s it had been shown that galantamine was present in other members of the Amaryllidaceae family and could be extracted from the leaves and bulbs of the common snowdrop, *G. nivalis*.

Chemistry
Role of acetylcholine in the brain
Acetylcholine is a chemical in the nervous system which functions as a neurotransmitter, a 'chemical messenger' that enables nerve cells (neurons) to communicate with each other. Processes such as moving, learning, thinking, and forming new memories all depend on nerve impulses being passed from one neuron to another using neurotransmitters such as acetylcholine. Neurons release neurotransmitter molecules at specialised structures, synapses, that bring part of one neuron very close to part of another. The neurotransmitter diffuses across the tiny gap and binds to specific receptors on the adjacent neuron, altering its electrical activity. Acetylcholinesterase plays a critical role in acetylcholine-mediated neurotransmission (cholinergic signalling) by rapidly destroying the liberated acetylcholine to keep the signal brief so that the neuron can respond to the next release of neurotransmitter. In 1951, the Russian pharmacologists Mashkovsky and Kruglikova-Lvova demonstrated that galantamine is an anticholinesterase, inhibiting the enzyme acetylcholinesterase thus preventing the breakdown of acetylcholine (Heinrich, 2010).

Acetylcholine in Alzheimer's disease
Alzheimer's disease, the commonest form of dementia, causes a progressive decline in brain function. The disease is associated with abnormalities of cholinergic signalling pathways as well as death of neurons. Galantamine, as an anticholinesterase, stops acetylcholinesterase from destroying acetylcholine so that levels of acetylcholine in the brain rise and neurotransmission improves. Galantamine has only a modest effect on mental functions such as learning and memory in Alzheimer's disease, but it and other acetylcholinesterase inhibitors may help behavioural symptoms by improving attention and concentration (O'Brien *et al*, 2017). In the UK, galantamine – as the pharmaceutical drug, galantamine – is licensed for use in mild to moderate Alzheimer's disease. Galantamine may also be helpful in Lewy body dementias and Parkinson's disease dementia, especially for associated neuropsychiatric symptoms such as hallucinations, apathy, anxiety and sleep disorders.

Homer, Odysseus and Circe
Homer, writing in the 8th or 7th century BCE, may have provided the earliest description of

the medicinal use of galantamine in the Greek epic poem The Odyssey. Homer describes how Odysseus and his crew visit the island of the sorceress Circe. Circe laid out a feast for the starving exhausted crew, but she had poisoned the food. Homer tells us that the crew forgot who they were and were transformed into swine. Bodily transformation is implausible, but swine-like behaviour is certainly possible if Circe's potion contained a plant in the deadly nightshade family (*q.v. Atropa*), the Solanaceae. These plants have high levels of alkaloids (atropine and scopolamine) that block the action of acetylcholine, leading to amnesia, vivid hallucinations and bizarre actions. Two millennia ago, Pliny [70 CE] wrote that such a hallucinogenic plant, the mandrake, *Mandragora officinarum* (*q.v.*) was called Circeium, so this belief is not new. The unfortunate victim might indeed believe he had become a pig. According to Homer, Odysseus did not suffer the same fate and was able to rescue his crew because the god Hermes had given him a herb 'Moly' that had a 'flower like milk' to protect him from Circe's magic. Was Moly a snowdrop? Could the action of galantamine have protected Odysseus from the effects of Circe's potion by increasing the levels of acetylcholine in his brain (Plaitakis & Duvoisin, 1983)? However, Homer's plant is regarded in 16th and 17th century herbals, and by Linnaeus, as wild garlic (*Allium ursinum*), named *Moly Homericum* in Gerard's *Herbal* (1633), but snowdrops would at least have worked.

Toxicity

The adverse effects of galantamine include dizziness, nausea, vomiting, abdominal pain, diarrhoea, muscle spasms and a slow heart rate. Snowdrop bulbs, sometimes eaten in mistake for spring onions, may be fatal. Snowdrops are equally toxic to dogs.

Other plant sources of galantamine

Although galanthamine/galantamine was originally isolated from *G. nivalis* and later from *G. woronowii*, the alkaloid has been obtained from *Narcissus* (daffodil) and *Leucojum*, especially *L. aestivum* (summer snowflake or Loddon lily) and, in China, from *Lycoris*, a relative of the daffodil. Since the 1990s, galantamine has been synthesized industrially (Heinrich, 2010; Keglevich *et al*, 2016) but it is expensive and difficult to make. Extracting it from cultivated daffodil bulbs and leaves is regarded as the only viable source of galantamine apart from *Lycoris* in China.

Extracts from many species of plant are being investigated in the search for new, clinically effective inhibitors of acetylcholinesterase (Dos Santos *et al*, 2018). Traditional medicines such as *Angelica gigas* (Korean angelica or female ginseng) are also being examined. The dried roots have been used in Korean traditional medicine for disorders ranging from constipation and irregular menstruation to joint pains and headache. Bioactive compounds in *Angelica gigas* have anticholinesterase activity, and extracts of the plant have been shown to improve spatial learning and memory in mice with mild cognitive impairment (Kim *et al*, 2019).

Conclusion

More work is needed to find treatments that will alleviate dementia, a life-changing disease with a huge psychological and emotional impact. The snowdrop family, Amaryllidaceae, has already played some part in offering hope for the future.

References

Brookes R. *The Natural History of Vegetables* vol 6, 2nd edn. London, T. Carnan & F. Newbery, jun., 1772

Dos Santos TC, Gomes TM, Pinto BAS, *et al*. Naturally occurring acetylcholinesterase inhibitors and their potential use for Alzheimer's disease. *Front Pharmacol* 2018;9:1-14

Heinrich M. Galanthamine from Galanthus and other Amaryllidaceae – Chemistry and biology based on traditional use. *The Alkaloids. Chemistry and Biology* 2010;68:157-165

Keglevich P, Szántay C, Hazai L. The chemistry of galanthamine. Classical synthetic methods and comprehensive study on its analogues. *Mini Rev Med Chem* 2016;16:1450-61

Kim M, Song M, Oh HJ, *et al*. Evaluating the memory enhancing effects of Angelica gigas in mouse models of mild cognitive impairments. *Nutrients* 2019;12:1-12

O'Brien JT, Holmes C, Jones M, *et al*. Clinical practice with anti-dementia drugs: A revised (third) consensus statement from the British Association for Psychopharmacology. *J Psychopharmacol* 2017;31:147-168

Plaitakis A, Duvoisin RC. Homer's moly identified as Galanthus nivalis L.: physiologic antidote to stramonium poisoning. *Clin Neuropharmacol* 1983;6:1-5

Angelica gigas contains bioactive compounds that improve memory in mice

Narcissus pseudonarcissus. Cultivated daffodil bulbs are the main source of galantamine for the pharmaceutical industry

CHAPTER 20 – MICHAEL DE SWIET

Galega officinalis
The source of metformin

Introduction

Galega officinalis, goat's rue, is a plant from which the medicine metformin has been derived; the latter is very widely used for the treatment of diabetes mellitus (diabetes), a metabolic disease associated with high blood glucose levels.

Plant profile

Galega officinalis is a fast-growing herbaceous perennial in the Fabaceae (pea) family. It is found throughout Central and Southern Europe across to Western Pakistan in dry grassland, scrub, meadows, open woodland and along roadsides.

The pretty flowers, clustered on upright racemes, have the characteristic pea-like form of this family and can be white, lilac or purple in colour. They bloom from mid to late summer and form a frothy mass above the large green pinnate leaves.

The plant is rather sprawling, adapted as it is to growing through and amongst scrubby vegetation. In a garden situation, staking the clump of floppy, metre-long stems is recommended in order to prevent it overwhelming its neighbours.

Nomenclature

Galega officinalis, a legume in the Fabaceae family, was formally named by Linnaeus (1753). The etymology of the genus name is unknown although it may derive from the Greek word for *milk*, *Gala* (Stearn, 1996), and was first used by Mattioli in the early 16th century. *Officinalis* is a name given to pharmaceutical plants.

Its common names include goat's rue, French lilac, holy hay, Italian fitch, Spanish sanfoin, false indigo and professor-weed and, historically, *Ruta capraria*. It is called goat's rue because *capraria* is the Latin for 'pertaining to goats', and was initially regarded as a variety of rue, *Ruta*, in the family Rutaceae.

Historical medical use

Galega officinalis first appears in the works of Mattioli as Galega sive [or] *Ruta capraria* [rue pertaining to a goat]. Dioscorides [70 CE], writing about the unrelated rue, *Ruta sylvestris* (now called *R. montana*) and *R. sylvatica*, says that that these plants are a cure for numerous different aches and pains and snake bite, but on the skin they cause blistering. The latter confirms its affinity to *R. graveolens*, rue, which has the same photosensitising effect.

In his Discorsi (1568), after describing the *Ruta* of Dioscorides, Mattioli says that recently he had found another species of *Ruta*, growing on the banks of ditches, which had been called *Capraria* or *Galega*, which was wonderful against the plague, and that many people had been saved by eating it regularly in salads or soups, and that it was also a treatment for epilepsy, intestinal worms and rashes. In his *De Plantis Epitome* (1581) he adds that it is an ingredient in theriacs (multi-ingredient cure-alls). The woodcuts of Mattioli clearly show the petalled flowers of *R. montana* and *R. sylvatica*, very similar to the well-known rue, *Ruta graveolens*. The very different leguminous flowers of his woodcut of *Galega* are a good representation of *G. officinalis*. It is not possible to trace *Galega* before Mattioli as neither *Galega* or *Capraria* appear in Dioscorides, Pliny or Avicenna and it is not to be found in Matthaeus Sylvaticus' massive compilation of medicinal works, the *Pandectarum Medicinae* (1524).

Later in the 16th century *G. officinalis* appears as *Galega* or *Ruta capraria*, with the English name of goat's rue, as a treatment against all venom and poison, and for intestinal worms, given as a spoonful of sap every day to children to prevent epilepsy, and, boiled in vinegar and taken with a theriac (any antidote to poison), used 'to heale the plague if taken within 24 hours' (Lyte, 1578; L'Escluse, 1557; Dodoens, 1583). Collectors of snakes in Piedmont used it to protect themselves against snake bites (L'Obel, 1576).

John Parkinson (1640) gives many more uses for '*Galega* or *Ruta capraria*', goat's rue, regarding it as a panacea against poisons that was better than the multi-ingredient theriacs. He adds that it was just as useful a medicine for goats (and sheep).

It is in the pharmacopoeia of the College of Physicians (1618), and its uses there are given as '...resists poison, kills worms, helps the falling sickness [epilepsy], resisteth the pestilence' by Culpeper (1649). Modern assertions that its use in diabetes was recommended by Culpeper are not to be found in his writing. It was still in the College pharmacopoeia in 1716 with the same uses. Elsewhere it was decried as being useless and most of its reputed medicinal properties mere fables (Quincy, 1718) and a 'useless ingredient in milk water' [a medicine – a galactagogue – for stimulating lactation] (Alleyne, 1733). However, it was still in Linnaeus's *Materia Medica* as *Galega*,

Nel terzo lib. di Dioscoride.

GALEGA OVERO RVTA CAPRARIA.

Galega officinalis – woodcut from Mattioli's *Discorsi*, 1568

a plant used by the apothecaries, with its properties given as being sudorific (to induce sweating) and to treat the plague, rashes and bites (Linnaeus, 1749, 1782).

No-one records any anti-diabetic action of *Galega* extracts until the 21st century when it is claimed that it was so used by Nicholas Culpeper (Bailey, 2017). Certainly, none of the editions of Nicholas Culpeper's publications recommend it, and the nearest he gets to diabetes and goats is claiming that the ashes of a burnt goat's bladder, if eaten, 'helped the diabetes or continued pissing'. This was a remedy for bed wetting according the book *On Simple Medicines* attributed to Dioscorides (Fitch, 2022). The story of Galega/goat's rue being used for treating diabetes comes from *Culpeper's Colour Herbal* (Potterton, 1983) where its modern use in the synthesis of metformin is being referred to and has been inserted into Culpeper's text. It is an important reminder that 'source' literature and not the compilations of other authors should always be sought.

Throughout the 16th to 20th centuries, the original uses as given by Mattioli are repeated with increasing scepticism, although occasionally – if mentioned at all – reporting historic use and saying that it was seldom used.

Treatment for diabetes – discovery of metformin

There are two main forms of diabetes, type 1 and type 2. In type 1 diabetes, which usually presents in young people, there is an absolute deficiency of insulin. Type 2 diabetes presents in older people and is strongly associated with obesity. Type 2 diabetes is caused by insulin resistance rather than by absence of the hormone. Insulin resistance occurs when cells in muscles, fat and liver do not respond well to insulin and cannot easily take up glucose from the blood.

Galega officinalis came back into medicine in the early part of the 20th century when it was shown that guanidine, a chemical found in *G. officinalis*, could improve the symptoms of diabetes mellitus (Watanabe, 1918; Sterne, 1969; Beckman, 1971) even though it was too toxic for human use.

Tanret (Simonnet & Tanret, 1927) identified another alkaloid from *G. officinalis*, galegine (isoamylene guanidine), that was less toxic (Muller 1927), but trials in the 1920s and 1930s in patients with diabetes were unsuccessful (Witters, 2001; Bailey *et al*, 2007). Two synthetic diguanides, decamethylene diguanide (Synthalin A) and dodecamethylene diguanide (Synthalin B) were used clinically in the 1920s, but concern about toxicity and lack of effect by comparison with insulin led to their discontinuation. Jean Sterne, a French diabetologist, also conducted studies with galegine and its derivatives. In 1956, at the Hôpital Laennec in Paris, he explored the anti-diabetic properties of the biguanides including dimethylbiguanide, metformin, that has made such a huge impact in the treatment of diabetes (Sterne 1957). However, the chemically similar phenethylbiguanide, phenformin, while effective was withdrawn because of side effects, including lactic acidosis, a potentially fatal metabolic complication.

Chemistry

Mode of action

Guanidine, the diguanides (Synthalin A & B) and the biguanides (metformin and phenformin) act by decreasing insulin resistance and by suppressing glucose production by the liver (hepatic gluconeogenesis) (Davis 2006). These actions are likely to be mediated at least in part by the activation of AMP-activated protein kinase (AMPK), an enzyme that plays an important role in insulin signalling.

Metformin, introduced to the UK in 1958, Canada in 1972, and the United States in 1995 is now believed to be the most widely prescribed anti-diabetic drug in the world; in the United States alone, more than 59 million prescriptions were filled in 2011 (Institute for Healthcare Informatics, 2011). In addition, metformin has particular application in women's health both for the treatment of diabetes in pregnancy and more generally for metabolic syndrome (see below).

Metformin is now synthesised entirely artificially. The final stage of this synthesis is the reaction of dimethylamine with cyanoguanidine. Both of these compounds can be made from a simple inorganic source of carbon, such as coke, together with water, limestone, and nitrogen (The Science Snail, 2000).

Diabetes and pregnancy

Type 1 diabetes is a dangerous complication of pregnancy and causes birth defects and the development of very large fragile babies which can die before or soon after birth. It must be treated with insulin. In addition, a form of type 2 diabetes, gestational diabetes, can develop specifically because of the insulin resistance induced by pregnancy. The main risk is that the babies will be very big. Metformin was originally used in South Africa for the treatment of pregnancy diabetes in the 'Cape Coloured' population *i.e.* persons of mixed race or Khoisan descent (Coetzee & Jackson, 1979). More recently a multi-centre international controlled trial has shown that metformin significantly reduces the pregnancy complications associated with gestational diabetes (Rowan *et al*, 2008). In view of the increasing prevalence of obesity and type 2 diabetes, this is a study of considerable importance and has led to increased use of metformin in pregnancy diabetes.

Insulin resistance and obesity have complications other than diabetes. Metabolic syndrome refers to a condition with obesity, diabetes, hyperlipidaemia, vascular disease and, in women, polycystic ovarian disease. In this condition, multiple large cysts form in the ovaries and can result in infertility because of failure of ovulation. Metformin treatment has been shown to reverse polycystic ovaries and restore fertility, though this effect is disputed (Lord *et al*, 2003; Tang *et al*, 2009).

Herbal use and toxicity

It is surprising that *G. officinalis* is said to be a galactagogue in animals, as it is listed as a Class A Federal Noxious Weed in 35 states of North America, and appears on the database of poisonous plants, causing pulmonary oedema, hypotension, paralysis and death in cattle. As little as 0.2% of body weight of the dried plant can be fatal to sheep (Colorado State University Guide to Poisonous Plants).

Galega officinalis does not have a licence as a herbal medicine from the MHRA (Medicines and Healthcare products Regulatory Agency) for sale in the UK due to its toxicity and lack of evidence that it stimulates lactation.

References

Alleyne J. *A New English Dispensatory*. London, Thos, Astley and S. Austen, 1733

Bailey CJ. Metformin: Historical overview. *Diabetologia* 2017;60:1566-76

Bailey CJ, Campbell IW, Chan JCN, *et al* (eds). *Metformin: the Gold Standard. A Scientific handbook*. Chichester: Wiley, 2007

Beckman R. *Biguanide (Expermenteller Teil)*. In: Maske H (ed) *Handbook of experimental pharmacology* 29. Berlin, Springer Verlag, pp439-596, 1971

Coetzee E, Jackson W. Metformin in management of pregnant insulin-dependent diabetics. *Diabetologica* 1979;16:241-5

Colorado State University. Guide to Poisonous Plants https://csuvth.colostate.edu/poisonous_plants/ accessed 29 December 2020

Davis SN. *Insulin, Oral Hypoglycemic Agents, and the Pharmacology of the Endocrine Pancreas*. In: Brunton L, Lazo J, Parker K. *Goodman & Gilman's The Pharmacological Basis of Therapeutics*, 11th edn. New York, McGraw-Hill, 2006

Fitch JG. *On Simples, attributed to Dioscorides. Introduction, Translation, Concordances*. Leiden, Koninklijke Brill, 2022

Institute for Healthcare Informatics. The use of medicines in the United States: Review of 2011. April 2012. https://www.kff.org/wp-content/uploads/sites/2/2012/10/ihii_medicines_in_u.s_report_2011.pdf accessed 15 January 2022

Lord JM, Flight IHK, Norman RJ. Metformin in polycystic ovary syndrome: systematic review and meta-analysis. *BMJ*. 2003;327(7421):951-3

Muller H, Rheinwein H. Pharmacology of galegin. *Arch Exp Path Pharmacol* 1927;125:212-28

Potterton D. (ed.) *Culpeper's Colour Herbal*. London, W. Foulsham & Co., 1983

Quincy J. *Pharmacopoeia Universalis Extemporanea or a Complete English Dispensatory*. London, A. Bell, 1718

Rowan JA, Hague WM, Gao W, *et al*. MiG Trial Investigators. Metformin versus insulin for the treatment of gestational diabetes. *N Engl J Med* 2008;258(19):2003-15

Simonnet H, Tanret G. Sur les propietes hypoglycemiantes du sulfate de galegine. *Bull Soc Chim Biol* Paris 1927;8

Stearn WT. *Stearn's Dictionary of Plant Names for Gardeners*. New York, Sterling Publishing Co. Inc., 1996

Sterne J. Du nouveau dans les antidiabetiques. La NN dimethylamine guanyl guanide (N.N.D.G.). *Maroc Med* 1957;36:1295-6

Sterne J. *Pharmacology and mode of action of the hypoglycemic guanidine derivatives*. In: Campbell GD (ed) *Oral hypoglycemic agents*. New York, Academic Press, pp193-245, 1969

Tang T, Lord JM, Norman RJ, *et al*. Insulin-sensitising drugs (metformin, rosiglitazone, pioglitazone, D-chiro-inositol) for women with polycystic ovary syndrome, oligo amenorrhoea and subfertility. *Cochrane Database Syst Rev* 2009;(4):CD003053. doi: 10.1002/14651858.CD003053.pub3

The Modern Materia Medica. New York, The Druggists Circular, 2nd edn revised and enlarged, 1911

The Science Snail. Total synthesis of metformin. 21 August 2020. https://www.sciencesnail.com/science/total-synthesis-of-metformin accessed 17 January 2022

USDA Natural Resources Conservation Service. *Galega officinalis* https://plants.usda.gov/home/plantProfile?symbol=GAOF accessed 15 July 2022

Watanabe CK. Studies in the metabolic changes induced by administration of guanidine bases. *J Biol Chem* 1918;33:253-65

Witters LA. The blooming of the French lilac. *Journal of Clinical Investigation* 2001;108:1105-7

Metformin, introduced to the United Kingdom in 1958, Canada in 1972, and the United States in 1995 is now believed to be the most widely prescribed anti-diabetic drug in the world; in the United States alone, more than 59 million prescriptions were filled in 2011.

CHAPTER 21 – MICHAEL DE SWIET

Glycyrrhiza glabra
The source of carbenoxolone

Introduction

The root of *Glycyrrhiza glabra*, liquorice, has been used as a medicine for many years. However, it is also the source of carbenoxolone, a medicine previously used for oesophageal, stomach and mouth ulcers. It has powerful, unwanted, potentially toxic endocrine effects.

Plant profile

Glycyrrhiza glabra is a hardy herbaceous perennial in the pea family, Fabaceae, which can grow very swiftly to 1.2m. It has a wide native distribution throughout Eurasia from the Mediterranean to Southwest Asia where it is found on dry open sites and, very often, on sandy soils near the sea.

The plant is cultivated for its roots and spreads by rhizomes or stolons, rooting from creeping stems. Liquorice roots are harvested in the autumn when the plant is 3 or 4 years old, and dried for later use. For good root production a deep, cultivated and fertile moisture-retentive but well-drained soil is needed. Like other members of the Fabaceae family this species has a symbiotic relationship with certain soil bacteria. These form nodules on the roots and fix atmospheric nitrogen which can be utilized not only by the plant itself but by other plants growing nearby.

Grown for use rather than decoration, the plant has pinnate leaves on tall gangly stems. From late June to July these are topped with pale purple/bluish flowers which are followed by clusters of seedpods. The species is monoecious (having both male and female organs on the same plant) and is pollinated by insects.

Glycyrrhiza can be propagated by dividing the root in spring or autumn.

Nomenclature and history

The name *Glycyrrhiza* has been used for at least 2000 years and comes from the Greek *glykys*, meaning sweet, and *rhiza*, meaning root (in Latin *dulcis radix*), appropriately as one of the chemicals it contains, glycyrrhizin, is up to 50 times sweeter than sugar. *Glabra* is Latin for

Dvlcis radix, Glycyrrhiza. Græcis, Γλυκυρρίζα
tia. Officinis, *Liquiritia.* Germanis, ℥eckritʒ/li
Gallis, *Recliſſe, ſiue Regaliſſe, ſeu Heculiſſe.*

D V L C I S   R A D I X.

Glycyrrhiza glabra from Mattioli's *De Plantis
Epitome*, 1586

'smooth' and refers to the texture of the leaves.
Liquorice is derived from the Latin, *Liquiritia* a
word which appeared in about 400 CE as a corrupt
transliteration of the Greek word *Glycorrhiza*.

Glycyrrhiza's history of medicinal use can be traced
back at least to Theophrastus (c. 350 BCE). He
called it 'Scythian root' and 'sweet root' and said
it was good for dry cough, chest problems, healing
wounds when mixed with honey, and quenching
thirst. He also reported that the Scythians, by
chewing the roots, and eating mares' milk cheese,
could go 11 or 12 days without drinking.

Its current medicinal uses were known by
Dioscorides [70 CE] who, calling it *Glycyrrhiza*
and *Dulcis radix* (sweet root), said that the roots
were good for heartburn, stomach problems
and tracheal soreness. Pliny [70 CE] also noted
its use for mouth ulcers (aphthous ulcers). They
attributed other virtues to it which – as with most
historic uses of plants – are no longer accepted
today, including curing whitlows, wounds, skin
ulcers, kidney stones and ocular pterygium
(a growth that occurs on the cornea). Paulus
Aeginetus, the 7th century Byzantine Greek physician, says it cured tracheal problems, bladder
ulcers and cystitis (Aegineta, 1542).

Liquorice

Liquorice cultivation came to England via the Cluniac Benedictine monks who had accompanied
the crusaders to the Middle East, where it grew profusely and was used to flavour drinks and
therapeutically as noted above. The Benedictines spread throughout Europe and, by the end of
the 12th century, they had about 30 monasteries in England including Pontefract in Yorkshire
where they grew *G. glabra* with great success. At the time of the dissolution of the monasteries
in the 16th century the local farmers bought the land and continued to grow liquorice. In
the 19th century the production of liquorice tablets, 'Pontefract cakes' was mechanised and
increased markedly. By the 1920s local supply of the plant could not meet the demand and the

manufacturers had to import the root from the original sources in the Middle East. These supplies were withdrawn in the Second World War, and the industry did not really recover afterwards, though some liquorice is still made in Yorkshire. Liquorice sweets are made from the root extract, sugar, and a binder, typically starch, gum arabic and gelatin.

The use of liquorice for stomach problems and mouth ulcers was documented again in the early 19th century when its use was noted for 'stomach complaints, which were thought to arise from a deficiency of the natural mucus which should defend the stomach against the acrimony [corrosiveness] of the flood and fluids secreted into it' (Duncan, 1819).

Liquorice root extract is a complex mixture of chemicals of which the more important are flavonoids, coumarins and especially certain terpenes, most notably the saponin glycosides, glycyrrhizin and glycyrrhizic acid, which are readily released from it after ingestion.

The role of liquorice for the treatment of peptic ulcer was thought at first to be due to its effect on the protective quality of the mucus lining the stomach making it more resistant to the hydrochloric acid produced in the stomach for digesting food. However, it is more likely to be due to the action of glycyrrhizic acid in preventing the breakdown of specific prostaglandins in the wall of the stomach. These potent chemical messengers, produced locally, make the lining cells more resistant to injury and help them to proliferate and so to heal small injuries to the lining. This results in further protection of the lining cells by increasing gastric mucus production.

Chemistry

Chemically glycyrrhizic acid is a steroid which is converted after ingestion by hydrolysis to glycyrrhetinic acid. This inhibits 11β-hydroxysteroid dehydrogenase, the main pathway for breaking down cortisol and aldosterone produced by the adrenal glands. In excess, the subsequent rise in cortisol and aldosterone causes water and salt retention and high blood pressure, a condition called hyperaldosteronism. In addition, aldosterone increases urinary potassium loss which causes low levels of potassium in the blood and subsequent muscular weakness and problems with heart rhythm. This may be fatal.

Carbenoxolone

Carbenoxolone is a hemisuccinate derivative of glycyrrhetinic acid, with a steroid-like structure, that was used for gastric, duodenal and oesophageal ulcers in the last century. It has been superseded because it too causes a condition similar to hyperaldosteronism, and by the development of other better drugs to counter hydrochloric acid production. In addition, it is now realised that the majority of peptic ulcers are caused by infection with a bacterium, *Helicobacter pylori*, which can be treated with antibiotics. Consequently, carbenoxolone treatment is no longer part of the current NICE guidelines for the treatment of gastric or duodenal ulceration.

Liquorice is registered as a traditional herbal medicine in Britain and Europe as a demulcent and expectorant. It carries warnings about the harmful effects of taking too much and about possible adverse effects on pregnancy.

Toxicity

There is concern that maternal liquorice consumption in pregnancy may cause impaired cognitive development (a reduced IQ) in the foetus. For example, in one recent Finnish study the children of women who had a high liquorice consumption in pregnancy scored seven points lower in IQ than a control group where the mothers consumed little or no liquorice in pregnancy. This could relate to liquorice affecting the transport of cortisol and cortisone across the placenta.

Liquorice poisoning can occur after eating daily more than 57g of black liquorice sweets for more than two weeks. It is particularly common in those abusing liquorice as a laxative. This may be fatal because of the cardiac dysrhythmias that it causes. For example, a construction worker died in Massachusetts after eating a bag and a half of black liquorice every day for a few weeks. When he was taken to hospital, he was found to have a very low blood potassium level which was the cause of the heart rhythm disorder from which he could not be resuscitated.

Summary

Carbenoxolone, derived from liquorice root, was a useful treatment for gastric, duodenal and oesophageal ulcers, but has now been superseded by better medication. Liquorice preparations made from the root of *G. glabra* are a pleasant form of sweetmeat and perfectly safe when taken in moderation. However, if they are used in excess for therapeutic purposes particularly as a laxative, they can cause dangerous, and occasionally fatal, side effects. This should be avoided.

References

Duncan A (Jun.). *The Edinburgh New Dispensatory*, 9th edn. Edinburgh, Bell and Bradfute, 1819

Gibson MR. Glycyrrhiza in old and new perspectives. *Lloydia* 1978;41:348-54

Lewis C, Short C. *A Latin Dictionary* (1st edn 1879). Oxford, Clarendon Press, 1927

Pinder RM, Brogden RN, Sawyer PR, *et al*. Carbenoxolone: A Review of its Pharmacological Properties and Therapeutic Efficacy in Peptic Ulcer Disease. *Drugs* 1976;11,245–307

Liquorice sweets made from *Glycerrhiza glabra* can be harmful if eaten to excess

Guaiacum officinale.

Published by Dr Woodville April. 1. 1790.

CHAPTER 22 – GRAHAM FOSTER

Guaiacum officinale
Roughbark lignum-vitae, guaiacwood – reducing deaths from bowel cancer

Guaiacum officinale in the family Zygophyllaceae is the national tree of Jamaica and was the main source of *Guaiacum* when it was used medicinally. *Guaiacum sanctum* (*palo santo*, holy wood, *lignum vitae*) is the national tree of the Bahamas and was mainly used for the hardness of its timber and not medicinally. Neither have ever been a 'Prescription Only' medicine or used therapeutically in modern times. However, they have a long history of herbal use for the past 500 years and have been included here because of the usefulness of extracts of the wood and bark (guaiac) for detecting blood in faeces.

Introduction

Chemicals from plants can treat diseases, but *Guaiacum officinale* is different. The resin, and α-guaiaconic acid extracted from it, were used to diagnose bowel cancers in early asymptomatic stages when treatment is more likely to be successful.

Cancer is a feared diagnosis throughout the world, immediately conjuring up images of a painful and inevitable demise. However, advances in therapeutics and diagnostics have transformed the picture for many common malignancies and the diagnosis of cancer is now often associated with an excellent chance of survival. For all cancers the key to effective therapy is early diagnosis – a tumour that has not spread to other organs is highly likely to be amenable to a surgical or chemotherapeutic cure but a cancer that is detected late has a greatly reduced chance of curative therapy.

For cancers of the large intestine (one of the commonest malignancies in the developed world) cure is almost assured if the cancer is identified whilst it is confined to the wall of the bowel. At this stage of its development the cancer is usually asymptomatic, but the abnormal tissue often leaks blood into the bowel where it appears in the stool. Blood mixed with faeces is often difficult to identify and much ingenuity has been deployed to find ways of encouraging people to inspect their bowel motions but, perhaps unsurprisingly, regular inspections have proved unpopular.

A major advance in early detection of bowel cancer was the development of easy-to-use testing kits that identify blood mixed in with faeces, the 'occult blood detection' systems. These systems were initially based on guaiac.

Plant profile

Guaiacum officinale is native to the Caribbean and tropical America, thriving where the temperature is 22-28°C, so it cannot be grown outdoors in the UK. It is slow-growing and makes a dome-shaped tree which eventually reaches a height of several metres. The wood is exceptionally hard and is used for pulleys and other devices requiring particularly hard timber. It has the fourth hardest wood in the world and is so dense that it does not float in water. It produces clusters of blue, five-petalled flowers followed by an orange outer seed capsule which splits open to reveal the scarlet fruit. This resemblance to the splitting seed capsules of *Euonymus alata* led Sir Hans Sloane (1696) to call it *Pruno vel evonymo affinis arbor* (*Prunus* or *Euonymus*-like tree).

History

The name *Guaiacum* is the first word from a native American language, being used by the Taino people of Santo Domingo in the Caribbean, to become an 'English' word. Then, as now, it meant the tree and the medicine which came from it.

Guaiacum in the 16th century

Santo Domingo was settled by the Spanish in 1493, bringing with them smallpox and measles. Syphilis was endemic in Santo Domingo and, almost in reciprocity, it was brought to Europe from there following the end of the war between the Catholic king of Naples [Pope Alexander VI] with Spanish help and King Charles of France of 1493-1495. People and soldiers from Santo Domingo who then came to Naples brought the disease with them. It spread through Italy and Spain and then to Germany and all of Europe and was estimated to have killed 10 million people. The disease was given different names (Mal de Naples, French pox, Spanish pox) depending on who was blaming whom.

The use of *Guaiacum* for treating syphilis was first described in European literature by Ulrich von Hutten in 1519, later translated into English by Thomas Paynell (von Hutten, 1533). von Hutten recognised that it was spread by sexual contact as it did not infect children or the elderly and was nothing to do with eclipses of the sun or other astrological phenomena or use of mercury as a treatment. von Hutten himself had a syphilitic ulcer which resisted all treatments until he used *Guaiacum*.

The utter ghastliness of untreated syphilis (before penicillin) is difficult to comprehend, but von Hutten is unsparing in the details of the disease and the multiplicity of worthless treatments that were current. He describes how *Guaiacum* came from Spagnola (Hispaniola, Santa Domingo) where syphilis was endemic and that the inhabitants called the tree 'Huaiacum' (pronounced 'why–a–cum' in English but transliterated into *Guaiacum* by the Spanish) and used it to treat their syphilis. He describes how *Guaiacum* is prepared by simmering the wood for some hours and using the froth to make a drink or mixing it with ceruse (lead oxide) to make an ointment. The resin, extracted by heating or burning the wood, was not used at that time. The patient needed to be kept in bed, in a warm room, starved and no other medication was allowed.

Guaiacum, its use, origins and preparation is well documented by Monardes (1569, 1580), Christopher Acosta (1593) and Fernel (1593). The latter adds its benefits in gout, arthritis, skin eruptions, ulcers, asthma, paralysis and more. It was seen as an alexipharmic, a universal cure-all.

Guaiacum in the 17th to 20th centuries

It was in the College's *Pharmacopoea Londinensis* (1618) where its use was described by Culpeper (1649) as 'Good for the French Poxe, as also for ulcers, scabs and leprosy'. Pemel (1652) reverted to Humoral theory, saying that it was hot and dry in the second degree, and expanded its uses, giving it an English name – pockwood. He gave the treatment of the French disease as its prime use, but also for dropsy, falling sickness, shortness of breath, catarrhs, coughs, consumption, gout, joint aches, disease of the bladder and kidneys, long lingering diseases, 'stoppings' of the spleen and liver, and for scabs and itch. It was made into a medicine by soaking a pound of wood and two ounces of bark in 12-14 pints of spring water for 24 hours and then boiling it down to seven or eight pints. The patient should drink a 'good draught' morning and evening and could add liquorice or aniseed to make it taste better.

John Quincy (1718) listed the resin from the tree as a treatment for genital discharges and ulcers, gonorrhoea and gout.

Linnaeus gave it its pharmaceutical name of *Lignum Sanctum* [Holy Wood] in his *Materia Medica*, where he reports its use for syphilis and leucorrhoea (1749), adding arthritis and scabies in the second edition of 1782.

However, by 1789, William Cullen, while still recommending it for arthritis and syphilis ('lues'), says that its efficacy for the latter is doubtful and advised mercury ointments. It was still available in apothecary shops in the mid-19th century as the wood or the resin for all its previously reported, putative uses (Bentley, 1861). A century later it had only vague uses in the pharmacopoeias including Wallers Mixture, a highly-diluted compound medicine '30 drops in a cup of milk … demanded by dry stone wall builders for back ache' and as 'Chelsea Pensioner' compounded with nutmeg, rhubarb, potassium tartrate and sulphur as a popular remedy for gout and rheumatism (Martindale, 1967).

Syphilis remained without an effective and tolerable therapy until penicillin was developed after the Second World War when it replaced the highly-toxic mercury-based treatments that had replaced *Guaiacum*. There appears to be no evidence that *Guaiacum* has any effect on treating syphilis, merely being a placebo for over 300 years in a condition in which the initial manifestations remit spontaneously (Eppenberger *et al*, 2017).

Practical uses

An interesting chemical property of the *Guaiacum* extract, α-guaiaconic acid, is that it changes colour in the presence of chemical peroxidases. As was first demonstrated by Van Deen in 1864, the oxygen-carrying compound in blood, haem, is a powerful pseudoperoxidase and if *Guaiacum* extracts are mixed with blood and hydrogen peroxidase a blue colouration develops. This change in colour upon contact with blood allows *Guaiacum* extracts, known as *guaiac*, to be used to detect small amounts of blood even when mixed with faeces.

The faecal occult blood test based on guaiac was widely used for many years to detect small amounts of blood in faeces when there was no obvious bleeding. Strips of guaiac paper were provided to patients, along with instructions to smear them with faeces before returning them to the laboratory where incubation with hydrogen peroxidase completed the reaction. If blood was present the paper developed an instantaneous blue discolouration. Patients with positive results could then to be targeted for more invasive testing. In a landmark paper in 1993, Mandel *et al* showed that screening in this way was associated with a 33% reduction in mortality over a 33-year period. It is hard to think of a better use for a piece of bark but, sadly for those of a romantic bent, the guaiac test has now been replaced by different methodologies using more convenient and sensitive antibody-based detection systems that can be used and analysed at home.

Editor's Note: Guaiac faecal blood testing kits can give false positives if the patient had eaten red meat or certain vegetables: lemons and plants with high levels of antioxidants, such as vitamin C, caused false negative tests. These are not found with the modern detection systems.

References

Bentley R. *A Manual of Botany*. London, John Churchill, 1861

Cullen W. *A Treatise of the Materia Medica*. Edinburgh, Charles Elliot, 1789

Eppenberger P, Galassi F, Rühli F. A brief pictorial and historical introduction to guaiacum – from a putative cure for syphilis to an actual screening method for colorectal cancer. *British Journal of Clinical Pharmacology* 2017;83:2118-9

Mandel *et al*. Reducing mortality from colorectal cancer by screening for fecal occult blood. *New England Journal of Medicine* 1993;328:1365-71

Quincy J. *Pharmacopoeia Universalis Extemporanea or a Complete English Dispensatory*. London, A. Bell, 1718

Woodville W. *Medical Botany* vol 1. London, James Phillips, 1790

CHAP. 110.

De Guajaco, of Guajacum, or Pockwood.

The Names.

IT is called in Latine *Guajacum*, *Lignum Indicum*, *Lignum Sanctum*, and *Lignum vitæ*, in English Pockwood and Indiall Pockwood.

The temperament.

It is hot and dry in the second degree, and hath a cleansing faculty.

The Duration.

It will keepe good many yeares.

The inward use.

The chiefe use of this Wood is against the French Disease, for it provoketh Sweate, resisteth contagion and putrifaction, and cleanseth the Bloud : It is good also in the Dropsy, Falling Sicknesse, Shortnesse of breath, in Catarrhes, Rheumes and cold distillations of the Lungs, or other parts, Coughes and Consumptions, the Gout and all other joyntaches , and for cold flegmatick humours, for the Diseases of the Bladder and Reines, and for all long and lingring Diseases proceeding from cold and moist causes ; it openeth the stoppings of the Liver and Spleene , warmes and comforts the stomack and entralls, and is good in Scabs, Itch, &c.

The manner of Administring it.

It is chiefly used in Decoction.

A Decoction of Lignum vitæ.

Take of *Lignum Vitæ*, or Pock wood a pound , of the barke thereof two Ounces , steepe them in twelve or fourteene Pints of spring water foure and twenty houres, then boyle them to seven or eight pints, straine it, and give thereof a good draught morning and Evening, and let the party sweate upon it. If you adde two Ounces of Licoris, or more, and some Anisseede, it will be much

H h 2   more

The uses of *Guaiacum*, known by the English name of 'pockwood', according to Dr Robert Pemel in 1652

CHAPTER 23 – ARJUN DEVANESAN

Hordeum jubatum
Arundo donax
The source of lidocaine

Introduction

Both a chlorophyll-deficient mutant of *Hordeum jubatum* and *Arundo donax* are sources of gramine, from which isogramine and then lidocaine were derived. *Arundo donax* was never used in the discovery of local anaesthetics. Lidocaine, also known as lignocaine, has become the most widely used local anaesthetic to this day.

Nomenclature

Hordeum jubatum, also called foxtail barley for its characteristic plume, is a hardy grass species from North America. It had previously been called *Hordeum aristis* by the Finnish botanist and explorer, Pehr Kalm (1716-1779), who discovered it in Canada, and renamed by his tutor, Linnaeus, as *H. jubatum*. It is not a source of cereals – that is *H. vulgare*, which we grow at the College. *Hordeum* is in the same family, Poaceae, as *Arundo donax*, a giant reed to which it bears little resemblance, which we do grow.

Plant profiles

Hordeum jubatum is a hardy, invasive, perennial, hybrid grass from the north of North America and Siberia, tolerant of a wide range of soils including those with high salinity. It grows to 100cm with a dense, nodding flower spike, similar to cultivated barley (*q.v. H. vulgare*). The barbed spines, called awns, on the ripe seed heads are damaging to herbivores when they lodge in their mouths, face and eyes causing ulcers and blindness.

Arundo donax, the giant reed, is a hardy, deciduous rhizomatous perennial which sends up stout stems, or culms, which race up to 6m in height during the summer months and from which hang long tapering leaves. Often mistaken for an enormous bamboo it is in fact a grass, native to West and Central Asia through to temperate East Asia. It can become invasive in its native habitat of ditches, riversides and marshland but in a temperate garden it seldom poses any such danger. It does, however, need a large space to look its best and to prevent it appearing out of scale beside its plant neighbours. In September long panicles of light green to purple spikelets appear at the tops of the canes and these are pollinated by the wind. The plant needs sun and prefers moist or wet soil that is poor and sandy. The best foliage effect is obtained if the plant is cut down to the base in late autumn.

Arundo donax has been cultivated as a source for biofuel because of its rapid growth and for centuries it has also been grown for the production of reeds for wind instruments such as oboes.

Local anaesthetics

Following the discovery of the local anaesthetic properties and chemical structure of cocaine from the South American shrub *Erythroxylum coca* (*q.v.*), and its chemical formula, reports of toxicity and addiction led to the systematic search for new local anaesthetic compounds. Knowing that cocaine was an ester of benzoic acid, over a hundred synthetic compounds were tested (often on the tongues of researchers) for local anaesthetic effects. Eventually, in 1904/5, the German chemist Alfred Einhorn discovered procaine, which, while it had a shorter duration of action than cocaine, was much less toxic and had no addictive effects.

Arundo donax var. *versicolor* growing in the Garden of Medicinal Plants at the Royal College of Physicians

Chemistry

Mode of action

Local anaesthetics are now classified into two main groups – the esters (of which the ultimate parent is cocaine) and the amides (a class of synthetic chemicals, the most famous of which is lidocaine). Amides store better, are more heat tolerant and cause fewer allergic reactions than esters and so are generally favoured.

Local anaesthetics act by disrupting the transmission of pain impulses to the brain along nerve fibres. They do this by entering nerve cells and blocking sodium ion channels which are needed to regulate electrical signals. Local anaesthetics are weak bases (alkaline) and so exist in partly ionised form. The degree of ionisation determines their speed of onset, with only the un-ionised fraction able to enter the cell. Different nerve fibres exhibit different sensitivities to local anaesthetics depending on the thickness of the myelin sheath (the fatty layer protecting the cell). Thin pain fibres are therefore more affected than thick touch fibres which is why people can often still feel touch even though they cannot feel pain under local anaesthesia.

In the early 20th century, a large proportion of surgery in Europe and America was done under local anaesthesia. It was much safer for the patient than general anaesthesia and could be administered with nothing more than a hypodermic syringe. With the increased interest in local anaesthetics Heinrich Braun, a prominent German surgeon and strong advocate of local anaesthesia, developed five criteria for the 'ideal local anaesthetic'. It should:

- have a better therapeutic ratio (relative safety) than cocaine
- have better tissue penetration
- not cause tissue irritation
- be chemically stable (and store well)
- be compatible with adrenaline (as adrenaline was added to local anaesthetics to prolong their action)

Procaine was irritant to tissues, had a short effect time and did not store well but was deemed the most successful of all other available local anaesthetics at the time. Procaine, marketed as Novocain, derived from cocaine, became the definitive local anaesthetic for decades until lidocaine, from barley, *H. jubatum*, superseded it.

Hordeum jubatum

The story moves from Austria and Germany to Stockholm University in Sweden. In the early 1930s, Hans von Euler-Chelpin worked on mapping out the inheritance of physical traits in chemical terms. He was specifically interested in how genes gave rise to enzymes. As such, he acquired some chlorophyll-deficient mutants of a common barley (*Hordeum jubatum*) which appeared to be resistant to certain pests compared to its non-mutant wild type. While investigating the chemical differences between these mutants and their non-mutant types, von Euler-Chelpin isolated an indole alkaloid they called gramine (after the grass family, Gramineae – the previous name for the Poaceae).

Arundo donax

In one of history's rare cases of synchronicity, Russian researchers at around the same time, investigating pest resistance in crops, heard of a grass, *Arundo donax*, which camels refused to eat. From this giant reed, the Russian researchers also isolated an indole alkaloid which they called donaxine. There is no mention, however, of the development of local anaesthetics from donaxine. That said, it was soon discovered that donaxine and gramine are identical.

History

Neither donaxine nor gramine have anaesthetic properties themselves. However, when Holger Erdtman, von Euler-Chelpin's student at the time, was asked to synthesize gramine he produced 2-dimethylaminomethylindole (an isomer of gramine) instead. On the usual taste test, Erdtman found that isogramine numbed the tongue. In what must have been great excitement, Erdtman

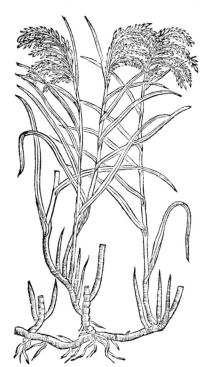

Arundo donax – woodcut from Parkinson's *Theatrum Botanicum*, 1640

and Nils Löfgren (then but a chemistry student) prepared several analogues with the starting material dimethylamino-o-acetotoluidide. These various compounds were systematically tested as local anaesthetics, but none were found superior to procaine and the project was abandoned. After graduating, and during the Second World War, Löfgren worked with Pharmacia on developing new local anaesthetics because wartime restricted the supply of procaine from Germany. The result, amoxecaine, was barely adequate but was occasionally used.

Nils Löfgren came back to von Euler-Chelpin's lab as a lecturer in 1941 and returned to his earlier experiments with an assistant Bengt Lundqvist in 1943. Taking a compound that he had synthesized as a PhD student and adding an extra methyl group in the 6 position of the benzene ring, they synthesized a new analogue and labelled it LL30 after the surnames of the two researchers. After the usual self-experimentation, LL30 was found to have a longer duration of action than procaine. Löfgren was an extremely energetic and eccentric character. It is said he tried spinal anaesthesia with LL30 on himself with the aid of a mirror. Lundqvist was an avid fencer with an international reputation and practiced with Tore Kornerup, a friend of Torsten Gordh. In the spring of 1943, Gordh had dinner with Kornerup at a restaurant, Stallmästaregården, after a staff meeting at the Karolinska Hospital. Kornerup mentioned that a friend of his, Lundqvist, had a new anaesthetic which was in need of clinical testing. Gordh was Sweden's first anaesthetist and was immediately interested. He suggested they contact Bengt Lagergréen (who had briefly studied under Gordh in the 1940s) and perform some tests with finger anaesthesia for a demonstration to Pharmacia in Uppsala later that year.

Gordh initially tested LL30 for toxicity with Leonard Goldberg as it was essential that they demonstrated its superiority to procaine in this respect. After that, Gordh began his clinical tests on himself, his wife, his students and volunteers. His wife was a medical student at the time and helped with the initial tests. Volunteers were offered 5 Swedish crowns and students were offered either a pack of cigarettes (Camel or Lucky Strike) or Gordh's thesis. Perhaps unsurprisingly, the students usually chose the cigarettes. American cigarettes were hard to come by in Sweden and so particularly attractive.

The evidence from clinical tests was astonishing and no statistical tests were needed because of the overwhelming superiority of LL30. It came to the attention of numerous pharmaceutical companies, but Lagergréen, who helped in the initial clinical testing, was related by marriage to the Wallenberg family who were closely connected with Astra AB. This may also explain the particularly lucrative deal for LL30, with Löfgren paid 4% of sales, giving Lundqvist a third of that. In 1949 Astra AB marketed it, named as lidokain (generic) and Xylocaine (commercial). Löfgren and Lundqvist both found themselves fabulously rich and retired to the tax haven Switzerland where, sadly, both met with a tragic end – the latter had an early stroke and the former committed suicide in 1966.

Lidocaine in local anaesthesia

Lidocaine is commonly given by injection and takes effect within minutes and lasts for up to three hours. It can be injected under the skin in the region where local anaesthesia is required – for example for suturing wounds – or around a nerve ('nerve block') for dental work. For these uses it is often mixed with adrenaline which causes constriction of surrounding blood vessels and reduces the rate that lidocaine is removed from the area, so prolonging the duration of action. It, and its derivatives, are also used for epidural anaesthesia. Applied topically it is used on the skin as a 'patch' for shingles pain, 'nerve pain' and jellyfish stings; to the throat by a spray before endoscopy or bronchoscopy, and as drops on the eye for minor ophthalmic operations. Intravenously, it has been used for treating chronic pain, such as phantom limb pain from an amputated limb, with occasional benefit.

Lidocaine and the heart

Intravenous lidocaine was also noted to be useful in reversing fatal rhythm disturbances of the heart. In a report published in 1950, James Southworth and colleagues described a case of a young woman who suffered from a life-threatening rhythm disturbance of the heart (ventricular fibrillation) while being investigated for vague breathlessness and chest pains. When she suddenly collapsed and was found to have a disordered heart rhythm and no pulse, she was resuscitated by manual compression of the heart through a surgical opening in her chest and had several electrical shocks but without success. She was then given a mixture of lidocaine and adrenaline and soon her heart returned to a normal rhythm and she subsequently (apparently) completely recovered. Lidocaine was then taken up by anaesthetists in the treatment of arrhythmias during anaesthesia for cardiac surgery and a large series of cases were reported in 1963. Since then, lidocaine gained in popularity as a treatment for rhythm disturbances, especially those originating from the ventricles, after heart attacks. It has several drawbacks, not least the fact that it reduces the strength of contraction of heart muscle, and so it is not as commonly used today as it once was. That said, lidocaine remains a trusted treatment for life-threatening rhythm disturbances which are refractory to other drug treatments.

Lidocaine remains one of the most commonly used local anaesthetics to this day. A number of other important alternatives were derived from it – bupivacaine, mepivacaine, ropivacaine – all seeking to increase its efficacy and reduce side effects.

References

Gordh T, Gordh TE, Lindqvist K, Warner DS. Lidocaine: The Origin of a Modern Local Anesthetic. *Anesthesiology* 2010;113:1433-7

Link WJ. Alfred Einhorn, Sc.D: Inventor of novocaine. *Dent Radiog Photog* 1959;32:1-20

Löfgren N, Lundquist B. Studies on local anaesthetics. *Svenks Kem Tidskr* 1946;58:206-17

Southworth JL, McKusick VA, Converse Pierce E, Rawson FL. Ventricular fibrillation precipitated by cardiac catheterisation: Complete recovery after forty-five minutes. *JAMA* 1950;14(8):717-720

Wildsmith JAW, Jansson JR. From cocaine to lidocaine: Great progress with a tragic ending. *European Journal of Anaesthesiology* (EJA) 2015;32(3):143-146

Further reading

Eriksson E (ed). *Illustrated handbook in local anaesthesia*. Munksgaard, A B Astra, 1969

Testing lidocaine for safety on his students (and his wife) in Sweden in the 1940s

Volunteers were offered 5 Swedish crowns and students were offered either a pack of cigarettes (Camel or Lucky Strike) or Gordh's thesis. Perhaps unsurprisingly, the students usually chose the cigarettes. American cigarettes were hard to come by in Sweden and so particularly attractive.

CHAPTER 24 – MICHAEL DE SWIET

Hordeum vulgare
Claviceps purpurea
The source of ergometrine and ergotamine

Introduction

Claviceps purpurea is a fungus that grows on the ears of rye (*Secale cereale*) and other cereal plants. Rye does not grow in the College Garden but barley, *Hordeum vulgare*, does and may host *C. purpurea*. *Claviceps purpurea* contains ergot alkaloids which are the cause of ergotism and the source of ergometrine which was previously of life-saving importance for stopping postpartum haemorrhage. *Claviceps* means club-headed and *purpurea* means purple, *i.e.* the purple club-headed fungus.

Plant profile

The annual grass *Hordeum vulgare*, or common barley, is a member of the Poaceae family. It has been so long in cultivation (about 10,000 years) that the cultivated variety is recognised as the species form. This domesticated form is hull-less or 'naked' which makes the seed both easier to harvest and a better crop to grow on a small scale. Originating in the Middle East in the area now known as Israel and Palestine, barley continues to be grown for its edible seed in temperate areas of the world. A field of barley is a beautiful thing: the awns, or needle shaped bristles, enclosing and extending above the flowering spikes, give the plant a distinctive and graceful form which sways in any breeze, causing ripples and waves to move across the crop. As the grain ripens the fields turn to gold.

The grass has pale green, linear leaves and an erect central stem with 2-5 lateral branches called 'tillers'. In two-row barley only the central spikelet is fertile, in six-row barley the lateral spikelets are fertile too. The two forms have different uses because of their varying sugar and protein content. Each fertile tiller carries a cylindrical grain-bearing inflorescence called a spike (ear or head) where the grains are arranged in a herringbone pattern. These are followed by golden to purple or black seeds.

Hordeum vulgare can grow in most soils and climates ranging from sub-arctic to sub-tropical but prefers a calcareous, moist and well-drained soil in a sunny position. Spring sowings will flower from June to August with later sowings flowering from September to early October. The species has both male and female organs and is pollinated by the wind.

Ergot alkaloids and ergotism

As early as 600 BCE an Assyrian tablet mentioned a 'noxious pustule in the ear of grain'. This is not very specific but, in the 4th century BCE, texts from the Parsees describe 'noxious grasses that cause pregnant women to drop the womb and die in childbed'. That is, if taken before delivery ergot can cause such severe contraction of the uterus that it will rupture, causing the woman to bleed to death.

The many active principles in ergot (ergot alkaloids) can also constrict the smooth muscle in blood vessels so severely as to obstruct blood flow and cause gangrene and convulsions. This combination of gangrene, convulsions and hallucinations is ergotism, sometimes called St Anthony's Fire, after the monks of the order of St Anthony who were said to be good at treating it. However, St Anthony's Fire is usually identified as erysipelas, a bacterial skin infection.

Hordeum vulgare – woodcut from Mattioli's *Discorsi*, 1568

Ergotism in the Middle Ages

Ergotism has occurred in outbreaks recorded since the Middle Ages and is caused by the fungus contaminating bread flours, most frequently rye flour but also others. The outbreaks were more common in countries to the east of the Rhine because that is where ryebread was commonly consumed. The symptoms are divided into gangrenous and convulsive forms. In the gangrenous form the hands and feet, which have relatively poor arterial blood flow, are most affected, with loss of sensation, painful swelling, skin loss and eventually loss of digits or limbs. The convulsive symptoms are convulsions, gastrointestinal upset and hallucinations. Both forms can be fatal.

It has been argued that the Dancing Manias of the Middle Ages, particularly the one that occurred in 1518 (said to be commemorated by the story of the Pied Piper of Hamelin), were due to ergotism, but the evidence is only circumstantial and often contradictory. The lack of a word for the disease hampers research into its occurrence. It was not until 1683 that black grains in the ears of corn were called ergot (OED, 1979). Ergot is a French word derived from Old French 'argot' that translates as 'cock's spur' as the fungal spores resemble a cock's spur and were so-called in England.

The mechanism of hallucinations is uncertain. They could be caused by reduced blood flow to the brain due to constriction of cerebral blood vessels or they could be caused by specific neurotoxins. Ergot alkaloids present in *C. purpurea* include lysergic acid which though not a hallucinogen itself can be changed to lysergic acid diethylamide (LSD) which is a potent hallucinogen as are others found in different fungi such as 'magic mushrooms'.

Epidemics occur when the grain is infected, typically in moist weather with cool temperatures: the cereal becomes infected at flowering time which is prolonged in cold wet weather. Ergotism is now very rare because of good farming practices which minimise the risk of infection, cleaning of the seed to remove the larger fungal spores (sclerotia) and rejection of any wheat samples intended for human consumption that contain any sclerotia. However, there was a relatively recent outbreak of ergotism in 1951 in Pont Saint Esprit in Southern France. Although ergotism seems likely, this cause has been disputed. Within one month of the onset of the episode about 150 cases following eating contaminated bread were reported, but eventually more than 250 people were affected including 50 interned in asylums, and seven deaths (Gabbai *et al*, 1951).

Obstetric use of ergot

In 1582 the German physician, Adam Lonicer in his *Kreuterbuch*, mentioned for the first time the use of what was to be later called ergot to stimulate the uterine contractions of labour by administering three sclerotia. The first accurate description of the ergot is also from his *Kreuterbuch*: 'long black hard narrow corn pegs, internally white, often protruding like long nails from between the grains in the ear'. Ergot appeared in the *Pharmacopoeia Londinensis* (Castle, 1836) as '*Ergota*, ergot, a species of fungus, feeding on the diseased seeds of some plants, especially rye'. It appeared in the *British Pharmacopoeia* (1864) as a tincture, infusion and liquid extract.

The contractions of the smooth muscle of the uterus provoked by ergot were so violent that the foetus could die because of inadequate maternal placental blood flow and the uterus could rupture, killing the mother from haemorrhage. This appears to have been first reported by the New York physician Dr David Hosack (1822). 'The ergot, has been called, in some of the books, from its effects in hastening labour, the *pulvis ad partum* [powder for childbirth]; as it regards the child, it may, with almost equal truth, be denominated the *pulvis ad mortem* [powder for death]; for I believe its operation, when sufficient to expel the child, in cases where nature is alone unequal to the task, is to produce so violent a contraction of the womb, and consequent convolution and compression of the uterine vessels as very much to impede, if not totally to interrupt, the circulation between the mother and child'. In the same article he recommended its use in postpartum haemorrhage (PPH).

By the later years of the 19th century the practice was to use ergot preparations only after the baby had been born and then to help with delivery of the placenta and to prevent or stop bleeding after delivery, postpartum haemorrhage.

Ergotamine and ergometrine

Arthur Stoll isolated the first pure ergot alkaloid, ergotamine, from it in 1920 (Spiro & Stoll, 1921). Ergometrine was isolated by Harold Dudley and was shown to be far more potent in causing uterine contraction by John Chassar Moir in 1935 (Dudley, 1935). Ergometrine was subsequently recognised as the best agent for encouraging the delivery of the placenta and for preventing and treating PPH. Before 1940 the maternal death rate from PPH was 3 per 10,000 deliveries. By 1952 when the use of ergometrine was well established, the maternal death rate from PPH had fallen fivefold to 0.6 per 10,000. Because ergometrine can cause nausea, vomiting and occasionally dangerous high blood pressure, it has been replaced by syntocinon for the management

Early 20th century bottle of ergot extract to treat postpartum bleeding

of labour after the mother has been delivered. Syntocinon is a synthesised form of the posterior pituitary hormone, oxytocin.

Migraine

Ergot alkaloids such as ergotamine have been used for the treatment of migraine since the 1920s. Longer-acting synthesised derivatives such as sumatriptan have also been used for migraine prophylaxis. A simple explanation of their action could be that ergotamine relieves the throbbing vascular headache of migraine by causing vasoconstriction in the cerebral blood vessels. However, it is disputed to what extent migraine is due to cerebral vasodilatation and to what extent ergotamine acts on cerebral blood vessels. Indeed, the pharmacology of ergot alkaloids is complex and variable between the different alkaloids. For example, they may act as both agonists mimicking the effect of adrenaline constricting blood vessels and antagonists inhibiting this effect.

Summary

Claviceps purpurea is an excellent example of a natural poison, in this case from a fungus, that causes gangrene, epilepsy and madness. It was known to cause uterine contraction for 500 years, and this property has been harnessed by pharmacists to make medicines to save lives and relieve pain. Thanks to these, mothers could have babies with greater safety and migraine sufferers can live a better life.

References

British Pharmacopoeia. London, Spottiswode & Co., 1864

Castle T. *A translation of the Pharmacopoeia Londinensis of 1836 with descriptive and explanatory notes on the Materia Medica etc*. London, E. Cox, 1836

Dudley HW, Moir C. The substance responsible for the traditional clinical effect of ergot. *BMJ* 1935;i:520-3

Gabbai, Lisbonne, Pourquier. Ergot Poisoning at Pont St Espirit, *BMJ* 1951;2(4732):650-1

Hosack D. On Ergot. *New York Medical and Physical Journal* 1822;1:205

OED. *Compact Edition of the Oxford English Dictionary*. London, Book Club Associates, 1979

Spiro K, Stoll A. Ueber die wirksamen Substanzen des Mutterkorns. *Schweiz Med Wschr* 1921;2:525-29

CHAPTER 25 – TIMOTHY CUTLER

Hydrangea febrifuga
The source of halofuginone

Hydrangea febrifuga is a plant from which a sedative, an antimalarial, treatments for protozoal infections in chickens, cattle and sheep, and treatments for abnormal fibrosis, autoimmune diseases and cancers have all been made.

Introduction

Hydrangea febrifuga was known until recently (2015) as *Dichroa febrifuga* and in herbal folklore as the fever bush. This is one of the 50 fundamental plants used in Chinese herbal medicine. Its roots are the original source of febrifugine, a powerful but very toxic antimalarial, from which methaqualone was synthesised in 1951, and which became a very popular sedative and hypnotic drug in the 1960s and 70s. A synthetic analogue, halofuginone (also known as HT100), was found to be a powerful treatment for the protozoal diseases coccidiosis and cryptosporidiosis in farm animals and for inhibiting and even enabling/stimulating the resorption of the excessive collagen production in diseases that cause fibrosis. It is a remarkable chemical which also inhibits the production of new blood vessels (angioneogenesis) in cancers. It also inhibits tumour growth, inflammation and autoimmune reactions, among several other properties.

Nomenclature

Hydrangea comes from the Greek for water+jar referring to its cup-shaped fruits. *Dichroa* means two-colours, alluding to its bi-coloured flowers, and *febrifuga*, meaning fever+flight, refers to its traditional use for treating fevers.

Plant profile

Growing to about 2m tall the plant has a sturdy look and rounded shape with large, obovate, deep green leaves with toothed margins and striking corymb-like panicles of flowers. The colour of the flowers varies according to the pH of the soil but they are always white on the outside of the petals and either pale pink or shades of blue on the inside. The berries are a reliably vibrant gentian blue: they mature in late summer and remain on the plant as a very ornamental feature well into the winter. They are also very attractive to birds.

Hydrangea febrifuga is a native of tropical and subtropical Asia, from India to China, Indonesia and New Guinea. It can be grown outside in the UK but, as it will not survive being frozen, it needs to be protected during severe winters. It prefers partial shade and moist, well-drained, acid soil, though it is also able to grow in alkaline soils. It can be propagated by semi-ripe cuttings in summer.

History
There is no European history regarding the medicinal use of hydrangeas.

Febrifugine
The root of this plant has been used as an antimalarial in traditional Chinese medicine for over 2000 years, but it was not until the mid-20th century that the active ingredient, a quinazolinone alkaloid initially called dichroine B and subsequently febrifugine, was isolated. When initially found in other hydrangeas it had been called hydragine. It was found to be more than 50-100 times more potent than quinine as an antimalarial, but its clinical use was dogged by severe gastrointestinal and hepatic toxicity, and it has never been put into regular use.

Methaqualone
In 1951, Kacker and Zaheer, in India, were researching new antimalarials derived from febrifugine and produced a compound called methaqualone. By 1955 its sedative and hypnotic effects were known and it was patented in the USA in 1962 and marketed as Quaalude.

In the 1960s, a methaqualone/antihistamine combination was marketed in Europe with the name Mandrax, a name alluding to mandrake (*q.v.*) and its properties (Drori, 2021). By 1965 it was the most commonly prescribed sedative in the UK. Its popularity was due to it replacing the long-term use and abuse of barbiturates which had been the primary sedative drugs of the 1950s. Although promoted as a sleeping pill, it quickly became a recreational drug of abuse (called 'Mandies' and 'disco biscuits') during the 1960s and 70s due to its effect of producing deep relaxation and reputedly as an aphrodisiac. It is a highly addictive drug to which tolerance develops so that regular users required ever larger doses to produce the same effect. It was found that taking one tablet a day for two weeks induced addiction, and withdrawal effects could be severe and sometimes fatal. Combined with alcohol 'people went to sleep and did not wake up'. It was also used as a 'date-rape' drug before Rohypnol.

The US Drug Enforcement Administration halted production of methaqualone in 1982 and it was soon withdrawn worldwide from legal manufacture. Illegal use continued from Mexican sources until the early 1990s when its popularity waned. It is still found in South Africa from clandestine manufacture where it is smoked in combination with marijuana (known as a 'white pipe'). Drug users also inject it, and dealers use it to 'cut' (adulterate) heroin.

This was a drug which one of the authors remembers being told when he was a junior doctor in 1966, by the drug company representative, that it was 'not addictive, not habit-forming, did not give a hangover after use, and was excellent in pregnancy' and that he could prescribe it safely to anyone!

Halofuginone for protozoal infections

Readers needing more detailed information on halofuginone and its mode of action should read the paper by Pines and Spector (2015), but the following is a summary of the main findings.

A synthetic halogenated derivative of febrifugine called halofuginone was synthesized in the late 1960s in the search for a less-toxic antimalarial drug by systematically altering parts of the febrifugine molecule. It was marketed under the brand name of Halocur as a potential antimalarial drug, effective in killing the malarial parasite *Plasmodium falciparum* in all three stages of its lifecycle. Adding bromide to the molecule made it less toxic to the recipients without lowering its effectiveness. Halofuginone bromide was found to prevent and treat coccidiosis in the poultry industry and halofuginone lactate treated cryptosporidiosis in sheep and cattle, both being protozoal diseases that cause significant morbidity and mortality.

Halofuginone and fibrosis

More recently, it has been identified as a specific inhibitor of collagen type 1 gene expression and, as a consequence, it inhibits the development of fibrosis – a progressive condition secondary to chronic inflammation that eventually leads to organ failure. This has opened a gateway into the treatment of many diseases where fibrosis due to collagen overgrowth is a significant problem, such as pulmonary fibrosis, liver cirrhosis, wound repair and fibrous adhesions after abdominal surgery. Not only does it inhibit the progression of the fibrosis, but it can also reverse established fibrosis without affecting normal collagen turnover, but clinical trials are still needed to demonstrate this.

Chemistry

Mode of action

A second (and perhaps more significant) mechanism of action is in the inhibition of transforming growth factor beta (TGF-β) signalling. This is a 'master pathway' in fibrosis which is responsible for fibroblast proliferation and transition to scar-forming myofibroblasts (Pines, 2015). This mechanism (together with an impact on collagen expression versus degradation) represents potential for a therapeutic medicine which would be different from current standards of care for (pulmonary) fibrosis for which pirfenidone (Esbriet) and nintedanib (Ofev) are currently used. If clinical trials confirm the multiple mechanisms of action assigned to it and show the impact on pathway and pathology-related biomarkers, this would be a great therapeutic advance.

Halofuginone and Duchenne muscular dystrophy (DMD)

Halofuginone has acquired Orphan Drug status for the treatment of DMD. This is the most advanced indication for the drug, aiming to reduce fibrosis and inflammation and promote healthy muscle fibre regeneration in DMD patients. Early DMD trial data indicated improvement in muscle strength for boys treated in the lowest three dose cohorts. These strength improvements, rather than just a slowed rate of decline, are a highly unusual finding. One importance of this finding is that halofuginone showed benefit in all types of DMD and not just in certain genetically determined groups. The follow-up phase two trial was terminated following an extreme adverse event case.

Halofuginone and cancer treatment

For a cancer to grow it needs to initiate the production of new blood vessels to 'nourish' it, a process called angioneogenesis. Halofuginone has been found to act at multiple points to inhibit the process of angioneogenesis and to reduce the size of existing blood vessels in cancers. It shows potential for treating gliomas, kidney, prostate and pancreatic tumours, and malignant melanomas, and in inhibiting secondary spread (metastases) of certain cancers.

By separate mechanisms, it is active against certain leukaemias, myeloma, breast cancer and pancreatic tumour, *in vitro* (*i.e.* in the laboratory). The National Cancer Institute has selected halofuginone for a rapid development programme for cancer therapy and halofuginone hydrobromide is being trialled in recurring bladder carcinoma in the UK.

Halofuginone and autoimmune disease

Halofuginone has been found to inhibit a multiplicity of genes responsible for development of inflammation and autoimmune diseases. Early clinical trials on the treatment of scleroderma (a chronic autoimmune disorder of connective tissue) have started, and the Food and Drug Administration (FDA) in the USA has granted halofuginone Orphan Drug status which will encourage further research into its use in this respect.

Chemistry

Interestingly, it is likely that an additional mechanism of action is at play for scleroderma, as interleukin 17 (IL-17)-mediated immune dysfunction is associated with autoimmune conditions and of skin in particular. Halofuginone has been shown to reduce T helper cell 17 (Th17) differentiation and subsequent IL-17 production preclinically.

Halofuginone and malaria

New data on the antimalarial properties of halofuginone both *in vivo* and *in vitro* strongly suggest it should be further developed as a new oral and intravenous drug, which will neatly return the interest in febrifugine back to its original use over the last two millennia.

Epilogue

The established drug pirfenidone (Esbriet) for treating interstitial pulmonary fibrosis was purchased by Roche recently for eight billion dollars ($8,000,000,000) so the potential for halofuginone, with a similar profile but as yet no confirmed pathway activity in patients, if it can be demonstrated in clinical trials to be safe and effective, is enormous.

References

Drori J. *Around the World in 80 Plants*. London, Laurence King, 2021

Kacker IK, Zaheer SH. Potential Analgesics. Part I. Synthesis of substituted 4-quinazolones. *J Ind Chem Soc* 1951;28:344-6

NicDaéid N, Savage KA. *Methaqualone and Meclaqualone*. In: Siegel J, Saukko P. *Encyclopedia of Forensic Sciences*, 2nd edn. Elsevier, 2013

Ningthoujam SS, Choudhury MD. Febrifugine and its analogs: Studies for their antimalarial and other therapeutic properties. In: *Studies in Natural Products Chemistry* vol 44. Elsevier, pp93-112, 2015

Pines M, Spector I. Halofuginone – the Multifaceted Molecule. *Molecules* 2015;20(1):573-594

Hydrangea quercifolia. Other hydrangeas also contain febrifugine, initially called hydragine

CHAPTER 26 – NOEL SNELL

Illicium anisatum and *I. verum*
The source of oseltamivir

Illicium verum (Chinese star anise) and *I. anisatum* (Japanese star anise) are closely-related species whose star-shaped seeds (that smell of anise) contain shikimic acid, which was used in the synthesis of the anti-influenza medication oseltamivir (Tamiflu). *Illicium anisatum* grows in the College Garden.

Plant profiles

Illicium anisatum and *I. verum* are members of the Schisandraceae family. They are similar in appearance but not in properties: *I. verum* yields the culinary spice star anise whereas *I. anisatum* contains dangerous neurotoxins and should not be eaten.

Both species have pale bark and glossy green leathery leaves of similar lanceolate shape but *I. verum* can grow to 18m whereas *I. anisatum* is usually only half that height. The flowers can be a useful distinguishing characteristic: while both tend to be yellow-green, those of *I. verum* can have a pinker tinge and are solitary while *I. anisatum* has larger ones with more tepals. The distinctive star-shaped seed heads are common to both. Both species require similar acid conditions to thrive. They are just about hardy in mild areas of the UK (*I. verum* slightly less so) where they should be grown in moist but well-drained lime-free soil in sun or semi-shade and sheltered from cold, drying winds.

The easiest way for the untrained botanist to distinguish the two is to note their native distribution.

Illicium verum, Chinese star anise, is native to the tropical forests of Southeast China and Vietnam where it has been cultivated since about 2000 BCE and highly prized for its aromatic fruits and foliage. Because of this long history of cultivation there is no certainty that the plants growing there are actually wild and not naturalised specimens.

Illicium anisatum, Japanese star anise, grows in woodland in Japan, Korea and Taiwan. It too is highly prized as an aromatic ornamental which is used to decorate shrines and temples in Japan (but not eaten!).

History

Illicium is Latin for 'an allurement' referring to the attractive fragrance of the seeds; *anisatum* meaning 'like anise' and *verum* meaning 'true'. *Illicium verum* was introduced to England from the Philippines in 1588 by the sailor Sir Thomas Candish (also known as Candi and Cavendish) after circumnavigating the world. He gave them to Queen Elizabeth's apothecary, Hugh Morgan (1530-1613) and James Geret (also Garret), a dealer in spices, according to Jacob Clusius, who calls it *Anisum Philippinarum insularum* [anise of the islands of the Philippines] (Clusius, 1601). Parkinson (1640) reported that one of the islanders who had come to England with Candish wrote that its name, 'in China characters, which as Clusius saith he could not imitate, was *Damor*, every letter being written under the other downewards' – referring to the vertical format of Chinese script. Gerard (1633) also retells Clusius, calling it *Anisum Indicum stellata* [Indian anise star]. The seeds were used to make sweets (Pomet, 1712); the distilled oil was used by the Dutch in Indonesia to flavour the alcoholic spirit arrack (or arak) (Hill, 1751) and as a carminative (to relieve flatulence), for kidney stones and coughs, and as a diuretic (Linnaeus, 1782). It is no longer used to make arrack but is an EU approved ingredient in the Italian liqueur sambuca.

Linnaeus (1759) described *I. anisatum* based on the seeds of Engelbert Kaempfer's '*Somo* vulgo *Skimi*' [*Somo* commonly called *Skimi*] from Japan that Kaempfer had described in *Amoenitatum exoticarum* (1712). Kaempfer recognised *Skimi* as being very poisonous as he wrote that it was mixed with puffer fish (*Tetraodonti ocellari*) to increase its poison. Linnaeus wrote that the 'ANISI STELLATA', star anise, of the apothecary shops, which came from China, Tartary and the Philippines (and would have been *I. verum*, the *Anisum Philippinarum insularum* of Clusius and the *Anisum Indicum stellata* of Gerard) was *I. anisatum* (Linnaeus, 1749). Consequently, the medicinal star anise was regarded as *I. anisatum* through most of the 19th century (Lindley, 1838; Bentley, 1861; *British Pharmacopoeia*, 1864; Waring, 1868). It was not until the 1880s when Dr Emil Bretschneider, Medical Officer to the Russian Embassy in Peking, pointed out that there were two star anise species, not one, and that *I. anisatum* from Japan was poisonous and the one from China was not, that *I. verum* from China was separated from the Japanese *I. anisatum* (which is only found there and in Korea and Taiwan) and described by Joseph Hooker (Hooker, 1888). Hooker's plant came from Henry Kopsch, a British Commissioner in the Chinese Customs Service who sent him seedlings, via Charles Ford, Superintendent of the Hong Kong Botanic Garden, which flowered at Kew in 1887 (Bretschneider, 1898).

The trade in what was called '*I. anisatum*' (fortunately, actually *I. verum*) in the 19th century in China was vast. Shanghai in 1872 handled 350 tons (320,000kg), most of it being reshipped to other Chinese ports or going on to India in addition to the overland route via the Silk Road through Yarkand (Flückiger & Hanbury, 1874). It had little place in the medicines of the 19th and 20th century in Britain. Oil of anise which was made by distillation of either the seeds of aniseed (*Pimpinella anisum*) or *I. verum*, was principally used as a flavouring for cough medicines and lozenges (Martindale, 1967).

The seeds of the Chinese star anise, *I. verum*, have a long history of use in traditional Chinese medicine and cookery, being used in many south-eastern Asian dishes for their distinctive flavour and aroma of aniseed and cardamom. Extracts from the seeds have been used as a topical antiseptic and as a tea to treat colic, backache, and toothache.

Chemistry

Toxicity of *I. anisatum*

The poisonous Japanese star anise, *I. anisatum*, contains neurotoxic sesquiterpene lactones including anisatin, which are irreversible inhibitors of the neurotransmitter gamma-aminobutyric acid, whose function is to reduce neuronal excitability throughout the nervous system.

Sadly, there have been many cases of accidental contamination of Chinese star anise by the toxic Japanese variety (which looks very similar). Consumption of the poisonous tea made from it leads to loss of consciousness, seizures, abnormal movements, and in many cases, death.

The seeds of the Japanese star anise have been used as incense, and their essential oil has been employed as a liniment for painful joints.

Oseltamivir

In 1993 Dr Norbert Bischofberger, then Director of Organic Chemistry at Gilead Sciences, started work on developing an orally-absorbed agent that was active against both influenza A and B. By 1996 clinical trials of oseltamivir had begun, and in 1999 patents on the drug were exclusively licensed to Roche, who completed the clinical development programme. Oseltamivir (Tamiflu) was approved by the US Food and Drug Administration for the treatment of influenza in adults in 1999, and by the European Medicines Agency in 2002.

The marketed product is oseltamivir phosphate, a pro-drug which is metabolised in the body to the active agent, oseltamivir carboxylate. Oseltamivir phosphate is synthesised from oseltamivir epoxide (Ro 64-0792) which itself can be synthesised in a number of steps from either shikimic acid or quinic acid. Shikimic acid was originally sourced from Chinese star anise but is now manufactured recombinantly in *Escherichia coli*. Quinic acid is derived from the bark of the *Cinchona* tree (*q.v.*), the source of quinine. A novel and efficient method of extracting and isolating shikimic acid from the leaves of the *Ginkgo biloba* tree using an ionic liquid was published in 2011 (Usuki *et al*, 2011).

Oseltamivir reduces replication of the influenza virus by inhibiting a viral enzyme, neuraminidase, which is necessary for the release of viral progeny from infected host cells.

Oseltamivir phosphate is rapidly absorbed from the gastrointestinal tract and converted by liver enzymes (esterases) to the active metabolite.

Chemistry

It is widely distributed throughout the body, including the upper and lower respiratory tract; the plasma half-life of the carboxylate is 6-10 hours. The parent drug and its metabolite are primarily excreted renally, with a small amount being found in the faeces. There is minimal potential for interactions with other medications. The predictable pharmacokinetics and relatively long half-life mean that it can be given once or twice daily, the recommended treatment course being for 5 days (Davies 2010).

The clinical studies submitted to the registration authorities showed that, if started within 36 hours of the start of symptoms, oseltamivir shortened the duration of fever and symptoms by about 30 hours compared with placebo, in patients with either influenza A or B. There was little effect on the duration of viral shedding. If given prophylactically to contacts of influenza patients, fewer subjects developed the disease. No major safety concerns were reported in any of the studies.

Ongoing concerns about a possible pandemic of avian influenza, together with the 2009 H1N1 swine flu pandemic prompted many governments to stockpile oseltamivir; the UK spent £424 million on this, and the USA $1.3 billion. In 2009 the Cochrane Collaboration were commissioned to update their systematic review of the efficacy and safety of oseltamivir. Despite promises by the licence-holder, Roche, to provide them with the complete patient data from their clinical trials, it took nearly four years for the complete data to be released, following a lengthy campaign by the *British Medical Journal* (Anon, undated). The updated review, published in 2014 (Jefferson 2014), found that, compared with placebo, oseltamivir reduced the time to symptom reduction in adults by 16.8 hours, and in previously healthy children by 29 hours (there was no effect in asthmatic children). Treatment had no effect on the risk of hospitalisation of adults or children, nor on serious complications of influenza.

In prophylaxis studies, oseltamivir reduced the risk of developing symptomatic, but not asymptomatic, infection. Oseltamivir was associated with an increased incidence of nausea and vomiting, and neuropsychiatric events.

Hence the benefits of oseltamivir were modest, the costs were high, and there were significant adverse effects. Post-marketing surveillance identified further uncommon side effects of treatment including hepatitis, cardiac arrhythmias and skin reactions (however, it seems safe to use during pregnancy) (Svensson, 2017). In 2017 the World Health Organisation, which had added oseltamivir to its list of essential medicines in 2010, downgraded its status.

References

Anonymous statement. Tamiflu campaign. https://www.bmj.com/tamiflu accessed 4 August 2022

Bentley R. *A Manual of Botany*. London, John Churchill, 1861

Bretschneider E. *History of European Botanical Discoveries in China*. Two volume reprint in 1981 from Zentral-Antiquariat Der Deutschen Demokratischen Repiblik, Leipsig, 1898

British Pharmacopoeia. London, Spottiswode & Co., 1864

Davies B. Pharmacokinetics of oseltamivir: an oral antiviral for the treatment and prophylaxis of influenza in diverse populations. *J Antimicrob Chemother* 2010;65 (Suppl 2): ii5–10. doi: 10.1093/jac/dkq015

Flückiger F, Hanbury D. *Pharmacographia. A History of the Principal Drugs of vegetable origin met with in Great Britain and British India*. London, Macmillan and Co., 1874

Graner S, Svensson T, Beau A, *et al.* Neuraminidase inhibitors in pregnancy and risk of adverse neonatal outcomes and congenital malformations. *BMJ* 2017; 356: j629. doi: 10.1136/bmj.j629

Hill J. *A History of the Materia Medica*. London, T. Longman, C. Hitch & L. Hawes (*et al*), 1751

Hooker J. Illicium verum. *Bot Mag* t.7005, 1888

Jefferson T, Jones M, Doshi P, *et al.* Neuraminidase inhibitors for preventing and treating influenza in adults and children. *Cochrane Database of Systematic Reviews* 2014, Issue 4. Art. No.: CD008965. doi: 10.1002/14651858.CD008965.pub4

Kaempfer E. *Amoenitatum exoticarum*. Lemgo, Heinrich Wilhelm Meyer, p880, 1712

Lindley, John. *Flora Medica*. London, Longman, Orme, Brown, Greene and Longmans, 1838

Martindale W. *Extra Pharmacopoeia*, 25th edn. London & Bradford, Percy Lund, Humphries & Co. Ltd., 1967

Pomet P. *A Compleat History of Druggs* (translated from *Histoire générale des drogues*). London, R. Bonwicke and others, 1712

Usuki T, Yasuda N, Yoshizawa-Fujita M, Rikukawa M. Extraction and isolation of shikimic acid from Ginkgo biloba leaves utilizing an ionic liquid that dissolves cellulose. *Chem Commun* 2011;47:10560-62. doi: 10.1039/c1cc13306c

Waring EJ. *Pharmacopoeia of India*. London, W.H. Allen & Co, 1868

CHAPTER 27 – ANTHONY DAYAN

Inula helenium
The source of inulin

The root of *Inula helenium* contains a polysaccharide carbohydrate, inulin. It has never been a 'Prescription Only' medicine or used therapeutically in modern times. However, it has a long history of herbal use in the past two millennia and is included here because of its usefulness in medicine for measuring kidney function.

Introduction

Inula helenium, elecampane, enula, horseheal, scabwort or wild sunflower, is a member of the Asteraceae family.

In the 1930s inulin became an important test for measuring kidney function, in particular its filtering capacity. Inulin is neither absorbed nor digested in the body but is excreted solely by filtration through the glomeruli in the kidney. Its clearance rate from the blood is called the glomerular filtration rate and is a measure of kidney function. The results of more convenient modern techniques are still compared with the inulin clearance test which is regarded as the 'gold standard'.

Nomenclature

The formal scientific name *Inula helenium* came from Linnaeus' systematisation of plant names in the mid-18th century but it had been known both as *Inula* and *Elenium* since Classical times. Perhaps named *helenium* from the mythology that it grew from the tears of the beautiful Helen of Greek mythology (whose abduction from her husband King Menelaus of Sparta by Paris of Troy led to the fabled Trojan war) according to Leonard Fuchs' *Historia Plantarum* (1542). Coles (1657) added that some say it was named for her as she discovered its properties 'against the biting and stingings of venomous beasts' – a use noted by Theophrastus (371-c. 287 BCE).

Elecampane is derived from Campania, the Roman province around Naples, where it was cultivated. Horseheal and scabwort come from use of an extract of its root in the Middle Ages to treat skin diseases of horses and scab in sheep, an infestation of the skin by mites. Wild sunflower refers to its large and glorious golden-yellow blooms.

Plant profile

Inula helenium is probably native to temperate Eurasia but is now naturalised throughout Europe and North America where it can be found growing in fields, waste places and on waysides. It is a herbaceous perennial: large, simple, felty leaves emerge in spring followed by sturdy branched stems up to 2m high, each tipped with several single, bright yellow flowerheads. In bloom between June and September, the whole plant has a striking presence and cheerful aspect. Leaving the flowerheads on to seed in the autumn prolongs the period of interest and also provides food for birds. *Inula helenium* is vigorous and easy to grow. It prefers deep, rich, moisture-retentive soil in full sun or partial shade, but it does not require much attention apart from staking of the tall flower stems. Over time the rootstocks form quite a solid mass from which the aromatic rhizomes spread. These can be lifted and divided in autumn or spring.

ELENIO.

G 4   na

Inula helenium – woodcut from Mattioli's *Discorsi*, 1568

Historical uses

There are many historical and current claims of the value of the roots or extracts of them in Ayurvedic and Chinese traditional medicine in Central and Western Europe and particularly in the Far East, as expectorants, anti-tussives, diaphoretics, anti-emetics, and bactericides and vermicides (Barnes *et al*, 2007; Seca *et al*, 2014). There are about 20 closely-related species of *Inula* with similar properties and uses.

Hippocrates (460-370 BCE) is alleged to have written 'Elecampane drives away anger and sorrow, strengthens the mouth of the stomach, clears the chest, expels the superfluities in the veins by the menses and urine, and more especially a wine made from it' according to Francis Adams (1847) quoting Serapion who attributes it to Hunain's (Hunayn Ibn Ishaq al-Ibadi (809-873)) translation of Hippocrates.

However, this quote is not to be found in Serapion's *De Simplicibus Medicinis* (1531) at the place that Adams indicated (or elsewhere), nor in *The Genuine Works of Hippocrates* (Adams, 1849). Evans' account appears to be the source of the often-reported statement that *Inula* was a medicinal plant known to Hippocrates.

Dioscorides [70 CE] attributed diuretic and emmenagogue properties to the root, good for cough, colic, shortness of breath, ruptures (which may mean dislocations which is what Galen says it is used for), convulsions, haemoptysis, and bites from serpents and venomous beasts, and that the leaves soaked in wine and applied were good for sciatica. Pliny [70 CE] said that anyone who used it would be amiable, gracious and loved, (like Helen of Troy); in wine it made one happy and took away sorrow and melancholy, cured orthopnoea, coughs, loin pain and snake bite. Chewing the root fastened loose teeth but, as a powder, killed mice, but Pliny's *Inula* may have been a different plant as he says it grows like wild thyme (which is rarely more than a few centimetres tall). Galen [200 CE] agreed with Dioscorides and Pliny and said that it cured hip pain called sciatica [*coxarum passiones, ischiadas vocant*].

Nearly fifteen hundred years later, Treveris in *The Grete Herball* (1529), says for asthma that *Inula* root was boiled with liquorice in barley water, and then dried, ground up and given as a powder. It was included in the *Pharmacopoea Londinensis* (1618), and the great English apothecary Culpeper wrote in 1649 that it was 'one of the most beneficial roots nature affords for the help of the consumptive', adding in 1650 that 'Elecampane, is ... wholesome for the stomach, resists poison, helps old coughs and shortness of breath, helps ruptures and provokes lust; in ointments it is good against scabs and itch' (Culpeper, 1650). Quincy (1718) says it makes good sweets but there were better medicines for those conditions for which it was recommended.

At the end of the 18th century, Woodville comments that 'we have no satisfactory evidence of its medicinal powers' (Woodville, 1792).

Despite this falling out of use in medicine, in more recent herbal medicine an extract of the root has been used as an expectorant, as a soothing medicine which brought on menstruation, and to treat coughs and mild respiratory-tract infections. An anti-fertility effect is claimed in some countries (Seca *et al*, 2014). In Britain it is recognised in complex mixtures for some of those purposes as a traditional herbal medicine, but not in the EU.

Inula helenium root extract as a medicine is in the official pharmacopoeias of Bulgaria and Poland (Seca *et al*, 2014). The extract has been used to kill common types of intestinal worms and it has sufficient bactericidal and fungicidal activity to have been used at times to treat skin infections.

Dietary uses

In the kitchen the grated, sliced and even dried aromatic roots and rhizome and the essential oil extracted from *Inula helenium* have been used to flavour many sorts of foods and sauces, wine, absinthe and vermouth. The inulin content of the root, which gives it a sweetish flavour (about 10-15% of sugar), has long led to use of the root itself or a water extract in sweets and confectionery since mediaeval times, and dried slices were sold as confectionery even in late 19ᵗʰ century London.

As inulin is not digested in the body it has been of particular interest in foods for diabetics as a non-calorific sugar replacement. An extract of the root has been used in France and Switzerland in the manufacture of absinthe.

In the EU inulin and its partial hydrolysis products, the oligofructans, are permitted food ingredients as sources of dietary fibre and non-nutritive sweetening agents to stabilise jellies and foams, to retain moisture in baked products and as a partial replacement for fats in dairy products such as yoghurts (Niness, 1999; Barnes *et al*, 2007; Franck, 2002; Nair *et al*, 2010). In the USA they have official GRAS status (Generally Recognised as Safe) for similar uses.

The longer chain inulin molecules are a significant type of dietary fibre and, together with their digestion products, may exert some cholesterol-lowering activity as well as promoting the activity of the gut.

Industrial production of inulin today is from the roots of chicory (*Cichorium intybus*; inulin content up to 20-30% by weight) and Jerusalem artichokes (*Helianthus tuberosus*), and synthetically from sucrose (Franck, 2002). Fructans of this type are found in storage roots of many other plants.

Chemistry

Phytochemistry

In addition to terpenoids, which includes several common sterols, the root contains up to 4% sesquiterpene lactones, most notably the isomers alantolactone and isoalantolactone, together known as Helenin or Elecampane camphor. This compound is present in the volatile oil that can be distilled from the root. It is probably the source of the anti-helminthic and bactericidal activity and perhaps of other claimed actions on blood pressure and blood sugar level. It may also be a skin sensitiser causing allergic reactions in some people. The lactone, isoalantodiene, is a potent plant growth inhibitor.

The inulin clearance test of kidney function

The classical inulin clearance test of kidney function involves constant intravenous infusion of inulin for several hours followed by serial measurements of the inulin concentration in blood and urine. The rate of fall in the blood level represents the inulin clearance by filtration through the kidney glomeruli: as a good measure of the glomerular filtration rate, it indicates the state of one of the most important functions of the kidney.

Measuring inulin clearance this way is arduous for the patient and technically demanding for those performing the test. It has largely been replaced by other, simpler and faster means to measure the kidney health.

Summary

Inula helenium is an attractive plant in the College Garden and has had recognised uses in human and veterinary herbal medicine since Classical times. The fructose polymer inulin extracted from its roots and those of other plants has many uses in foods but its major importance in conventional medicine was in establishing inulin clearance as a basic measure of kidney function.

References

Barnes J, Anderson LA, Phillipson JD. *Herbal Medicines*, 3rd edn. London, Pharmaceutical Press, pp240-2, 2007

Franck A. Technological functionality of Inulin and oligofructose. *Br J Nutr* 2002;87(2):S287-S291

Nair KK, Kharb S, Thompkinson DK. Inulin Dietary Fiber with Functional and Health Attributes – A Review. *Food Rev Internat* 2010;256:189-203

Niness KR. Inulin and Oligofructose: What Are They? *J Nutr* 1999;129:1402S-1406S

Quincy J. *Pharmacopoeia Universalis Extemporanea or a Complete English Dispensatory*. London, A. Bell, 1718

Seca AML, Grigore A, Pinto DCGA, Silva AMS. The genus Inula and their metabolites: From ethnopharmacological to medicinal uses. *J Ethnopharmacol* 2014;154:286-310

Shannon JA, Smith H. The Excretion of Inulin, Xylose and Urea by Normal and Phlorinized Man. *J Clin Invest* 1935;14:393-401

Woodville W. *Medical Botany* vol 2. London, James Phillips, 1792

CHAPTER 28 – TIMOTHY CUTLER

Melilotus officinalis
The source of dicoumarol which gave rise to warfarin

Introduction

The effect on cattle that ate mouldy *Melilotus officinalis*, melilot, known also as yellow sweet clover, king's clover (also king's claver) and yellow melilot, was the source of the research that led to the production of warfarin. In the process of becoming mouldy, a bitter-tasting, sweet smelling compound in melilot called coumarin – which is not an anticoagulant – was converted by fermentation to an anticoagulant, dicoumarol. Warfarin, an entirely synthetic coumarin-based anticoagulant, is not produced directly from *Melilotus*. However, it is chemically related to dicoumarol and has better clinical properties. It works by depleting body stores of vitamin K, an essential compound needed for blood to clot.

Warfarin has been the primary anticoagulant drug to prevent and treat thrombo-embolic (blood clotting) disease from the mid-1950s until 2011 when the first Direct Oral AntiCoagulant (DOAC), called dabigatran, which inhibits thrombin (followed by others like rivaroxaban which inhibit clotting factor Xa) was introduced as a possibly safer alternative.

Plant profile

Melilotus is a small genus comprising 22 species in the Fabaceae family. *M. officinalis* is an upright, spreading, hardy annual or biennial herb. Given the right conditions it can persist as a short-lived perennial growing up to 1.8m high.

Its native distribution remains unclear as sources differ on whether it is native to Europe as well as parts of Asia, or if it was introduced to Europe at a later date. It has become naturalised and is even an invasive weed in many temperate and tropical regions. This is due to its extensive use as a forage and nitrogen-fixing crop allied to the qualities detailed below.

The stems are smooth and branched with trifoliate pale green leaves. The long, slender, spike-like clusters of yellow flowers are sweetly scented, and their nectar attracts many kinds of insects, including bees, wasps, flies, butterflies and beetles.

The seeds are produced in large numbers and can remain viable in the soil for many years. Drought-tolerant and fast-growing, it thrives in a wide range of environmental conditions as long as it is in a sunny position and the soil is not acid. Sow seed from spring to mid-summer *in situ* or in pots. As for other legumes, pre-soaking the seed for 12 hours in warm water will speed up the germination process, particularly in dry weather. Germination will usually take place within 2 weeks.

MELILOTO.

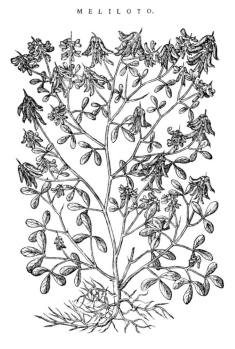

Melilotus officinalis – woodcut from Mattioli's *Discorsi*, 1568

History

Melilotus's use in medicine has been documented for centuries and can be traced to the *Materia Medica* of Dioscorides in 70 CE. He knew it from around Athens and Istanbul and from Campania in southern Italy where it was used to make garlands and called *Sertula* [Latin for a little garland]. Boiled with grape syrup and mixed with ingredients such as baked egg yolk, fenugreek flour, linseed, meal, poppy capsules or chicory, it was plastered on inflammations, especially around the eyes and genitalia. When soaked in water it was used for impetigo and, with other ingredients, for 'scurf, stomach-ache, earache and headaches'; it was not taken orally. Dioscorides' contemporary Pliny concurred and added that it was also good for cancerous sores and treating sebaceous cysts. Galen [c. 200 CE] in his book on simples, described its properties (which governed its use) as being mixed 'astringent and that it digested and concocted, and abundantly supplied with hot rather than cold properties' – concepts that pertained to Humoral medicine and have little relationship to the use of these words today. However, inflammation was attributed to an excess of hot Humors and 'cold' plants were usually used for this. The Arabian physicians, from Paulus Aegineta (625-690) to Rhazes (865-925), Avicenna (980-1037), Averrhoes (1126-1198) and Serapion (11th century), made no changes to the recommendations of the Graeco-Romans. Serapion called it king's crown (*corona regia*) alluding to its use in garlands, but (still perplexingly) both he and Avicenna said it had hot and dry properties and the hotness was greater than its coldness; that it had a bitter taste and a strong smell. Its uses continued almost unaltered and in the *Herbario volgare* (1522) it was 'hot and dry' and recommended for softening hardness of the liver and spleen, and for fistulas and haemorrhoids in addition to those noted before, including abscesses on the testicles and posterior. The writings of Dioscorides continued to be propounded into the 16th and 17th century through the books of Gerard (1597, 1633) and Parkinson (1640) which gave numerous varieties of *Melilotus* and the opinions of authorities throughout Europe, but with no validation of any of the hypothetical uses.

It was traditionally used topically in a poultice to soothe inflammations. The flowers of *Melilotus* were used by the physicians attending King Henry VIII (among whom was Thomas Linacre (1460-1524), founder of the Royal College of Physicians), as an ingredient of the King's Grace's oyntement designed to 'coole and dry and comfort the member', no doubt as a result of his strenuous sexual and sporting activities.

John Gerard (1597) also followed Dioscorides, writing 'melilot boiled in sweet wine untill it be soft, if you adde thereto the yolke of a rosted egge, the meale of fenugreeke and linseed, the roots of marsh mallowes and hog's grease stamped together, and used as a pultis [poultice], doth assuage and soften all manner of swellings'.

Culpeper (1649) more or less repeated Gerard's advice 'melilot boiled in wine and applied, mollifieth all hard tumours and inflammations that happen in the eyes, or other parts of the bodies, as the fundament, or privy parts of men and women'.

By the beginning of the 18th century *M. officinalis* was only valued for its strong and rather unpleasant ('pungent') fragrance in medication (Quincy, 1718) which was thought to be due to the presence of coumarin or benzoic acid (Lindley, 1838) in the dried plant (fresh it lacks fragrance). By then it had disappeared from pharmacopoeias. Coumarin itself was isolated from tonka beans (the seed of *Dipteryx odorata*) in 1820 independently by Vogel and Guibourt, the name derived from the French word for the bean, coumarou. It was synthesized in 1835 by Guillemettè who confirmed that Vogel and Guibourt had both isolated coumarin independently. The plant is thought to produce coumarin as a defence against predators.

Discovery of warfarin and its use in medicine

Melilotus officinalis, sweet clover, was planted as a fodder crop on the Alberta and Dakota plains at the end of the 19th century as it flourished on poor soil. In 1924, Schofield reported from the North Dakota Agricultural Experiment Station (NDAES) a hitherto undescribed fatal haemorrhagic disorder in cattle which seemed to be caused by the ingestion of spoiled mouldy silage containing sweet clover. The haemorrhage was often spontaneous and internal but procedures such as de-horning and castration were especially lethal. By 1931 Roderick had identified the cause of the haemorrhage as a reduction in prothrombin levels (an essential part of the coagulation process). The story of the discovery of the causative haemorrhagic agent is told by Karl Link (1959). In February 1933 he and his senior student, Eugen Schoeffel, were working in the biochemistry building of the NDAES, when a farmer arrived having driven nearly 200 miles through a snowstorm with a dead cow which had bled to death, a milk can full of blood which would not clot and 100 pounds of rotting sweet clover which his animals had been eating. This was the beginning of a prolonged research project and in June 1939, after staying up all night, Harold Campbell in Link's laboratory isolated 6mg of the anticoagulant chemical which was to be called dicoumarol. This was the anticoagulant released from the naturally-occurring coumarins (which are not anticoagulants) in sweet clover silage by the action of certain fungi which had caused the death of countless cows.

Dicoumarol was tried as a possible rodenticide but proved unreliable as it was too slow to cause toxicity in target rodents.

Many derivatives of dicoumarol were synthesised in the early 1940s and 'No. 42' proved to be a much more effective rodenticide and five to ten times more potent than dicoumarol. It was introduced in 1948 under the name warfarin which was derived from the initials of its patent holder the Wisconsin Alumni Research Foundation plus the coumarin-derived suffix. The use of warfarin as a rat poison eventually resulted in the natural selection of warfarin-resistant rats. Second generation and 'superwarfarins' have now been produced which have greater toxicity and speed of onset with subsequently less chance of the development of resistance.

Chemistry

Mode of action

Warfarin, like dicoumarol, blocks the production and availability of vitamin K, without which the clotting factors II, VII, IX and X become less effective. The formation of fibrin from fibrinogen is inhibited so the production of blood clots (thrombi) in the arteries and veins is much reduced, and bleeding from injuries is prolonged. A thrombus can completely block an artery leading to a heart attack or stroke, or bits can break off and travel as emboli in the bloodstream to the lungs (a pulmonary embolus) and brain (a stroke).

Potential for warfarin as a therapeutic agent for thrombo-embolic disease in humans was initially resisted, but in 1951 an army inductee was admitted to a naval hospital in Philadelphia after taking a massive overdose for attempted suicide. He was treated with a blood transfusion and large doses of vitamin K and made a full recovery. This acted as a catalyst for its introduction into clinical use for thrombo-embolic disease which was pioneered by Shapiro and Meyer (Link, 1959). By 1955, it was even used on the President of the United States, Dwight Eisenhower, who had been hospitalised in Denver, Colorado after a heart attack.

Since then, warfarin became the standard oral anticoagulant for all varieties of thrombo-embolic disease, including deep vein thrombosis and pulmonary embolism, as well as the prevention of embolic strokes in atrial fibrillation (an abnormality in heart rhythm) and in heart valve disease and in patients with implanted artificial heart valves.

Treatment with warfarin requires regular monitoring with a blood test to prevent overdosage which could lead to bleeding. The result is known as the International Normalised Ratio, which is a measure of the delay in blood clotting caused by the warfarin. Careful consideration needs to be given to possible impairment of its anticoagulant effect by numerous other drugs. St John's wort (*Hypericum perforatum*) may reduce its anticoagulant action; aspirin, anti-platelet agents, conazoles, statins and possibly cranberry juice all increase its activity. Vitamin K is used to treat warfarin overdose, and foods rich in vitamin K (leafy green vegetables and herbs) may also inactivate the anticoagulant property. For this reason, the introduction of the new class of direct oral anticoagulants (the DOACs) has replaced warfarin in a proportion of patients because there is no need for constant monitoring.

Side effects

Apart from unintended haemorrhage, side effects from warfarin are uncommon. Rarely, one can experience tissue necrosis (skin ulceration and gangrene) if it is given to patients with a concomitant protein C deficiency (a low level of one of the body's innate anticoagulants). Long-term use may be associated with an increased risk of osteoporosis (bone thinning) mainly in men and not women. Genetic factors are partially responsible for variations in warfarin activity which explains the different dosage required in different individuals to maintain a stable anticoagulant effect. Polymorphism in two particular genes explains much of the variation which is seen in African American and Asian ethnic groups.

Other plant sources of coumarins

Coumarins are found widely in the plant kingdom, *e.g. Galium odoratum*, sweet woodruff; *Dipteryx odorata*, tonka beans; *Hierochloe odorata*, sweet grass, but only dicoumarol from mouldy sweet clover has been developed into a registered medication. Warfarin is now produced by organic synthesis.

References

Guillemette A. Recherches sur la matière exicana ee du exicant. *J. de Pharmacie* 1835;21:172-8

Link KP. The discovery of Dicumarol and its sequels. *Circulation* 1959;XIX:97-107

Lindley J. *Flora Medica*. London, Longman, Orme, Brown, Greene and Longmans, 1838

Quincy J. *Pharmacopoeia Universalis Extemporanea or a Complete English Dispensatory*. London, A. Bell, 1718

Roderick LM. The pathology of sweet clover disease in cattle. *J Am Vet Med Assn* 1929;74:14-325

Schofield FW. Damaged sweet clover: The cause of a new disease in cattle simulating hemorrhagic septicaemia and blackleg. *J Am Vet Med Assn* LXIV. 1924;New series 17:553-575

CHAPTER 29 – ANTHONY DAYAN

Morus alba
The source of miglustat, migalastat and miglitol

Introduction

The iminosugars such as miglustat and its derivative, migalastat, were initially extracted from the leaves of *Morus alba*. They are used to treat rare congenital lipid storage disorders such as Gaucher's disease, Niemann Pick's disease and Fabry's disease. Miglustat, also known as deoxynojirimycin (trade name Zavesca), inhibits the production of glucosylceramide, a lipid that accumulates in the organs of people with these diseases who lack the enzyme for metabolising it. Another derivative, miglitol, is used to treat type 2 diabetes.

Miglustat, a registered drug in some countries, is an iminosugar, one of a family of compounds in which a nitrogen atom replaces one of the carbon atoms in a sugar molecule. Miglustat has an imino group in a glucose derivative.

Mode of action

Its medical importance comes from its ability to inhibit an enzyme, glucosylceramide synthase, in humans that synthesises a particular lipid molecule called glucosylceramide (a cerebroside), which is important in the normal structure and function of nerve and many other cells in the body.

Gaucher's disease and miglustat

In an unusual, serious genetic disorder called Gaucher's disease, recognised by Phillipe Gaucher in Paris in 1882, and defined and named by the American pathologist Nathan Brill (Gaucher, 1882; Brill & Mandlebaum, 1913) the patient lacks another enzyme, glucocerebrosidase, which normally metabolises (removes) that type of cerebroside resulting in the build-up and storage of excessive amounts of glucocerebroside and failure of the cells which contain it. The effects include enlargement and severe impairment of many organs, including the liver, spleen, kidneys and, in Gaucher's disease type II, epileptic fits and failure of brain development, often resulting in death of affected babies.

Morus nigra, the black mulberry (the white mulberry *Morus alba* was used for the extraction of miglustat)

Basic biochemical studies in Oxford in the 1990s showed that miglustat could inhibit the synthesis of glucocerebroside reducing its production and storage and retarding or even arresting the disease. It became a registered medicine in the early 2000s under the proprietary name Zavesca.

Various iminosugars occur in other plants, including *Commelina communis*, but the initial source of miglustat was an extract of mulberry leaves (*M. alba*), using fermentation with a fungus, *Zygosaccharomyces rouxii*. Attempts were made at industrial cultivation of different strains of the tree to provide larger quantities of the drug. That was soon replaced by extraction from large-scale fermentation of *Bacillus subtilis* and nowadays by a more convenient semi-synthetic process.

Plant profile

Morus alba is a hardy deciduous tree in the family Moraceae, native to China and India, which grows to 20m. It has been cultivated for nearly 5000 years because the large, heart-shaped leaves are the preferred food of silkworms and this has led to it being introduced across the world. The fruit is edible but not as juicy and delicious as that of *M. nigra* which was introduced in error to Britain by King James I to initiate a silk industry. It can be propagated from large cuttings. It is tolerant of many soil types but prefers a well-drained loam.

Further research into iminosugars showed that other derivatives could inhibit other related enzymes enabling treatment of similar storage diseases in which other types of cerebrosides accumulate due to genetic defects, particularly Niemann-Pick disease type C. This illness is characterised by progressive neurological impairment and miglustat slows the progress of the deterioration.

Fabry's disease and migalastat

Iminosugars as a group are being studied because they may have therapeutic value in other diseases. A derivative of miglustat, migalastat, is being used to treat Fabry's disease, an analogous genetic disorder in which a galactocerebroside called globotriaosylceramide accumulates. Sufferers from this condition experience kidney, skin, heart and neurological problems, particularly pain and strokes, due to a lack of the enzyme alpha-galactosidase A which removes globotriaosylceramide. Migalastat inhibits the production of globotriaosylceramide and slows the progress of the disease, at a cost per patient of over £200,000 per annum (NICE, 2016).

Type 2 diabetes and miglitol

Another analogue of miglustat, miglitol, an alpha-glucosidase inhibitor, has been registered for the treatment of type 2 diabetes. The enzyme alpha-glucosidase breaks down certain carbohydrates into glucose in the gut, and miglitol inhibits this so less sugar is produced, less is absorbed and the peaks of blood glucose after a meal are avoided.

Antiviral potential

They have been of interest, too, as potential anti-infective agents against HIV, HPV, hepatitis C, bovine diarrhoea, Ebola, Marburg, Zika, and dengue viruses and influenza (Zamoner *et al*, 2019). They work by interfering with viral replication and while not yet approved as antiviral medicines their potential is exciting.

References

Brill NE, Mandlebaum FS. Large-cell splenomegaly (Gaucher's disease). *The American Journal of the Medical Sciences* 1913;146(6):863-82. doi: 10.1097/00000441-191312000-00008

Gaucher PCE. De l'epithélioma primitiv de la rate; hypertrophie idiopathique de la rate sans leucémie. Paris, Medical School Thesis, 1882

National Institute for Health and Care Excellence. *Final evaluation determination – migalastat for treating Fabry disease*. December 2016. https://www.nice.org.uk/guidance/hst4/documents/final-evaluation-determination-document accessed 23 March 2022

Zamoner LOB, Aragão-Leoneti V, Carvalho I. Iminosugars: Effects of Stereochemistry, Ring Size, and N-Substituents on Glucosidase Activities. *Pharmaceuticals* 2019;12(3):108 doi: 10.3390/ph12030108

Further Reading

https://www.ema.europa.eu/en/medicines/human/EPAR/zavesca accessed 22 March 2022

Commelina communis contains iminosugars, potential sources of miglustat, but has not been used commercially

CHAPTER 30 – JOHN NEWTON

Nicotiana tabacum
Source of nicotine as an aid to stopping smoking, with a note on *Lobelia*, and *Laburnum anagyroides*

Introduction

Tobacco, from *Nicotiana tabacum* and *N. rustica*, is probably the most widely-consumed pharmaceutically-active plant in the world and also the most harmful. Popularised in Europe in the 16th century, demand for tobacco in Europe drove early expansion of the transatlantic slave trade. By the time the many health harms of tobacco were established in the later 20th century the use of this highly addictive plant had already exploded worldwide. Mass production of cigarettes from the late 19th century onwards played a part, but so too did aggressive marketing by the global tobacco industry. Even now there are high levels of tobacco consumption in many countries causing widespread harm.

Tobacco contains nicotine, responsible for causing addiction to it. *Nicotiana rustica* contains around 9% nicotine (*N. tabacum* between 1 and 3%). Some 70 or so other chemicals released from burning tobacco damage DNA and cause cancers and severe lung disease.

There are however some benefits that emerge from the otherwise tragic story of tobacco.

- Careful studies of tobacco as a cause of cancer in the 1960s required, and in turn stimulated, the development of the new science of epidemiology.

Nicotiana rustica, Aztec tobacco

- The challenge of addressing the worldwide epidemic of tobacco-related harm led to the formation of international partnerships and public health infrastructures with much wider benefits.
- The study of the effects of nicotine on the nervous system led to many pharmacological discoveries and the development of new medicines.
- Nicotine itself is now widely used in various forms as a smoking cessation aid in countries that pursue harm-reduction policies.
- Tobacco plants have been bio-engineered to mass produce drugs made by other plants.
- So-called 'bio-pharming' techniques have been used with tobacco plants to produce completely new beneficial proteins such as recombinant antibodies, vaccines, enzymes, and other regulatory proteins.

Plant profile

Nicotiana tabacum is an annual or short-lived perennial native to Bolivia but it has been so long cultivated across the world that it is no longer found in truly wild situations.

The lower leaves of the plant are large, hairy and pale green in colour, forming a rosette at the base of the stem which can grow to a height of 1.4m. The leaves are arranged alternately up the stem and diminish in size towards the top. All parts of the plant are sticky and covered with short viscid-glandular hairs which exude a yellow secretion containing nicotine.

From July to September, pink, fragrant flowers appear arranged on a terminal, branched panicle, and the seeds, which are numerous and tiny, ripen in light brown capsules from August to October. The species is hermaphrodite (monoecious, having both male and female organs) and is pollinated by bees, moths and butterflies.

Nicotiana tabacum prefers sun and moist soil and is frost-tender. Although we have some sturdy plants at the Royal College of Physicians which have survived two or three mild winters outside in the Garden, we sow new seed annually and raise the seedlings in the greenhouse until the risk of frost has passed.

History of tobacco use

There is archaeological and ethnographic evidence that tobacco has been smoked for some 8000 years in both North and South America. *Nicotiana tabacum* is associated with North America and the West Indies, as an introduced plant which was heavily cultivated there, and *N. rustica* with Latin America where it was known as Aztec tobacco. Christopher Columbus recorded the use of smoked tobacco in 1492, and other early explorers noted its medicinal use in countries such as Mexico. The common name tobacco comes from the Spanish word for the pipe (a long tube) with which the indigenous people of America smoked the plant, but it quickly came to be used for the plant itself.

The popularity of tobacco in Europe took off in 1560 when Jean Nicot, the French ambassador in Lisbon, introduced it to Queen Catherine de Medici, the mother of the King of France. Nicot had been sent to Lisbon to arrange a royal marriage, but he had failed in this task and was presumably looking for other opportunities to find favour at court. He had been impressed by the use of tobacco in Lisbon as a painkiller and treatment for cancer. He sent back seeds to France and, on his return, encouraged the Queen to try it for her migraine. His preparation of powdered leaves inhaled through the nose clearly worked and

Nicotiana tabacum from l'Obel and Pena's *Stirpium Adversaria Nova*, 1571, with a man smoking a 'tabac'

she was impressed. Nicot prospered and what the English later called snuff became both fashionable and a lucrative international trading commodity. Promoted as a medicine, early uptake of tobacco was more likely to have been driven by its addictive nature than by any medicinal benefit. In 1586, the French physician and botanist Jacques Daléchamps called tobacco '*Herba Nicotiana*', and subsequently Linnaeus (1753) named the tobacco genus *Nicotiana* in recognition of Nicot's role.

In 1597, Gerard described tobacco plants as 'Henbane of Trinidad' and 'Henbane of Peru' and attributed many virtues to them. For example, he notes that Nicolas Monardes (a Spanish physician from Seville) recommends tobacco applied to the affected part as a treatment for migraine, toothache and gout. Gerard also recommends it taken by mouth for fits in childbirth and against the poison of venomous beasts. He eloquently describes smoking tobacco 'taken in a pipe set on fire and sucked into the stomach and thrust forth against the nostrils' and recommends it as a general panacea for aches and pains of any origin. He warns however that the effect is short lived and only palliative.

Tobacco was first imported commercially by the Portuguese and grown in their colonies in Brazil using slave labour. It is also thought that Portuguese sailors spread the use of tobacco beyond Europe and America to countries such as India and Japan as they navigated the globe. Demand for tobacco grew in the 16th and 17th centuries, and this resulted in many more slaves being brought over from Africa. The gruesome efficiency of the slave trade meant that the poorest-quality tobacco could be used to buy slaves in Africa who were then delivered to America on the same ships that would bring the better-quality tobacco back to Europe for sale.

The English also played their part in the early exploitation of tobacco as a commercial commodity. John Rolfe (husband of Pocahontas) planted the first cultivated crop of tobacco in Virginia in 1612, and within ten years it was that county's biggest export. King James I of England in 1604 wrote *A Counterblaste to Tobacco* attacking the 'vile use of taking tobacco in this country', introduced a 30% import tax and warned of its danger to the lungs. It took a further 350 years 'before anyone noticed' that around 90% of deaths from lung cancer were associated with smoking. It remains one of the commonest causes of fatal disease in the world, with 8 million deaths attributable to smoking each year (that is one death every four seconds).

Nicholas Culpeper (1616-1654) frequently smoked tobacco having been introduced to it as a student in Cambridge. Tobacco was in the *Pharmacopoea Londinensis* (1618), and in Culpeper's translation of it (1649, 1650) he described numerous uses including as a cure for plague sores (buboes). However, in the precursor to his Herbal (*The English Physitian*, 1652) it is clear that he is writing about 'English tobacco' (*N. minor*) with its yellow-green flowers which grew wild in England, identified by a woodcut in Parkinson (1640) as *N. rustica*, and as henbane, *Hyoscyamus luteus*, by Quincy (1753). Throughout the 17th century tobacco was thought to prevent plague, and the boys of Eton College were apparently compelled to smoke it for that purpose or be thrashed. By the 18th century its role as a panacea had disappeared (Brookes, 1753), and the great Leipzig physician Rivinus (1652-1723), the Luxembourger Adam Chenot (1722-1789) and others, had shown it to be useless in treating the plague (Cullen, 1789). There was widespread acceptance of its toxicity, its induction of addiction and recognition of its physiological functions and absorption by every route. Excessive smoking caused vomiting and purgation, and a forcible enema of tobacco smoke (using a special bellows) was recommended for severe colic ['iliac passion'] secondary to constipation (James, 1752; Sydenham, 1809). Dr Richard Mead recommended that tobacco smoak (sic) should be blown into the intestines for those who had 'been drowned many hours' in a chapter on the treatment of tetanus by immersing the patient in cold water which occasionally resulted in death by drowning (Mead, 1756). Jacob Robinson (1746) also supplies several case histories of the use of tobacco enemas for drowning. Their use 'for reviving the vital powers [signs of life] in some kinds of asphyxia' were still recommended in the first American pharmacopoeia (Coxe, 1806) and through to 1895 (King, 1895).

For centuries, the way tobacco was consumed differed greatly between countries. The French generally sniffed it, the Americans chewed it, the Spanish smoked it in cigars while the British used a pipe. However, in the 19th century cigarettes as a development of the cigar gradually became the norm everywhere. The story behind the invention of the first cigarette is disputed but their use seems to have spread from Spain to France during the Peninsular war and from

France to Britain during the Crimean war. In 1856, a factory was set up in Walworth (London) to manufacture the new cigarettes. Rolling by hand was slow but, in 1880, James Bonsack from Virginia patented the first cigarette rolling machine – it could produce 12,000 cigarettes per hour – reducing costs and increasing volumes, paving the way for market expansion and huge profits for producers.

It is currently estimated that 1.3 billion people smoke across the world. The popularity of smoked tobacco has never been due to its modest and, as Gerard warned, transitory medicinal benefits but has always been due to the highly addictive nicotine content, a feature still exploited for profit on a massive scale by the tobacco industry.

Pharmaceutical uses of nicotine

Nicotine is the alkaloid responsible for many of the effects of tobacco, including its addictive nature, but is not one of the many carcinogens in the smoke which cause cancers. Nicotine has been used extensively in physiology laboratories to understand human and animal nervous systems. This understanding has led to the discovery of whole classes of drugs such as the beta-blockers and beta-stimulants (*e.g.* salbutamol), drugs which have dramatically improved the treatment of heart disease, high blood pressure and asthma.

The psychopharmacological effects of nicotine include mild euphoria, increased energy and arousal, and reduced stress, anxiety and appetite (Koob *et al*, 2014).

Nicotine also reduces pain when used locally, and raises pain thresholds. Its physiological effects are due to activation of the autonomic nervous system including the adrenal medulla. Nicotine increases blood pressure, heart rate and gut activity, dilates the bronchi and increases salivation.

But activation is short lived and is followed by longer-lasting depression of the same systems. In a habitual smoker, withdrawal of nicotine causes classic symptoms of anxiety, restlessness and irritability beginning in 6-12 hours, peaking at 1-3 days and lasting from 7-30 days. The psychological craving for nicotine can last 6 months. Smokers effectively titrate their intake of nicotine to maximise the positive effects and keep these withdrawal symptoms at bay.

Soon after Doll and Hill demonstrated the harmful effects of smoking (Doll & Hill, 1952) other scientists began tentatively to establish the theory and practice of managing nicotine addiction. In 1967, Lichtneck and Lundgren, two doctors from the University of Lund, proposed the development of a nicotine-containing chewing gum (Elam, 2015). It took 20 years before nicotine replacement therapy became an established therapeutic strategy (Russell, 1991). It seems extraordinary now, but there was real resistance among the medical establishment in the 1970s to the concept of smoking as an addictive disorder. Since smoking was not considered to be a disease, a treatment for it could not be classed as a medicine, and nicotine-containing gum was initially refused a pharmaceutical licence.

Nicotine replacement therapy is now widely used in smoking cessation practice and is available in many forms including electronic cigarettes and other vaping devices. The topic remains controversial, and some authors still argue against nicotine substitution as an approach. Electronic cigarettes have been criticised for simply replacing one form of nicotine addiction with another and potentially recruiting new addicts who have never smoked. However, smokers are likely to benefit greatly from switching entirely from smoking to using an electronic cigarette (Newton, Dockrell & Marczylo, 2018). The evidence is that, with strong regulation, electronic cigarettes are helping tens of thousands of smokers to quit and are almost entirely used by smokers or ex-smokers.

The tobacco plant is finally playing one entirely positive role in human health. Bioengineering technologies are being used to programme tobacco plants to mass produce drugs and other beneficial compounds. For example, transgenic specimens of the *N. tabacum* plant have been bio-engineered to manufacture artemisinic acid the precursor of the antimalarial artemisinin and a compound made much more sparingly by *Artemisia annua* (*q.v.*) (Malhotra *et al*, 2016).

Chemistry

Another approach with even wider potential is agroinfiltration (Chen *et al*, 2013) in which *N. benthamiana* has been used to produce a vaccine for Ebola using a bacterium, *Agrobacterium tumefaciens*, that naturally infects tobacco plants and produces tumours of meristematic/stem cell tissue. A mouse infected with Ebola recombines the DNA in its antibody-producing white blood cells to encode anti-Ebola antibodies. The genes for these are extracted, cloned and inserted into a Ti plasmid in *A. tumefaciens*. The bacterium is then introduced into a tobacco plant leaf by direct injection or by vacuum infiltration. The meristematic tumours that grow in the leaf contain the bacterium and the allele. The tumours are cut out and the allele-containing cells are treated as meristems, multiplied up on nutrient agar plates, and repeatedly divided. When enough meristematic tissue has been produced, the meristems are treated with auxins to make them differentiate into roots and leaves – seedling tobacco plants. All the cells of these full-grown genetically-engineered tobacco plants have the allele, producing antibodies to Ebola in substantial amounts, which can then be extracted and used. This technique could also be used for producing COVID-19 antibodies for therapeutic use.

Epilogue

Nicotine – an insecticide

Nicotine is produced by the plant for its own benefit as a deterrent to sap-sucking insects, and it is an effective insecticide. In the College Garden the ants deposit aphids on the stems of *N. rustica*, but as soon as the aphids start to feed on its sap they die, and the stems are littered with their corpses. It is a very powerful, indiscriminate insecticide, as toxic to bees as it is to aphids, so is no longer commercially available for agricultural use. It is absorbed through the skin, so agricultural workers in tobacco fields wear protective clothing to avoid being poisoned.

Stems of *Nicotiana rustica* with dead blackfly

NOEL SNELL

Lobelia and *Laburnum*

Lobelia spp. the source of lobeline, and *Laburnum anagyroides* the source of cytisine, to aid smoking cessation

The genus *Lobelia* was named after the botanist (and physician to King James I) Matthias de l'Obel, or Lobel (1538-1616) (Oakeley, 2012). Lobeline is a piperidine alkaloid found in several species of *Lobelia*, particularly *L. inflata* (Indian tobacco). It was first isolated by Nobel Prize winner Heinrich Wieland; he and his brother Hermann, a pharmacologist, were cousins of the owner of Boehringer and both had consultancy contracts. Lobeline has actions similar to nicotine but is only 1-2% as potent, and hence attempts were made to use it as an aid to quitting smoking. It is no longer used, but is included here because of its historical relationship with tobacco and smoking.

Lobelia siphilitica, one of many *Lobelia* species with allegedly curative properties

Plant profile

Lobelia inflata, commonly known as Indian tobacco, is an annual or biennial native to East Canada and to North-Central and East USA. It can grow in part shade and prefers dry situations such as roadsides and waste ground. In common with other opportunistic plants which grow in this sort of habitat, the size varies according to conditions but when happy it has a spreading habit and reaches up to 90cm tall. The leaves are pale green with blue, purple or pink-tinged tubular flowers that are loosely held on upright racemes from summer through to autumn.

History

Smoking the dried leaves of *Lobelia* is a traditional Native American remedy for respiratory ailments. The medicinal use of *Lobelia* was taken up and promoted in the USA by Samuel Thomson (1769-1843), a herbal practitioner who devised his own system of 'natural medicine' (he was subsequently charged with murder after one of his patients died after taking lobeline).

Chemistry

The structure of lobeline was first described in 1925. Its pharmacological effects are complex but generally similar to those of nicotine, although it is structurally dissimilar. In low doses it is a short-acting respiratory stimulant, in moderate doses it induces cough and expectoration, and as the dose is increased it can cause nausea, vomiting, tremor, dizziness and sweating. Overdosage leads to tachycardia, hypertension, Cheyne-Stokes respiration, convulsions, hypothermia, coma and death. It can interfere with the 'reward pathway' of endogenous dopamine release by addictive substances including nicotine and amphetamines. It has been tried in the treatment of drug addictions, but with limited success.

Lobeline was one of the earliest drugs introduced (in Germany) by Boehringer Ingelheim (in 1921) as 'the respiratory analeptic, Lobelin' – the first respiratory stimulant. It was first used to aid smoking cessation in the 1930s (Dorsey 1936). It became quite widely used in proprietary smoking remedies (usually in tablet form), but in 1993 the FDA banned all over-the-counter smoking cessation products in the USA, including lobeline, due to a lack of acceptable clinical efficacy

data. A systematic review in 2012 of all the clinical data concluded that there was no evidence that lobeline had any benefit (Stead, 2012). It is no longer available as a licensed medicinal product.

Laburnum and cytisine

Cytisine is an alkaloid that occurs naturally in several plant genera and is obtained from the seeds of *Laburnum anagyroides* (the golden rain acacia). Its pharmacological actions are similar to nicotine, and is available on prescription, in many countries, as an aid to smoking cessation. However, it is not listed in the British National Formulary as being available in the UK.

Plant profile

Laburnum anagyroides, golden rain acacia, is a small, short-lived tree from the limestone regions of the mountains of Southern Europe. It is in the Fabaceae, pea family. It has hanging racemes of yellow flowers followed by poisonous seeds and is a popular garden plant.

During the Second World War it was smoked by both German and Russian troops as a tobacco substitute; in the 1960s it was marketed in Eastern Europe as a smoking cessation aid. A 2013 meta-analysis of clinical trial data (which included a rigorous study published in 2011 in the *New England Journal of Medicine*) showed significant clinical benefits which were comparable to other established treatments (Hajek *et al*, 2013). A 2014 systematic review and economic evaluation concluded that cytisine was more likely to be cost-effective for smoking cessation than the established agent, varenicline (Leaviss, 2014). Because it is cheap, appears to be effective, and is relatively safe, there have been calls for it to be made available more widely in the West as an approved medication (Prochaska 2013).

Reported side effects of cytisine at therapeutic doses include dry mouth, nausea, dyspepsia and intestinal cramps. In overdosage it can interfere with breathing and eventually lead to convulsions and death due to respiratory failure. It is possibly teratogenic and its use is contra-indicated during pregnancy and breastfeeding.

References

Brookes R. *The General Dispensatory containing a Translation of the Pharmacopoeia of the Royal College of Physicians of London and Edinburgh*. London, J. Newbery and W. Owen, 1753

Chen Q, Lai H, Hurtado J, *et al*. Agroinfiltration as an Effective and Scalable Strategy of Gene Delivery for Production of Pharmaceutical Proteins. *Adv Tech Biol Med* 2013;1(1):103

Coxe JR. *The American Dispensatory*. Philadelphia, printed by A. Bartram for Thomas Dobson, 1806

Cullen W. *A Treatise of the Materia Medica*. Edinburgh, Charles Elliot, 1789

Doll R, Hill AB. A study of the aetiology of carcinoma of the lung. *BMJ* 1952;2(4797):1271-86

Dorsey JL. Control of the tobacco habit. *Annals of Internal Medicine* 1936;10:628-31

Elam MJ. Nicorette reborn? E-cigarettes in light of the history of nicotine replacement technology. *International Journal of Drug Policy* 2015;26(6):536-42

Hajek P, McRobbie H, Myers K. Efficacy of cytisine in helping smokers quit: systematic review and meta-analysis. *Thorax* 2013;68:1037-42

James R. *Pharmacopoeia Universalis*, 2nd edn. London, J. Hodges, 1752

King J. *The American Dispensatory*. Cincinnati, The Ohio Valley Company, 1895

Koob GF, Arends MA, Le Moal M. *Nicotine*. In: Koob GF, Arends MA, Le Moal M (eds) *Drugs, Addiction, and the Brain*. San Diego, Academic Press, pp221-59, 2014

Leaviss J, Sullivan W, Ren S, *et al*. What is the clinical effectiveness and cost-effectiveness of cytisine compared with varenicline for smoking cessation? A systematic review and economic evaluation. *Health Technol Assess* 2014;18(33) doi: 10.3310/hta18330

Malhotra K, Subramaniyan M, Rawat K, *et al*. Compartmentalized Metabolic Engineering for Artemisinin Biosynthesis and Effective Malaria Treatment by Oral Delivery of Plant Cells. *Molecular Plant* 2016;9(11):1464-77

Mead R. *A Mechanical Account of Poisons*, 5th edn. London, J. Brindley, p174, 1756

Newton JN, Dockrell M, Marczylo T. Making sense of the latest evidence on electronic cigarettes. *The Lancet* 2018;391(10121):639-42

Oakeley HF. *Doctors in the Medicinal Garden. Plants named after Physicians*. London, Royal College of Physicians, 2012

Perkins EE. *The Elements of Botany*. London, Thomas Hurst, 1837

Prochaska J, Das S, Benowitz N. Cytisine, the world's oldest smoking cessation aid. *BMJ* 2013;347:f5198

[Robinson J] Anon. *A Physical Dissertation on Drowning*. London, Jacob Robinson, 1746

Russell MA. The future of nicotine replacement. *Br J Addict* 1991;86(5):653-8

Stead LF, Hughes JR. Lobeline for smoking cessation. *Cochrane Database of Systematic Reviews* 2012, Issue 2. doi: 10.1002/14651858.CD000124.pub2

Sydenham T *The Works of Thomas Sydenham on Acute and Chronic Diseases*. Philadelphia, Benjamin and Thomas Kite, 1809

Tobacco's successful symbiosis with mankind

It is the euphoriant effect of nicotine that has been the great evolutionary advantage for tobacco as it is cultivated on four million hectares of land worldwide in a symbiotic relationship with mankind. Smokers need it, the tars do not kill them for 20 years by which time they have reproduced, and their children have copied their habit into addiction; the farmers prosper; the plants prosper: a classical arrangement. That more people die from smoking than almost any other single cause, emphasises the importance for our health of removing it from our lives.

Nicotiana tabacum from Perkins *Elements of Botany*, 1837

CHAPTER 31 – GRAHAM FOSTER

Papaver rhoeas
The source of rhoeadine, thebaine and powerful opioids, with a note on other Papaveraceae

Introduction

The corn or Flanders poppy, *Papaver rhoeas*, is a common sight in European fields and roadside verges where its bright red petals lift the spirits. Like the opium poppy, *P. somniferum (q.v.)*, it contains opiate alkaloids but those of *P. rhoeas* are of low potency, and their chief medical use is as the starting point for more potent opioids.

Plant profile

Papaver rhoeas, the Flanders or field poppy, is a hardy annual thought to be native to the eastern Mediterranean region but is widely naturalised throughout Europe where it is found on disturbed soil, wasteland and roadsides. It is smaller (90cm) than *P. somniferum* and is covered in fine hairs, both on its stems and on its pinnate green leaves. The petals look crumpled as they emerge from their buds in June and July, becoming delicate, papery, scarlet flowers. Ornamental cultivars can be pink or white, some with a dark centre. The wine glass-shaped seed pod has the same 'crown' as *P. somniferum*. It contains numerous small brown seeds which are released through the pores that open under the 'crown', rather like a pepper pot. These seeds can remain dormant in the soil for 80 years or more and will spring up easily when the soil surface is disturbed and conditions are right, hence the familiar sight of swathes of red running across arable land. This is also why they are associated with the First World War, for they emerged *en masse* from the churned-up earth of the battlefields of northern France. *Papaver rhoeas* is often found in wildflower seed mixtures designed for habitat restoration and annual displays: it has the added benefit of attracting wildlife.

It should be sown in the spring on bare earth in full sun for flowering in June and July in the same year: a single plant may produce a succession of up to 400 short-lived flowers.

From earliest times, poppies have been used to treat pain, as a sedative, and as a cough suppressant, either by ingestion or by absorption of their opiates through the skin. Because *P. rhoeas* grew in northern Europe and *P. somniferum*, the opium poppy, did not, its use was better known in Britain.

Papaver rhoeas from Egenolph's *Plantarum Arborum*, 1562

History

Despite the reputation for having low potency, six seed heads of *P. rhoeas* in wine were said to be sufficient to give a good night's sleep; the leaves and seeds applied as a poultice treated inflammation, and even a decoction sprinkled on the skin induced sleep, according to Dioscorides [70 CE]. He also wrote that smelling the plant would induce sleep, a claim repeated by Quincy (1718) who had found a whole family put to sleep for several days by having corn poppies hanging up to dry in every room in their house. They woke up when the poppies were removed. Culpeper (1650) wrote 'Syrup of Red, or Erratick Poppies: by many called Corn-Roses [that] some are of the opinion that these Poppies are the coldest [*i.e.* most dangerous] of all other … I know no danger in this syrup, so it be taken in moderation; and bread immoderately taken hurts; the syrup cools the blood, helps surfets and may safely be given in Frenzies, Feavers and hot Agues'. Culpeper (1650) also used it for coughs, sore throats, pleurisy and toothache, and as a poultice to treat erysipelas. Regarding the seeds he writes 'Poppy seeds ease pain and provoke sleep. Your best way is to make an emulsion of them with barley water'. The petals were used to make a sedative syrup for children (Smith, 1818).

Poppies, as herbal medicines, are one of the few plants whose effects and uses were correctly known for the past two millennia (and more). However, in the recent past our understanding of their constituent chemicals, their function, and the medicines that can be made from them, has advanced beyond any comprehension of our predecessors.

The medicines

Papaver rhoeas contains thebaine and rhoeadine which are both mild analgesics and sedatives. They are readily converted to more potent opioids. The medical properties and effects of these compounds are discussed in the chapter on *P. somniferum*. The chief pharmaceutical interest in *P. rhoeas* is in the conversion of the plant's constituents to highly-active opioids. Thebaine is used to manufacture oxycodone and oxymorphone, the former being widely used as an analgesic in the USA where its overpromotion has led to an escalation in drug addiction and associated consequences. It is also used to make naloxone, diprenorphine and buprenorphine, potent opiate antagonists.

The second opiate found in *P. rhoeas* is rhoeadine which can be converted to etorphine, although initially this compound was derived from other members of the poppy family, specifically *P. orientale* and *P. bracteatum*. Etorphine is a highly-active opioid that is several thousand-fold more potent than morphine and has very rapid effects. Its action is readily reversed by the antidote, diprenorphine. The combination of high potency, rapid onset of action with a ready antidote makes etorphine an ideal veterinary sedative – large animals, including elephants, can be quickly sedated with an etorphine-filled dart and equally rapidly restored to full vigour by timely use of diprenorphine. The opiate-filled darts are usually dispensed with the antidote so that if the user inadvertently scratches themselves with the dart, death from respiratory failure can be avoided by administration of the antidote. It is wise to select the friend to administer the antidote with some care, those prone to panic when colleagues stop breathing may not be ideal partners for this particular venture. Its derivative, dihydroetorphine, is up to 12,000 times more powerful than morphine, and is used in China for pain relief, usually sublingually or in a dermal patch as it is poorly absorbed when taken by mouth. It is not available as a prescription medicine in the UK.

The development of potent opioids has been further expanded by the development of fully-synthetic compounds such as fentanyl, whose high potency and ease of absorption from skin patches has led to its widespread use as an easy-to-administer, long-acting analgesic that can be taken even by those temporarily unable to swallow. Unfortunately, these highly-potent opioids are very attractive to those wishing to smuggle drugs as they can be readily hidden and then reconstituted. The reconstitution process of the super-active opioids involves the massive dilution of tiny amounts of the drug, and a minor miscalculation leads to highly-potent agents which, when ingested, are rapidly fatal. The widespread use of potent narcotics in the USA has exacerbated the problems of an inadequate harm reduction programme for those addicted to drugs and led to a surge in drug overdose deaths and a reduction in the life expectancy of US citizens for the first time since the Second World War.

Footnote

Other members of the Papaveraceae, *Argemone mexicana* (Mexican poppy), *Chelidonium majus* (greater celandine), *Fumaria officinalis* (common fumitory) and *Corydalis species* also contain protopine which is an antihistamine and an analgesic. *Glaucium flavum* (horned poppy) contains glaucine, a PDE4 inhibitor which is a calcium channel blocker so dilates arteries and bronchi, and has been used to treat cough, depression, and Alzheimer's and Parkinson's diseases, but can cause delirium and sedation.

References

Quincy J. *Pharmacopoeia Universalis Extemporanea or a Complete English Dispensatory*. London, A. Bell, 1718

Smith JE. *English Botany or Coloured Figures of British Plants … the figures by James Sowerby* vol 5. London, C.E. Sowerby, 1818

Further reading

Advisory Council on the Misuse of Drugs (ACMD). *Misuse of Fentanyl and Fentanyl Analogues*. 2020. https://www.gov.uk/government/publications/misuse-of-fentanyl-and-fentanyl-analogues accessed 16 July 2022

The Mexican poppy, *Argemone mexicana* contains the antihistamine, protopine. However, the seeds contain toxic alkaloids and as a contaminant of mustard seed oil cause outbreaks of Epidemic Dropsy, right sided heart failure with oedema.

Glaucium flavum, the horned poppy, contains glaucine a PDE4 inhibitor which is a calcium channel blocker so dilates arteries and bronchi, used to treat cough, depression, Alzheimer's and Parkinson's disease

CHAPTER 32 – GRAHAM FOSTER

Papaver somniferum
The source of morphine, codeine, noscapine, protopine, papaverine and verapamil

Introduction

The opium poppy (*Papaver somniferum*) is a commonly-grown ornamental flower whose sap serves as the source of opiates, morphine and codeine, as well as verapamil which is used in a variety of heart conditions including angina and dangerously fast heart rhythm disturbances. *Papaver*'s use as an analgesic and intoxicant is ancient, with pictures of poppies being found in pharaonic tombs suggesting that the ancient Egyptians were aware of its benefits.

Plant profile

Papaver somniferum is thought to have originated in the western Mediterranean region. It is a hardy annual which grows to about 120cm with straight stems covered in tiny hairs. The glaucous grey-green leaves have a distinctive powdery sheen with undulate, serrated margins which give them a frilly look. Drooping buds open into beautiful soft-textured flowers which appear to be made

of tissue paper. Sometimes variegated and often double-flowered, they vary in colour from pale white to mauve and purple, often with dramatic dark spots at the base of the petals. The smooth, spherical seed pods are lime-green, creating a striking contrast to the petals, turning brown as the seeds mature, with a fluted cap which looks a little like a crown. The ripe seeds are liberated through holes that appear under this cap. *Papaver somniferum* is often cultivated as an ornamental and there are many named varieties. It is pollinated by bees, and will self-seed once established; otherwise, seed can be sown in the spring. It should be grown in full sun and will flower from June to September.

Papaver somniferum with incised seed head, opium sap leaking out

Papaver somniferum is one of the best examples of multiple medicinal chemicals made from a single plant; of pharmaceutical manipulation; of multiple uses and the evolutionary benefit to it from its interrelationship with mankind. Because of its usefulness to us both medicinally and as an item of trade, it is grown in vast quantities. Few plants not grown for food have had such an impact on our civilisation, with perhaps the exception of *Nicotiana tabacum*. In 2006 the world production of opium from *P. somniferum* was 6610 metric tons. In 1906 it was over 30,000 tons when 25% of Chinese males were regular users, caused by Britain selling huge quantities of opium to China to reverse our balance-of-payments deficit.

Opium poppies flower from June to September and after flowering the large bulbous seed pods can be used as a source of opioids. The pods are scarified with a curved blade and the sticky, milky sap dries to form a scab over the wound from where it is easily scraped off and used to form small balls of dried sap which contain a variety of alkaloids of medicinal benefit. Alternatively, the whole plant is harvested and crushed to extract the sap. Clinically, the most important of the secreted alkaloids are morphine and codeine which have potent neurological effects. In addition, poppy sap contains the alkaloid papaverine which has minimal neurological effects but in animals relaxes smooth muscle, including that in the arterial wall. This activity led to attempts to modify papaverine for human use and in the 1950s German chemists derived a potent vasodilator, verapamil, from papaverine. This drug is now used to treat angina and disturbances of the heart rhythm (ventricular tachycardia). It is also used for intestinal colic, bile duct and ureteric spasm, cerebral artery spasm after subarachnoid haemorrhage. Papaverine is used by local intracavernosal injection in erectile dysfunction due to neurological causes. Unwanted side effects such as prolonged priapism may occur.

In addition to morphine (12%), codeine (3%) and papaverine, the sap contains protopine an analgesic antihistamine which relieves the pain of inflammation and also inhibits platelet aggregation; noscapine, which is a cough suppressant, and thebaine which is mildly analgesic and made into oxycodone and oxymorphone which are (addictive) analgesics (*q.v.* thebaine in *P. rhoeas*). Varieties of *P. somniferum* have been raised which predominantly produce thebaine or the highly toxic oripavine instead of morphine – the ingenuity of the commercial production of addictive chemicals is almost limitless.

History

Papaver somniferum seeds have been found, dated to 4200 BCE, in Spanish caves and evidence from the Neolithic era indicates the use of opium since 5000 BCE. It is the oldest medicine in continuous use, mentioned in the Ebers papyrus from 1550 BCE (Bryan, 1974) and it was known to Theophrastus in the 3rd century BCE. It would have been swallowed, smoked, given as a suppository and applied topically.

Dioscorides [70 CE] writes that the seed is baked with bread instead of sesame seed, that the leaves and seed heads boiled together induce sleep or, ground up to a paste with barley, make a poultice for treating erysipelas. Boiled with honey until it thickens it made lozenges for coughs, chest and abdominal problems. He knew that its active ingredients were absorbed through the skin, and as a tea applied to the forehead it induced sleep, and that even small amounts of the sap were analgesic, soporific and treated diarrhoea, but too much induced coma and was fatal. It could be given by suppository. The seed head and leaves could be crushed, strained, dried and stored as tablets or the sap could be harvested from the seed head by making incisions into it and collecting it with a spoon. Pliny [70 CE]

Papaver somniferum from Mattioli's *Discorsi*, 1568

recognised three types of poppy (as did Dioscorides), one whose seeds were used on the surface of loaves of bread, the black poppy whose sap could be harvested by incising the seed heads, and the scarlet corn poppy which we know as *Papaver rhoeas*. Neither mentioned the problem of addiction, but both noted death, sometimes deliberately for unendurable illness, if taken to excess. Pliny warned of the dangers of adding it to compound medicines because of the risk of overdosage.

These basic understandings of the uses and dangers of opium sap remained unchanged through the Golden Age of Islam, and, for example, Avicenna [980-1037] mostly follows Dioscorides.

In the 16th century in England, there was no first-hand knowledge of the opium poppy, and Dioscorides was repeated verbatim. The corn poppy, *P. rhoeas* (*q.v.*), was familiar, and Turner (1568) discusses putting the flowers into butter to colour it, and the sap into cheese, adding that it would make a man go to sleep. He personally experienced the effects (including difficulty in breathing) when he inadvertently swallowed an opium-containing mouthwash he had been using for a sore tooth. He cured himself by taking 'pellitory of Spain', (*Anacyclus pyrethrum*), but recommends inducing vomiting as the best treatment. Pemel (1652) recommended a decoction of *Artemisia*. That opium was a 'cold' medicine in the Hippocratic Humoral tradition was clearly understood, for if sufficient was ingested the user died and turned cold. That cold medicines were not necessarily suitable for treating fevers was also recognised, for opium was not so used, whereas pomegranate juice and cucumbers were.

Culpeper (1650) writes that meconium is 'the juyce of English Poppies boyled till it be thick' and 'that Opium is nothing else but the juyce of poppies growing in hotter countries, for such Opium as Authors talk of comes from Utopia [*Ed:* he means an imaginary wonderland, I suspect]'. He cautions 'Syrups of Poppies provoke sleep, but in that I desire they may be used with a great deal of caution and wariness…' and warns in particular against giving syrup of poppies to children to get them to sleep.

Opium-containing medicines were marketed under names such as laudanum, and paregoric and Dover's powders. Dover's powder contained opium and ipecacuanha, the latter to induce vomiting if an overdose was taken. Up to 60mg of morphine was permitted in Gee's Cough Linctus, J. Collis Browne's Chlorodyne (which also contained cannabis) and 'Kaolin and Morph' for diarrhoea, and these were available 'across the counter' – without prescription – until the latter part of the 20th century. All came as syrups, loaded with sugar to counteract the extremely bitter taste of opiates.

It has frequently been stated that the seeds of *P. somniferum* do not contain morphine. However, in reality 1g of poppy seeds can contain 0.25mg of morphine, and while one poppy seed bagel can make a urine test positive for morphine for a week, one would need a lot of bagels to have any discernible effect.

Discovery

Morphine was extracted from opium in 1805 by the 22-year-old Friedrich Sertürner while still a pharmacy apprentice – it was the first alkaloid to be isolated from any plant, and its discovery revolutionised pharmacy. It took another 20 years before its formula was deduced and it was only synthesised for the first time in 1952 (Krishnamurti & Chakra Rao, 2016). Like other researchers on addictive drugs, he too became an addict. While morphine is still sourced by a complex extraction process from opium sap from poppies, the search is on to engineer yeast DNA so it can synthesise morphine from sugar – something which could be done in a domestic kitchen – a thought to engender anxiety in health professionals and law-enforcement officers.

Pain management

Pain is a common symptom and has presumably evolved to protect damaged body parts from further injury and encourage healing. However, there are circumstances where stopping an activity due to pain would be counter-productive – Bert Trautman famously broke his neck playing in goal in the 1956 FA cup final but continued to play, allowing Manchester City to win the trophy. To allow goalkeepers to collect cup-winners' medals evolution has developed an 'anti-pain' system in which chemical transmitters in the brain, endorphins, bind to receptors that activate a variety of pleasurable sensations and reduce pain. These receptors, the opioid receptors, are also engaged by opioids inducing a sense of euphoria, somnolence and pleasure, and reducing both pain itself and the associated anxiety. As powerful analgesics, opiates are widely used in medicine (as oral medicaments, intramuscular and intravenous injections, or as patches applied to the skin) and have been the mainstay of pain relief for millennia. Unfortunately, the sleepiness induced by opiates can be profound and lead to a reduction in breathing that, in extreme cases, may be fatal.

Morphine derivatives

Drug companies have long sought ways to retain the benefits of plant-derived opiates, chiefly morphine, without their addictive effects. The English chemist Alder Wright synthesised an acetylated form of morphine in the late 19th century, and this was later developed by the pharmaceutical company Bayer who named the drug 'heroin' meaning heroic and strong. They claimed the drug was non-addictive to boost sales, but this bogus claim led to widespread use of the drug with associated addiction. A similar marketing ploy was used recently by Purdue Pharma who marketed a modified version of codeine – oxycodone – as having little risk of addiction potential when used to treat pain. They ran an aggressive marketing campaign that persuaded many US physicians to prescribe larger and larger doses of this 'non-addictive' drug leading to a surge in people becoming addicted to opiates and a six-fold increase in deaths from overdose, which in 2017 was a greater cause of deaths than HIV/AIDS. These premature deaths have been associated with a reduction in the average life expectancy of American citizens, the first such reduction since the Second World War. The exploitation of gullible doctors, who might have been expected to question why a drug marked as 'non-addictive' was required by patients in larger and larger doses, that increased the profits of the pharmaceutical company, remains a cause of ongoing litigation in many jurisdictions (DeWeerdt, 2019)

Addiction

The pleasurable sensations induced by opiates lead rapidly to the desire to repeat the experience and as an individual becomes accustomed to the effects, tolerance develops leading to the need to take ever-increasing amounts to induce the same effects. Regular intake of opioids leads to dependence upon them and if the drugs are rapidly withdrawn a painful state of anxiety, sweating and arousal of body hairs develops, the so-called 'cold-turkey' syndrome, that is most unpleasant. Opioids are highly addictive and opiate craving, combined with the risk of accidental overdose leading to respiratory failure, may have lethal consequences as addicts take ever greater risks purchasing illegal opiates of unknown purity and strength.

Opiate addiction before 1920 was mainly with oral use. However, in Asia since the 1950s, 'chasing the dragon', where heroin is heated on tin foil and the fumes inhaled through a small tube, has become the primary recreational use (Strang *et al*, 1997). Smoking is a relatively safe way to take the drug as induction of sleep prevents further intake and protects against respiratory failure but causes severe lung damage. A form of severe brain damage, spongiform leukoencephalopathy, is a rare consequence. The introduction of the hypodermic syringe and the usage of injectable opiates has led to an increase in deaths from accidental overdose and the transmission of blood-borne viruses, such as HIV, hepatitis C, as well as sepsis among addicts who often share injecting equipment. The hazards associated with opiate addiction can be greatly mitigated by the provision of long-acting, safer opiates – such as methadone or buprenorphine – to reduce the need to buy illicit, untested drugs as well as providing drug users with opiate antagonists such as naloxone (which very rapidly reverses the effects of opiates) and by providing supplies of clean injecting paraphernalia to prevent the spread of infections. Some countries have expanded these harm-reduction measures and introduced 'safe injection areas' where people who use drugs can inject in safety with trained medical staff available to treat any overdoses. Other countries have adopted a puritanical approach to illicit drug use and discouraged the promotion of harm-reduction activities on the grounds that they encourage illegal behaviour. The USA has adopted this policy with a spectacular lack of success – over 70,000 Americans died in 2019 of drug overdoses, over half being due to synthetic opioids (NIDA). The USA leads the world in the spread of hepatitis C, with a surge in new infections over the last few years.

The opiate receptors in the brain that induce pleasure are also used in the cough-suppressant pathway and, surprisingly, they play a role in the gut where they reduce motility. Opiates are therefore powerful cough suppressants as well as extremely constipating. These two properties were initially used pharmaceutically with morphine being marked to treat coughs and a mixture 'kaolin and morphine' being widely available to treat diarrhoea, as noted at the beginning. More enlightened marketing legislation has removed these products; more selective anti-diarrhoeal products are now used that lack the pleasurable side effects of the morphine-based compounds.

Vasodilators

Vasodilators are compounds that reduce the contraction of the smooth muscle surrounding blood vessels and this relaxation reduces blood pressure. Papaverine, derived directly from *P. somniferum* has vasodilatory effects in animals but has minimal impact in humans. However, modification of papaverine led to the development of verapamil, a potent vasodilator in humans. Initially verapamil was used to treat hypertension and angina – high blood pressure is caused by excess contraction of blood vessels and when this is eased there is a reduction in blood pressure and an associated decrease in the output of the heart leading to both lower blood pressure and reduced cardiac strain that reduces angina. Verapamil is now known to work by blocking channels which allow calcium to enter smooth muscle cells that provokes their contraction. Calcium channels are also found in heart-muscle cells where they play a role in the electrical currents that regulate the heart rhythm. Blocking these channels reduces cardiac muscle excitability and verapamil is used to manage some of the associated disturbances of rhythm. The mechanism of action of verapamil in the heart was for many years controversial and the claims of American physicians who conducted complex analyses of statistical models to suggest that calcium channel blockade was not responsible for its cardiac effects were disproven by the Oxford scientist Miles Vaughan-Williams who conducted elegant experiments that showed when calcium was excluded from the experiment verapamil had no effect (*pers. comm.*).

Summary

Opium, from *P. somniferum*, and its constituents such as morphine, and its derivatives such as heroin, have been the drugs in longest usage and of greatest importance for 7000 years, principally for the control of pain. Their euphoriant action causes widespread addiction and no opiate derivatives have been synthesised that do not have this problem. For the relief of pain they have no equal but the social, health and addiction problems from their use has become a pandemic, for which no solution has been found.

References

Bryan CP (translator). *Ancient Egyptian Medicine, The Ebers Papyrus*. Chicago, Ares Publishers Inc. (translated from the German version), 1974

DeWeerdt S. Tracing the US opioid crisis to its roots. *Nature* 2019;573:S10-S12 doi: 10.1038/d41586-019-02686-2

Krishnamurti C, Chakra Rao SSC. The isolation of morphine by Serturner. *Indian J Anaesth* 2016;60(11):861-2

National Institute of Drug Abuse *Overdose death rates*. https://nida.nih.gov/research-topics/trends-statistics/overdose-death-rates accessed 12 September 2021

Strang J, Griffiths P, Gossop M. Heroin smoking by 'chasing the dragon': origins and history. *Addiction* 1997;92(6):673-83. doi: 10.1046/j.1360-0443.1997.9266734

Leguminosae

Physostigma venenosum Balfour

CHAPTER 33 – ARJUN DEVANESAN

Physostigma venenosum
The source of physostigmine and the basis for the synthetic analogue neostigmine

Introduction

Physostigma venenosum, also known as the Calabar or ordeal bean, is a legume endemic to West Africa. It was found there by Scottish missionaries in the early 19th century and its extremely poisonous seeds brought back to Scotland where it was investigated scientifically. It contains a number of alkaloids, predominantly physostigmine which is now known as a reversible acetylcholinesterase inhibitor. It has numerous important uses, being the first chemical known to constrict the pupil, reverse the toxicity of atropine and reverse the effects of the paralytic agent curare. It has, therefore, important uses in ophthalmology, toxicology and anaesthesia and is still in limited use today. No part of the plant except the seed has any toxic or medicinal property (Fraser, 1863).

Chemistry

Acetylcholinesterase inhibitors increase acetylcholine levels at the neuromuscular junction (where nerve impulses are transmitted to muscle fibres) and in cerebral synapses where acetylcholine is the neurotransmitter. It does this by preventing the acetylcholinesterase enzymes from destroying the acetylcholine liberated from the nerve endings (*cf* galantamine from *Galanthus q.v.*). Its central effects, both toxic and therapeutic, are due to its ability to pass through the blood-brain barrier.

Plant profiles

An evergreen woody vine, in the family Fabaceae, *P. venenosum* is native to tropical rainforests in West Africa from Sierra Leone to the Congo, growing to a height of 15m. It carries pendent clusters of purple, kidney-shaped flowers. The fruit is a yellow-brown pod 15cm long containing two to three oblong or slightly curved dark brown to maroon 3cm-long seeds. The Calabar bean requires a minimum temperature of 15-18°C, a sunny position and well-drained soil and in the wild is often found on riverbanks. It cannot be grown outdoors in the UK.

Medicinal use

Neostigmine (an entirely synthetic analogue of physostigmine developed by Aeschlimann and Reinert in 1931) has a similar but simpler structure. It does not pass through the blood-brain barrier so is more widely used, particularly as an antidote when the action of acetylcholine on muscles has been blocked peripherally, *e.g.* by anticholinergic drugs such as curare, or where the acetylcholine receptor sites (the neuromuscular junction) are impaired, *e.g.* in myasthenia gravis. When poisoning by anticholinergic substances (such as atropine, nicotine, strychnine, organophosphate poisons and nerve gases) causes central nervous system effects, physostigmine is the preferred antidote because it penetrates the blood-brain barrier.

Myasthenia gravis

Myasthenia gravis is a rare autoimmune condition where the body produces antibodies to the acetylcholine receptors at the neuromuscular junction, so the muscles do not receive enough acetylcholine to function normally. It is characterized by easy muscular fatiguability, either generally or just the eye muscles. The use of physostigmine to treat myasthenia gravis was discovered by Dr Mary Walker MRCP (1888-1974), working at St. Alfege's Hospital, Greenwich, UK (Walker, 1934). She demonstrated that the muscular fatigue in that condition could be abolished for a few hours by injections of physostigmine. However, the neurological side effects are unpleasant because physostigmine passes through the blood-brain barrier and can cause vomiting, dizziness, ataxia and, in bigger doses, convulsions, coma and death. In 1935 Dr Walker successfully treated a patient with myasthenia using oral neostigmine. Neostigmine remains a treatment of choice for the symptoms of myasthenia as it does not pass through the blood-brain barrier.

Physostigmine has been used to treat glaucoma, a condition where intraocular pressure of the aqueous humour is raised. Its effect is by increasing acetylcholine levels available for the ciliary muscle of the eye to contract, which allows a better outflow of aqueous humour (Pinheiro *et al*, 2018). It has been used to increase intestinal mobility if it is impaired after surgery.

Physostigmine's use in anaesthesia for reversing the neuromuscular block and muscle paralysis caused by curare has problems because of its ability to cross the blood-brain barrier, causing neurological side effects, and neostigmine has been used instead since the 1950s. This ability, however, makes it potentially useful for treating memory loss due to dementia, where there is a deficiency of acetylcholine in the brain, particularly in Alzheimer's disease. However, its use is limited by its short duration of action and its side effects. As such, it is not currently licensed for treating dementia. Acetylcholinesterase inhibitors such as galantamine (*q.v. Galanthus*), which have a longer duration of action are used.

History

Physostigma venenosum was called the Calabar bean by European Missionaries to Old Calabar, an eastern province of Nigeria, where they first heard of the use of a seed as a test to see if a person was guilty of witchcraft. The ordeal that the accused was put through by virtue of ingesting the bean led to the name 'the ordeal bean'. If the accused vomited and survived, he or she was deemed innocent while those that died were considered guilty. One such missionary, Rev. Hope Masterton Waddell, describes first encountering the mysterious beans lying on a table when meeting a tribal chief at his house. When he asked what they were he was told that they were 'the poison bean, ésere, commonly called 'chop nut', used as an ordeal in cases of imputed witchcraft, to discover the innocence or guilt of the party accused' (Waddell, 1863).

The use of the 'ordeal bean' was also undertaken to prove the innocence or guilt in other matters or as a trial of worthiness to become a warrior. The accused (or potential warrior) would take a quantity of the bean, either by chewing it or by an infusion. Often an enema of the powdered bean was administered as well. There would be no effect for about 10 minutes, then 'thirst developed with abdominal cramps, flushing and swelling of the face, protrusion of the eyes, trembling, lacrimation, perspiration, salivation, foaming at the mouth, purgation, constricted pupils … progressive paralysis beginning in the lower extremities, inability to stand; and cardiac and respiratory failure in about one-half an hour' (Morton, 1977). Those familiar with the ordeal would take a large quantity of the bean rapidly to induce vomiting before its poison could take effect, the guilty would sip it slowly so absorbing sufficient to be killed by it. The accuser of a person who 'proved' himself not guilty by surviving was often obliged to take the 'ordeal bean' afterwards, a policy which may well have reduced the number of accusations made.

Waddell was a Scottish missionary, and this provides some explanation as to why much of the early research on Calabar beans has Scottish connections (particularly Edinburgh). Indeed, the first serious account of the toxicology of the bean was documented in 1855 by the famous Scottish toxicologist Robert Christison. Waddell gave Christison some Calabar beans and recounted a trial by ordeal that he had witnessed on his missionary journeys. At the time, Christison was not aware of any European botanical description and so he gave some seeds to his friend John Hutton Balfour to cultivate. It was six years before Balfour reported his botanical findings and named the plant *P. venenosum*.

At first, Christison had some doubts as to whether the beans he received were indeed the Calabar bean and so he proceeded to ingest small quantities of the bean and examine the effects on himself. In small doses (an eighth of a bean) he only observed a slight numbness in his limbs and possible drowsiness. A slightly larger dose caused giddiness and 'the peculiar indescribable torpidity over the whole frame which attends the action of opium or Indian hemp in medicinal doses' followed by other effects that he was lucky to survive (Christison 1855). He then swallowed shaving water as a purgative, vomiting up the poison, convinced that he did indeed have the Calabar bean in his possession. Christison then tested it on a number of animals, examining the effects of the poison and noting that it acted on many vital bodily systems, including the circulation, breathing and the nervous system, but did not seem to cause tetany (like strychnine) or paralysis (like curare).

It was not long until the first report of a potential medicinal use for an extract of the Calabar bean was first described. It would come to be called physostigmine after the botanical name of the plant. Argyll Robertson, a famous ophthalmologist, published in 1863 a paper which announced the first agent capable of constricting the pupil of the eye, an extract of the Calabar bean. Argyll Robertson openly admitted that he was alerted to the potential effects of the extract by a friend, Thomas Fraser, who was completing his MD on the systemic effects of physostigmine. In this paper, Argyll Robertson also noted that the effects of the Calabar bean were antagonistic towards the effects of *Atropa belladonna* (*q.v.* from which we get atropine) on the eye – which is to cause dilatation of the pupil.

Chemistry

Fraser's thesis, submitted in 1862, detailed the systemic effects of physostigmine and soon a number of studies examining the effects on a wide variety of animals were published. While Fraser is often credited as developing the theory that physostigmine and atropine are antagonistic, a number of earlier articles had already mentioned this possibility. In particular, the ophthalmologist Niemetschek noted the antagonism between atropine and physostigmine in the eye and mentioned it to his colleague Kleinwächter who successfully used physostigmine to treat a patient suffering from atropine poisoning. However, at the time, pharmaceutical antagonism was not a well-established theory. Fraser's contribution in this area was the systematic and precise investigation of this antagonism.

In 1900, the physiologist J Pal was the first to discover that physostigmine reverses the effect of curare in an experiment on dogs. Having given a dog curare as part of the anaesthetic, Pal administered physostigmine to observe the effect on peristalsis in the gut (the muscle contractions that propel food along the digestive tract). Surprisingly, not only did he observe an increase in peristalsis but the dog also started breathing (Nickalls & Nickalls, 1985). This heralded the use of physostigmine as a reversal agent for the prolonged muscle paralysis caused by curare when used in anaesthesia (*q.v. Chrondrodendron tomentosum*). So, the ability to quickly reverse the effect of curare with physostigmine not only contributed to scientific understanding of the physiology of the neuromuscular junction, but also allowed the widespread use of curare in anaesthesia.

End note

This is a short summary of the long and fascinating history of this plant and its uses over the years and across continents. For a more detailed review, see Alex Proudfoot's 'The Early Toxicology of Physostigmine: A Tale of Beans, Great Men and Egos'.

References

Aeschlimann JA, Reinert M. Pharmacological Action of Some Analogues of Physostigmine. *J Pharmacol & Exper Therap* 1931;43(3):413-444

Argyll Robertson D. On the Calabar bean as a new agent in ophthalmic medicine. *Edinb Med J* 1862;8:815-20

Balfour JH. Description of the plant which produces the ordeal bean of Calabar. *Trans R Soc Edinb* 1861;22:305-14

Christison R. On the properties of the ordeal bean of Old Calabar, Western Africa. *Edinb Med J* 1855;1:193-204

Fraser TR. On the characters, actions, and therapeutical uses of the ordeal bean of Calabar (*Physostigma venenosum*, Balfour). Section I: history, employment as an ordeal, botanical characters etc. *Edinb Med J* 1863;8:36-56

Kleinwächter. Beobachtung über die Wirkung des Calabar-Extracts gegen Atropin-Vergiftung. *Berl Klin Wochenschr* 1864;1:369-71

Morton JF. *Major Medicinal Plants. Botany Culture and Uses.* Springfield, Illinois, Charles C Thomas, 1977

Nickalls RWD, Nickalls EA. The first reversal of curare. A translation of Pal's original paper 'physostigmine, an antidote to curare'. *Anaesthesia* 1985;40:572-5

Pal J. Physostigmin ein Gegengift des Curare. *Zentralblatt fur Physiologie* 1900;14:255-8 (noted in Nickalls & Nickalls, 1985)

Pinheiro GKLO, Araújo Filho I, Araújo Neto I, *et al*. Nature as a source of drugs for ophthalmology. *Arq Bras Oftalmol* 2018;81(5):443-454

Proudfoot A. The Early Toxicology of Physostigmine: A Tale of Beans, Great Men and Egos. *Toxicology reviews* 2012;25:99-138

Waddell HM. *Twenty-nine years in the West Indies and Central Africa: a review of missionary work and adventure.* 1829–1858. London, T Nelson & Sons, 1863

Walker MB. Treatment of myasthenia gravis with physostigmine. *The Lancet* 1934;i:1200-1

CHAPTER 34 – NOEL SNELL

Pilocarpus microphyllus
The source of the drug pilocarpine

Plants of the genus *Pilocarpus*, called jaborandi, native to South America, are the source of pilocarpine, which was isolated from the leaves in the 1890s. Pilocarpine is a medication used for the treatment of dry mouth (xerostomia) and raised intra-ocular pressure (glaucoma).

Plant profile

Pilocarpus species are tender neotropical shrubs or small trees from 3-7.5m high that can still be found in the wild, but now grown in plantations for harvesting the leaf and to protect the wild population which was being damaged by overcollection.

Pilocarpus pennatifolius cultivated at Palazzo Orengo (the Hanbury Botanic Gardens, Ventimiglia, Italy) in Thomas Hanbury's *herbarium*, 1895, at the Royal College of Physicians

Pilocarpus microphyllus from a commercial sample in 1905

In 1873 a Dr Coutinho from Brazil took samples of jaborandi leaves to France and demonstrated that extracts were powerful inducers of sweating and salivation. Within two years the active principle, an alkaloid subsequently named pilocarpine, was identified almost simultaneously by Hardy (in France) and Gerrard (in England). Initially it was used mainly to induce fluid and electrolyte loss (by sweating) in the management of renal failure, then for the treatment of glaucoma, and more recently for the treatment of the dry mouth that occurs in Sjögren's syndrome and after radiotherapy for head and neck cancers. It has also been used as an unapproved treatment for hair loss (although hair loss has been reported as a possible side effect of pilocarpine use).

Jaborandi is the local name for several species of *Pilocarpus* (and possibly plants of other genera). Pilocarpine is now produced from the leaves of *Pilocarpus microphyllus*, which is native to Northern Brazil.

Chemistry

Pilocarpine acts as an agonist at muscarinic receptors, in other words it stimulates the parasympathetic nervous system. This accounts for its medicinal effects on the sweat glands and on intra-ocular pressure (by constricting the pupil and increasing the rate of exit of aqueous humour from within the eye). It can also lead to side effects such as bronchospasm, a slow heart rate, and diarrhoea. Its effects are therefore the opposite of muscarinic antagonists such as atropine (*q.v.*).

Damage to the parasympathetic innervation of the eye causes dilatation of the pupil. This is called Adie's syndrome and is usually due to inflammation infection in the ciliary ganglion, behind the eye. The dilated pupil constricts very slowly to light. Administering a small dose of pilocarpine will cause an Adie pupil to dilate, whereas a normal pupil needs a greater quantity of pilocarpine to react. Dilated pupils that do not react to light. following a head injury or cerebral bleed, are an important sign of rapidly progressing brain stem herniation and terminal damage. The use of very dilute pilocarpine to constrict the pupils can be a useful diagnostic tool.

Pilocarpine eye drops can cause irritation of the cornea and increased tear production.

Further reading

Holmstedt B, *et al.* Jaborandi: an interdisciplinary appraisal. *J Ethnopharmacol* 1979;1:3-21

CHAPTER 35 – JOHN NEWTON

Podophyllum peltatum
Podophyllum hexandrum
The source of podophyllotoxin, etoposide and teniposide

Introduction

Podophyllum peltatum and *P. hexandrum* are traditional herbal purgatives and emetics used historically in North America and the Himalayas, respectively. They contain similar chemicals known as lignans which, although highly toxic when ingested, also have potential therapeutic value. The most effective of these is podophyllotoxin which has been used for many decades as a topical treatment for sexually-transmitted genital warts (condyloma acuminata caused by a papilloma viral infection). A chemically-modified derivative of podophyllotoxin, etoposide, is widely used as a systemic anti-cancer drug and bone marrow suppressant. Another, teniposide, is principally used in the treatment of recurrent lymphocytic leukaemia and Hodgkin lymphoma.

Plant profiles

It is the foliage of *Podophyllum* which is its most striking feature: it has large, round (sometimes lobed and palmate) leaves in different shades of green, sometimes mottled with different hues, held like open umbrellas.

Podophyllum peltatum is a relatively small, hardy herbaceous perennial whose native range is Southeast Canada through to Central and Eastern USA. It is found in species-rich woodland, thickets and pastures where it spreads by underground rhizomes.

The deeply-lobed, glossy, palmate leaves, about 20-30cm in diameter, emerge first like small,

Podophyllum 'Spotty Dotty' with flowers

Flower of *Podophyllum peltatum*

furled umbrellas pushing up out of the ground. Below these, later in the spring, hang single white to pale pink, slightly scented flowers which are pollinated by flies. From these develop fleshy green fruit which ripens to yellow or red. The plant should be grown in moist soil in full or semi-shade.

Podophyllum hexandrum, commonly known as the Himalayan mayapple, has a native range from Northeast Afghanistan to Central China, where it is found in perennial scrub forests and alpine meadows. Like *P. peltatum* it has deeply lobed leaves 25cm across but mottled with purple. Late frosts may damage young leaves but otherwise the plants are quite hardy. In early spring on stems of about 30cm high they produce pinkish or white cup-shaped flowers with yellow stamens which face upwards, followed by the appearance below them of furled, slowly-opening, red-brown leaves which become mottled green when fully expanded. These are followed in late summer by scarlet, egg-shaped berries which hang below the leaves on short drooping peduncles. *Podophyllum hexandrum* benefits from added organic matter in the soil as it likes more moisture than *P. peltatum* and takes some years to become established. It is very long-lived given a suitable habitat. Overcollection for medicinal purposes has meant that wild populations are being endangered.

Both *Podophyllums* can be propagated by division in spring or late summer.

Nomenclature and history

Podophyllum peltatum, of the family Berberidaceae, was described by Linnaeus (1738, 1753) from a plant previously called *Aconitifolia humile bifolium, sive flore albo unico Leucoii campanulato, fructu cynosbati* (with an engraving showing its rhizomatous habit) by the German physician, Christian Mentzel (1682). It was rediscovered and named *Anapodophyllum canadense* by Mark Catesby in his *Natural History of Carolina, Florida and the Bahama Islands* (1731). It is felicitous that Dr Richard Mead (1673-1754), a famous Fellow of our College, was one of the sponsors of Catesby's travels. Although the fruit may be edible, the roots contain toxic compounds that make it a powerful purgative and emetic (Kelly & Hartwell, 1954). *Podophyllum* was used by indigenous Americans both as a medicine and a poison. It was a treatment for intestinal worms for the Cherokees and several Native American tribes, and a decoction of the roots or whole plant as an insecticide for corn and potatoes by the Cherokee, Iroquois and Menominee (Moerman, 2009).

Catesby writes that 'The Root is said to be an excellent Emetic and is used as such in Carolina, which has given it there the name of *Ipecacuana* (sic), the stringy Roots of which it resembles'. The early settlers adopted the root as a remedy, and it appears in the first *American Dispensatory* (Coxe, 1806) as a purgative, confirming that it was used by the Cherokee as an anti-helminthic and noting that the leaves are poisonous, the fruit edible. It is a purgative ('cathartic and a cholagogue') in the *United States Pharmacopoeia* published in 1820, where it stayed until 1942.

It enters the *British Pharmacopoeia* (1864) as 'Root imported from North America' and in the *Pharmacopoeia of India* (Waring, 1868) as the resin obtained by distillation.

Podophyllin

In 1835, Dr John King prepared podophyllin resin as an alcoholic extract of *Podophyllum* root. This was more reliable pharmacologically than preparations of the plant itself but was much more potent. The first time King used it the patient very nearly died; was severely ill for days and suffered chronic illness for years afterwards (Kelly & Hartwell, 1954). Despite this dramatic start, and in a suitably reduced dose, podophyllin resin went on to became popular as a cathartic, purgative, vermifuge and cholagogue. It was especially popular with the so-called 'eclectic practitioners', a branch of 19th century American physicians whose treatments were generally based on native North American herbs and the traditions of indigenous Americans.

From about 1850, podophyllin resin was extracted commercially and became the normal preparation of *Podophyllum*. Its use spread to England and Europe where it was widely used until the early 20th century, for example as an ingredient in common proprietary 'liver pills' – laxatives, with no effect on the liver but for which many benefits were claimed by the manufacturers.

There are historical accounts of podophyllin being used to treat skin growths and fungal infections, but by the end of the 19th century it was also being used in typhoid, chronic hepatitis, tuberculosis, syphilis, rheumatism, menstrual problems, intestinal worms and urinary incontinence – on the basis of the violent purging and vomiting it caused (King, 1895). In 1942, Kaplan showed that it was a highly effective treatment for condyloma acuminata (Culp & Kaplan, 1944).

Cytological analysis showed that podophyllin (like colchicine) stops cells dividing by causing mitotic arrest, an effect which explains both its curative effect on warts and its toxicity on gut mucosa. Podophyllin and podophyllotoxin are not used as a treatment for other forms of warts partly because of their toxicity and because they are

Podophyllin resin for treating genital condylomata

relatively ineffective in penetrating the thick keratin layer overlying most cutaneous warts. They will, however, penetrate normal skin.

Toxicity

A nurseryman in Surrey, visiting the College Garden, related that he had been handling plants of *P. peltatum* and must have absorbed sufficient through the skin of his palms as it induced vomiting for several hours, and caused sensitisation, so even a minor contact with the leaf now makes him ill. In Martindale (1967), podophyllin (the crude extract) was reported as being used for lichen simplex, seborrhoeic warts and senile keratoses, but principally for genital warts as a topical application, and as a violent purgative. It reports death from renal failure in a girl treated with podophyllin cream for a large condyloma and deaths in others who had oral podophyllin; a baby born with multiple deformities to a mother who took podophyllin as a laxative in the first trimester of pregnancy, and another young woman who endured 11 days of coma followed by 10 months of peripheral neuropathy after taking an alcoholic extract of the root. There is no doubt that this is a poisonous plant, to be treated with care (and gloves).

Chemistry

Podophyllotoxin

The active chemicals in *Podophyllum* and podophyllin resin do not contain nitrogen and are therefore not alkaloids. They are, instead, lignans – natural products based on a 2,3,dibenzylbutane skeleton (Stähelin & von Wartburg, 1991). The most effective of these anti-mitotic lignans is podophyllotoxin which was first isolated in crystalline form in 1880.

Regarding the use of podophyllotoxin for cutaneous warts, the British Association of Dermatologists' guidelines state: 'Podophyllotoxin can inhibit cell division by interfering with the mitotic spindle and will affect normal skin as well as warts. It can have dangerous systemic effects if used in high concentrations or over large areas, and its use is contraindicated in pregnancy. Although podophyllotoxin (and previously the cruder podophyllin) is a standard treatment for anogenital warts, its evaluation in cutaneous warts has been limited. The assumption is that penetration of the thick, cornified layer of cutaneous warts is poor compared with that achieved at mucosal sites. A very small open study of 40 patients with plantar warts treated with podophyllin 25% in liquid paraffin under prolonged adhesive plaster occlusion reported a 67% clearance rate of patients at 3 months. However, the side-effects of this treatment include an intense inflammatory reaction with blistering, which can be very painful. There are no recent studies using podophyllotoxin for cutaneous warts, except as a 1% component of a combination therapy with cantharidin and salicylic acid, so the contribution of podophyllotoxin alone is impossible to evaluate' (Sterling *et al*, 2014).

Podophyllotoxin

Purified podophyllotoxin is now available in modern pharmaceutical products and can even be self-administered. It is so much more reliable, less toxic, and more effective than the herbal extract (podophyllin) that the latter can only be preferred as a treatment for condyloma acuminata on cost grounds (von Krogh & Longstaff, 2001). Both podophyllin resin and podophyllotoxin are included in the WHO's list of essential medicines, those that meet the priority health needs of the world's population. This is partly because of the continuing public health importance of condyloma acuminata as a complication of HIV and AIDS. In view of podophyllotoxin's toxicity, it is a relief to know that vaccination against the specific papilloma viruses that cause condylomata is being successfully deployed worldwide, as is vaccination against the variety which causes cervical cancer.

Etoposide and teniposide

A systematic effort began in the 1950s to produce anti-cancer therapies by modifying these naturally-occurring lignans in the knowledge that *Podophyllum*-derived compounds can stop cells dividing. It was not until the 1970s, and after testing around 500 compounds, that clinical studies showed promising results (Stähelin & von Wartburg, 1991). Two glycosides of podophyllotoxin, teniposide and etoposide, were found to be effective in animal tumour models. Etoposide was approved for use in 1983 and is now widely used in the treatment of many cancers including testicular, bladder, prostate, lung, stomach, and uterine cancer as well as lymphoma and leukaemia, Kaposi's sarcoma, Ewing's sarcoma, and glioblastoma multiforme. It is also used in bone marrow transplantation to suppress leukaemic marrow prior to recolonising it with healthy marrow cells. Teniposide is approved for the treatment of cancer in children including Hodgkin lymphoma and acute lymphocytic leukaemia.

References

Austin DF. *Florida Ethnobotany*. Boca Raton, Florida, CRC Press, 2004

British Pharmacopoeia. London, Spottiswode & Co., 1864

Catesby M. *Natural History of Carolina, Florida and the Bahama Islands* vol 1: t 24. London, W. Innys and R. Manby, 1731

Coxe JR. *The American Dispensatory*. Philadelphia, A. Bartram for Thomas Dobson, 1806

Culp OS, Kaplan IW. Condylomata Acuminata: Two Hundred Cases Treated With Podophyllin. *Ann Surg* 1944;120(2):251-6

Kelly M, Hartwell JL. The biological effects and the chemical composition of podophyllin: a review. *Journal of the National Cancer Institute* 1954;14(4):967-1010

King J. *The American Dispensatory*, 17th edn. Cincinnati, The Ohio Valley Company, 1895

Mentzel C. *Pinax. Nominum plantarum. Universalis … adjectus est Pugillus Rariorum Plantarum*, t.11. Berlin, Berolini, 1682

Martindale W. *Extra Pharmacopoeia*, 25th edn. London & Bradford, Percy Lund, Humphries & Co. Ltd, 1967

Moerman DE. *Native American Ethnobotany.* Portland, London, Timber Press, 2009

Stähelin HF, von Wartburg A. The chemical and biological route from podophyllotoxin glucoside to etoposide: ninth Cain Memorial Award lecture. *Cancer Research* 1991;51(1):5-15

Sterling JC, Gibbs S, Haque Hussain SS, *et al*. British Association of Dermatologists' guidelines for the management of cutaneous warts. *Br J Dermatol* 2014;171:696-712

von Krogh G, Longstaff E. Podophyllin office therapy against condyloma should be abandoned. *Sex Transm Infect* 2001;77(6):409-12

Waring EJ. *Pharmacopoeia of India*. London, W.H. Allen & Co, 1868

Podophyllum hexandrum from Asia also contains podophyllin

Display card, possibly from the early 1910s, advertising the multiple benefits of podophyllin-containing 'liver pills'
Source: Wellcome Images (a website operated by Wellcome Trust) reproduced under the Creative Commons
Attribution 4.0 International license

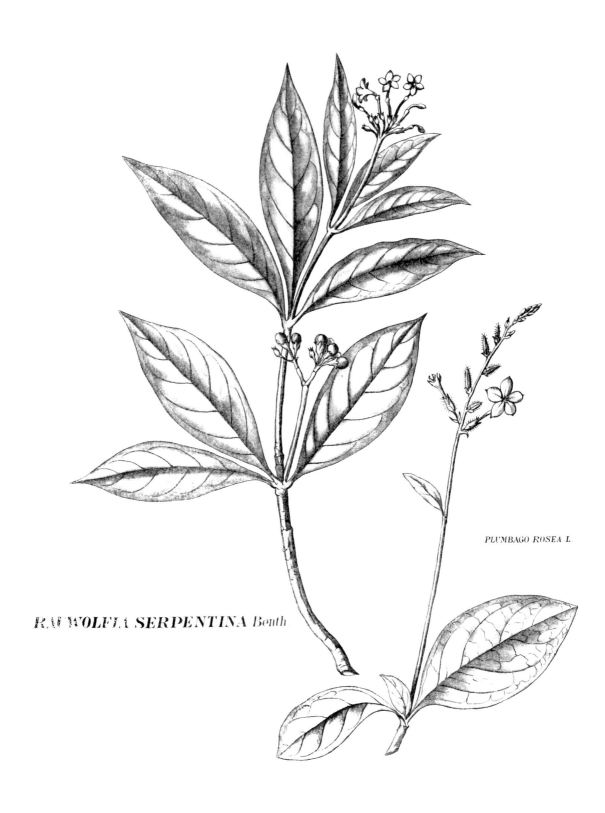

RAUWOLFIA SERPENTINA Benth.

PLUMBAGO ROSEA L.

Engraving of *Rauwolfia serpentina* from the Pharmaceutical Society's herbarium
at the Royal College of Physicians

Rauvolfia serpentina
Rauvolfia vomitoria
The source of reserpine, ajmaline and ajmalicine

Introduction

The root of *Rauvolfia serpentina* (known also as Indian snakeroot, devil pepper and serpentine wood) is the source of over 50 alkaloids of which reserpine, ajmaline and ajmalicine are the most important. *Rauvolfia* is a genus of 75 species of shrubs and climbers in the family Apocynaceae. Nearly all members of this family (*q.v.* also *Catharanthus roseus*) are poisonous but are an important source of alkaloids used both medicinally and as tools for assessing many biochemical and physiological functions of the body (Kumari *et al*, 2013).

The principal ingredient of snakeroot, the alkaloid reserpine, was identified in 1952 at the laboratories of Ciba in Basel, Switzerland (Gawade & Fegade, 2012).

Reserpine has been an important drug for the treatment of hypertension (raised blood pressure) for much of the second half of the 20th century, although its use has now declined dramatically, and it is no longer available in the USA and UK.

Ajmaline, named by its discoverer in 1931, Professor Siddiqui, after the preeminent Unani practitioner Hakim Ajmal Khan, has been used in certain cardiac arrhythmias as an alternative to digitalis and quinidine in continental Europe and Japan, but has never entered UK pharmacopoeias for this purpose.

Chemistry, ajmaline

Ajmaline is a potassium- and sodium-channel blocker and is used intravenously as a provocation test for diagnosing Brugada syndrome – a congenital autosomal dominant disorder that causes episodes of ventricular fibrillation, resulting in fainting or sudden cardiac death. These diagnostic cardiac arrhythmias appear on intravenous ajmaline and require immediate cardiac resuscitation. Treatment depends on the seriousness of the condition and includes medication and/or an implanted cardiac defibrillator. Genetic screening for all family members is advised.

Ajmalicine, another chemically-related alkaloid sourced from *Rauvolfia* is an α-receptor blocker which was found to treat hypertension. It was available under many trade names but is no longer used.

In India, chewing the root of *R. serpentina*, or using extracts, is a traditional remedy in Unani medicine as a tranquilizer. It is alleged that Mahatma Gandhi was a regular user for its calming effect (Court, 2012). It is from this traditional use (since at least the 1930s) that reserpine, extracted from its roots, became used for treating the thousands of disturbed and distressed patients with schizophrenic illnesses who were incarcerated in huge mental hospitals, once its sedative properties became known and accepted in the West.

Plant profile

Rauvolfia serpentina is a tropical, perennial, evergreen, erect, woody shrub that grows to 60-100cm. The genus is native to Africa, India, Pakistan and southern Asia. It requires tropical to subtropical conditions (Zone 10) and grows in damp forests, either in shade or in the open. It is tolerant of different soil types. Frequently described as having a 'twining' habit in pharmaceutical articles which appears to be due to confusion with *Aristolochia* species, also called 'snake root'.

History and nomenclature

Powdered *R. serpentina* root is said to have an ancient history, and was reputedly used in the *Ayurveda* from about 600 BCE. It was then known as either chandrá or sarpagandha. Chandrá means 'moon' and refers to its former use in treating 'moon disease' or lunacy. Sarpagandha means 'snake's smell or repellent' and refers to its other former use for snake bites and scorpion stings (Monachino, 1954; Somers, 1958). However, in an excellent review of reserpine, Dr Pradipto Roy (2018) quoted from *Pharmacopoeia Indica* (Bose, 1932) that there was no evidence for this from the Ayurvedic Pharmacopoeia, that the names are misnomers and that no drug with that name having similar properties can be found in the Ayurvedic literature. Historic source literature in Sanskrit to support its use for psychiatric illness is unobtainable and the literature from the Early Modern Era, detailed below, does not support claims for treating insanity.

1593: One of the earliest descriptions of *R. serpentina* comes from Garcia ab Horto (1593), saying the root of the tree called *Lignum columbrinum* (serpent-like wood) as a powder kills intestinal worms, treats 'spots' rashes and impetigo, cures jaundice and circuitous fevers, and cures the bites of poisonous snakes. He adds that the root is more powerful than the wood.

1650: The unrelated *Serpentaria virginianae*, Virginian snake root, which appeared in the College's pharmacopoeia of 1650 and in other pharmacopoeias for the next 200 years is *Aristolochia serpentaria*.

1703: The genus was named *Rauvolfia* by Charles Plumier (1646-1704) in 1704 to commemorate Leonard Rauwolf (1535-1596), a German physician and botanist who had explored Syria and Mesopotamia in 1574-1575. Plumier was a French botanical explorer from the discalced monks of the Roman Catholic Order of the Minims who discovered *R. tetraphylla* in the West Indies. The genus is pan-equatorial and is found as far south as South Africa. As genus names must be in Latin it was named *Rauvolfia*, rather than *Rauwolfia*, because there is no 'W' in classical Latin, and Plumier specifically indicates this change in spelling, so it was not a typographical error.

1712: Engelbert Kaempfer, in his *Amoenitatum Exoticarum* (1712), called it *Radix mungo*, the most important antidote from the vegetable kingdom, named in honour of *Mustela*, the mongoose, the animal 'genius' which is the natural enemy of serpents. He recounts the legend that the mongoose eats the roots or leaves if it is going to attack a serpent or if it has been bitten by one, and the local people regard the plant as a cure-all against all venoms and poisons.

1737: Johannes Burman in his *Thesaurus Zeylanicus* (1737) has an elegant engraving of the plant, as *Lignum colubrinum* (serpent-like wood) and *Radix serpentum*, describing the sinuous roots which give it its name. He says the inhabitants of Sri Lanka praise it as a cure for snake bite and all poisons. It has a useful bibliography of other writers about this plant.

1753: Linnaeus named it *Ophioxylon* (from Greek, *Ophios* + *Xylon*, serpent + wood) *serpentinum*, from Sri Lanka, with the pharmaceutical name of 'Serpentinum lignum' [snake wood] from the shape of the roots (Linnaeus, 1753). In his *Materia Medica* (1749, 1782) he says it has a bitter taste and is used to induce sweating and to treat quartan fevers [malaria], bites and stings. He did not recognise it to be in the same genus as Plumier's *R. tetraphylla*.

1755: Georg Everhard Rumphius in his *Herbarii Amboinense* (1755), gives an early reference to its sedative properties in Batavia (now Jakarta, Indonesia) where the roots chewed with 'Pinanga' were used for treating anxiety and abdominal pain due to flatus. Its principal use was to treat snake bite and poisoning. He called it *Radix mustelae*, and *Mongo*, and repeats the story from Kaempfer that the mongoose (the Ratto de Mongo of Portugal, the perpetual enemy of serpents), chews the leaves if it is about to attack, or has been bitten by, a poisonous snake. It was also used for fevers and headache, and the sap from the leaves dropped in the eyes cured white marks, and dissolved in alcohol and anointed on the umbilicus, cured jaundice and impetigo. He too has an engraving which confirms its identity with *R. serpentina*. Rumphius notes its previous names of *Lignum colubrinum* and Linnaeus' *Ophyxylum* (sic).

1799: Hipólito Ruiz and Joseph Pavón incorrectly described a sweet-smelling Peruvian species of *Citharexylum*, a genus in the Verbenaceae, as *R. flexuosum*.

1868: *Ophioxylon serpentinum* is described as an erect or twining shrub in the *Pharmacopoeia of India* (Waring, 1868), used for snakebites, fevers, to increase uterine contractions in labour, and as an antihelminthic. Both the description as 'twining' and its use as for uterine contractions suggests it may have been confused here with *Aristolochia* which is a 'birthwort' for speeding up labour, and its description taken from other sources.

Rauvolfia serpentina did not appear in the pharmacopoeias of the West for nearly 200 years after Rumphius, although modern writers believe that the root was widely used in India for centuries for treating insanity. The published reports noted above, with the exception of Rumphius noting its use for anxiety, are primarily concerned with its use for snake bite and poisoning, and to a lesser extent as a febrifuge and to aid uterine contractions.

Little interest has been shown in the West for traditional herbal remedies from Asia, although the use of the roots of *R. serpentina* as a treatment for high blood pressure (hypertension) in India had been noted in the 1940s. This was investigated and the careful work of Vakil (1949) over ten years using tablets of the root, led to the introduction and widespread use across India and then the world, of reserpine as a primary treatment for hypertension.

Gananath and Bose (1931) reported that 20-30 grains of powdered *R. serpentina* abolished violent maniacal symptoms and lowered blood pressure in a patient. It took 25 years for their findings to be taken up by the Western pharmaceutical and medical establishment.

Chemistry, reserpine

Peripheral effects: reserpine is a monoterpene indole alkaloid which acts on the sympathetic nervous system by depleting the monoamine neurotransmitters noradrenaline, serotonin and dopamine from central and peripheral axon terminals. This results in interruption of synaptic transmission, producing relaxation of the smooth muscle in peripheral blood vessel walls and a reduction in cardiac output leading to a lowering both of systolic and diastolic blood pressure.

Central effects: A similar effect in reducing serotonin and dopamine levels in the brain probably explains its sedative and anti-psychotic actions, which have been widely employed medicinally for humans. In the veterinary world it has been used to calm over-excitable horses. Its principal shortcoming in clinical use is the risk of it inducing a severe, deep and sometimes permanent depression with suicidal thoughts. This effect was an important pointer to the importance of brain monoamines in the genesis of psychiatric disorders and contributed to the later introduction of monoamine oxidase inhibitors such as phenelzine (Nardil) and selective serotonin reuptake inhibitors such as fluoxetine (Prozac) for treating depression.

Reserpine was marketed in the USA for the treatment of hypertension with FDA approval in 1955, under the brand name of Serpasil (Ciba Pharmaceutical Products Inc.). It was the first potent drug widely used for long-term treatment of hypertension. In the 1970s annual sales exceeded $30 million.

It was a particularly effective anti-hypertensive drug when combined with a thiazide diuretic, and is one of the very few medications for hypertension that have been shown in clinical trials to specifically reduce mortality. Its principal shortcoming in clinical use is the risk of it inducing a severe, deep and sometimes permanent depression with suicidal thoughts. This is the main reason its use has declined with the advent of more modern therapy for raised blood pressure. It is no longer available in the UK and USA.

Reserpine started to be used for treating patients with schizophrenia in the 1940s in India, initially because of the tranquilizing reputation of *Rauvolfia* extracts. It was found to reduce paranoid delusions and hallucinations as well as being calming. The treatment of schizophrenia and other psychoses with reserpine or *Rauvolfia* was mostly ignored in the West but became established with the work of Nathan Kline at the Rockland State Hospital in the USA, following the presentation of a paper by Dr RA Hakim of Ahmabad, India on the cure of schizophrenia with *R. serpentina* (Kline, 1954). This spurred an explosion of interest in the pharmaceutical treatment of psychiatric disease which revolutionised care and the concept of cure.

With the increased popularity of reserpine in clinical use, both for hypertension and schizophrenia, *R. serpentina* became over-harvested in the wild in India and an alternative commercial source was found in *R. vomitoria* from West Africa (van Wyk & Wink, 2004). Local collection for traditional herbal remedies is currently the biggest threat to its survival and this is reducing the genetic diversity of the remaining populations. It is included in the IUCN list as 'Endangered'.

The use of reserpine to treat schizophrenia and other psychoses was effective but superseded with the advent of chlorpromazine (a medicine in the phenothiazine group) in the mid-1950s. This and other phenothiazines signalled the end of the large mental hospitals which, at that time, had around a third of all NHS beds in the UK. By the 1970s, the introduction of these drugs had reduced the average length of stay of a patient admitted with a diagnosis of 'schizophrenia' from one year to a matter of weeks.

Rauvolfia vomitoria (photo by Lucinda Lachelin)

In the 1970s, *Rauvolfia*, reserpine and its other alkaloids were widely prescribed, and the pharmacopoeias of that time, *e.g.* Martindale (1967), are a wonderful source of information about the clinical uses and results. The commoner side effects of reserpine, largely due to its blockade of the sympathetic nervous system, were reported as nasal congestion, nausea, vomiting and diarrhoea, gastric ulceration, erectile dysfunction, nightmares and Parkinsonism. More important side effects, especially depression, contributed to its demise in psychiatric use, as the unwanted effects of the common alternative, chlorpromazine, while not trivial, were less distressing. The newer anti-hypertensive medications also had fewer side effects and replaced reserpine and its related drugs.

The extrapyramidal, Parkinson-like movements which were observed in rabbits treated with reserpine led to the work by Carlsson in 1959 which uncovered the role of dopamine deficiency in Parkinson's disease and its eventual treatment with levodopa, for which he was awarded the 2000 Nobel Prize for Medicine/Physiology (Hollman, 1992).

Summary

Rauvolfia serpentina is an extraordinarily important medicinal plant whose (putative) centuries-old traditional use as a tranquillizer and for mental illness led to the isolation of reserpine and other important alkaloids after it came into use in India in the 1930s. These alkaloids gave us significant advances in the treatment and understanding of blood pressure, psychiatric illnesses and Parkinson's disease. Reserpine is no longer a 'modern' drug, but it led to huge advances in our knowledge of how the central and peripheral nervous systems worked and so to better medicines for the treatment of a wide range of diseases.

References

Burman J. *Thesaurus Zeylanica exhibiens plantas in insula Zeylana nascentes*. Amsterdam, Ianssonio-Waesbergios & Salomonem Schouten, tab64, p141-2, 1737

Bose KC. *Pharmacopoeia Indica: Being a Collection of Vegetable Mineral and Animal Drugs in Common use in India*. Calcutta: Bishen Singh Mahendra Pal Singh, p153, 1932

Court WE. The Rauvolfia story. *Pharmaceutical Historian* 1998;28(3):43-48

Gananath S, Bose KC. *Rauwolfia serpentina*: A new Indian drug for insanity and high blood pressure. *Indian Med World* 1931;11:194-200

Gawade BV, Fegade SA. Rauwolfia (Reserpine) as a potential anti-hypertensive agent: A Review. *Int J Pharm Phytopharmacol Res* 2012;2(1):46-49

Hollman A. Plants in Cardiology. *British Medical Journal* 1992;25

Kaempfer E. *Amoenitatum Exoticarum Politico-Physico Medicarum*. Fasciculus 3. Gemina Indorum Antidota. Lemgoviae, Henrici Wilhelmi Meyeri, pp573-8, 1712

Kline NS. Use of Rauwolfia serpentina Benth. in neuropsychiatric conditions. *Annals of the New York Academy of Sciences* 1954;59(1):107-132

Kumari R, Rathi R, Rani A, *et al. Rauwolfia serpentina* L. Benth. ex Kurz.: Phytochemical, Pharmacological and Therapeutic Aspects. *Int J Pharm Sci Rev Res* 2013;23(2)n° 56:348-50

Lewis WH, Elvin-Lewis MPF. *Medical Botany: Plants affecting human health*, 2nd edn. Hoboken, New Jersey, John Wiley and Sons, p286, 2003

Martindale. *Extra Pharmacopoeia*, 25th edn. London & Bradford, Percy Lund, Humphries & Co. Ltd., 1967

Monachino J. *Rauwolfia serpentina*: its history, botany and medicinal use. *Economic Botany* 1954;8(4):349-365

Plumier C. *Nova Plantarum Americanarum Genera*. Paris, Joannem Boudet, 1703

Roy P. Global Pharma and Local Science: The Untold Tale of Reserpine. *Indian J Psychiatry* 2018;60(Suppl 2):S277-S283

Ruiz H, Pavón J. *Flora Peruviana et Chilensis*. Madrid, Typis Gabrielis de Sancha, 2:26, lamina 152, 1799

Rumphius GE. *Herbarii Amboinense Auctuarium* vol 6. Amsterdam, Mynardum Uytwerf, & Viduam ac Filium S. Schouten, t16, pp29-32, 1755

Somers K. Notes on Rauwolfia and ancient medical writings of India. *Medical History* 1958;2(2):87-91

Vakil RJ. A clinical trial of *Rauwolfia serpentina* in essential hypertension. *British Heart Journal* 1949;11(4):350-5

van Wyk B-E, Wink W. *Medicinal Plants of the World*. Timber Press, London, p266, 2004

Waring EJ. *Pharmacopoeia of India*. London, W.H. Allen & Co, 1868

CHAPTER 37 – SUSAN BURGE

Salix alba
The source of salicylic acid

Filipendula ulmaria
The source of aspirin

Gaultheria procumbens
The source of methyl salicylate

Introduction

Salicin is a chemical defence for plants, reducing the likelihood of it being eaten, although interestingly some insects have developed strategies to ingest and store salicin so that the salicin can be used for their own defence. Salicin also helps a plant to resist viral infections. Some medicinal properties of willow bark have been recognised since ancient times, but now we use two derivatives of salicin, known as salicylates – salicylic acid, which is used for skin conditions, and acetylsalicylic acid, which is aspirin.

Filipendula ulmaria (Photo by Michael de Swiet)

Gaultheria procumbens

Acetylsalicylic acid (aspirin) was originally derived from salicin in *Filipendula ulmaria*, meadowsweet, a herbaceous plant in the family Rosaceae.

The bark of *Salix alba*, the white willow, was the source of salicylic acid, used (topically) for warts and corns. It is a tree in the Salicaceae, a diverse family that includes *Populus* (poplars) and over 50 other genera. Its bark and leaves were traditionally used for extracting salicin, a bitter-tasting chemical, which is converted to the analgesic substance salicylic acid in the body.

Gaultheria procumbens, the wintergreen or eastern teaberry, from the family Ericaceae, contains methyl salicylate, the active ingredient in Oil of Wintergreen. This naturally-occurring salicylate is found in other plants including some species of birch, particularly *Betula lenta* (black birch). It is used topically for joint and muscle pains because of its mild analgesic activity when absorbed through the skin; taken orally it is extremely poisonous but in low concentrations it is used as an antiseptic in mouth washes, and for flavouring in sweets. Excessive topical use can cause salicylate poisoning. Methyl salicylate is now produced artificially. It is not a 'Prescription Only' medicine and is mentioned here only for its part in the story of salicylates.

Plant profiles

Salix alba, the white willow, has a native range from Europe to North China and Northwest Africa. It is the largest species of willow reaching 25m x 10m in size with narrow silvery leaves on dark grey stems that develop deep fissures. When young, the stems are light grey-pink to olive-brown and are covered with silky hairs. Male and female catkins develop on separate trees in spring and are an important source of early nectar and pollen for bees and other insects. After pollination, the female catkins lengthen and develop small capsules, each containing minute seeds which are covered in a white down which helps them to be dispersed by the wind. This plant tolerates maritime exposure and atmospheric pollution and is a valuable wildlife habitat for the many species of insect which are associated with it. The catkins and the branches make good nesting and roosting sites for birds. There are many subspecies and cultivars.

Filipendula ulmaria, meadowsweet, is a herbaceous perennial in the Rosaceae family native to Eurasia. It is found growing in wet conditions such as swamps, marshes, fens, wet woods and meadows, wet rock ledges and by rivers. The plant has tall, reddish stems up to 2m high with dark green, aromatic leaves. The leaves are pinnately divided and are a whitish colour on the undersides. In summer, large frothy sprays of creamy white flowers with a sweet perfume appear on leafy stems high above the foliage. They are pollinated by various insects, particularly houseflies. In the garden this plant grows best in a humus-rich moist soil in semi-shade. As it is rhizomatous it can be propagated by division in autumn or winter.

Gaultheria procumbens is a slow growing, dwarf, evergreen shrub native to Central and East Canada through to North Central and East USA. It prefers a position in semi-shade as it grows naturally in the woodland understorey and on forest margins. Attractive pink-tinged, white, bell-shaped flowers appear in summer amongst the small, dark green, shiny leaves. As this species is dioecious both male and female plants are needed to produce the shiny, red, aromatic berries that are very ornamental in the winter months. Like other members of the Ericaceae family the plant requires humus-rich, acidic soil. It is fully hardy but prefers a sheltered site and does not like to dry out.

Manufacture

These salicylates are no longer made from salicin in plants. Acetylsalicylic acid (aspirin) was originally made from salicin extracted from *F. ulmaria*. The name aspirin comes from the old name for *F. ulmaria*, *Spiraea ulmaria* – a[cetyl]SPIRin (the suffix 'in' being added to indicate it was a medicine).

Chemistry

Synthesis of salicylic acid and acetylsalicylic acid (aspirin)

Salicylic acid is synthesised by reacting phenol with sodium hydroxide to make sodium phenoxide. A reaction of this with carbon dioxide forms sodium salicylate which is then acidified to make salicylic acid. Salicylic acid is combined with acetic anhydride to make acetylsalicylic acid.

History

Willow, *Salix* (also Itia and Itea, in Greek), has a long tradition as an analgesic and for treating skin conditions, but it is difficult to find any early recommendation for its use in fevers.

Hippocrates [350 BCE] taught that the use of extracts of the leaves and bark of *Salix* could be used for treating warts and corns. The *Materia Medica* of Dioscorides [70 CE] agreed and additionally recommended the leaves for intestinal obstruction; to prevent conception; and in various topical preparations for the pain of gout and earache, dandruff and clouding of vision – but makes no mention of fevers. Pliny [70 CE],

Salix from Mattioli's *Discorsi*, 1568

agreed with Dioscorides, adding that it was good for treating 'freckles' (although this may refer to senile keratoses), for cooling 'lascivious lust' and would disable the 'act of generation'. Galen [200 CE] recommended preparations of the leaves and flowers for healing bleeding wounds.

None of the great physicians of Classical Greece and Rome, or of the Islamic Golden Age, not Galen, Paulus Aegineta [625-690], nor the great Arabians like Avicenna [980-1037] and Al-Rhazes [865-925], nor Serapion [11th century] mention it being used to treat fevers, even though they gave its (Humoral) properties as drying and cooling. Many of the uses mentioned by Dioscorides were still recommended by William Turner in the 16th century: he added that the juice of the leaves and bark applied to the skin treated scaly skin, scurf, earache and gout. More surprisingly, he said that, by mouth, it was good for stomach-ache and coughing up blood (Turner, 1568).

Nicholas Culpeper (1652) echoed Hippocrates, writing, 'the Burnt ashes of the Bark, being mixed with vinegar taketh away Warts, Corns and Superfluous Flesh, being applied to the place'. The doctrine from Hippocrates and Galen, that the hypothetical Humoral properties in a plant might be used medicinally, was invoked by Culpeper (1650) who wrote, 'Willow leaves; are cold, dry and binding … the boughs stuck about a chamber wonderfully cool the air and refresh such as have the feavers, the leaves applied to the head, help hot diseases there, and frenzies'. He had copied this verbatim from John Parkinson (1640), who had probably sourced this from Pietro Mattioli who wrote (1586) that the leaves were useful for strewing around the beds of the feverish. No mention is made about ingesting it, but while topical application for headaches might have been effective, cooling the air would not treat fevers. The ancient pseudoscience, the Doctrine of Signatures, which stated that features or 'signatures' of the plant such as the shape of leaves or seed, or the conditions in which the plant grew, indicated how a plant might be used, was favoured by William Coles who said of willow, 'any stick thereof, although almost withered, being fixed in the earth, groweth which Signature doth duly declare that a bath made of the decoction of the leaves and bark of willow, restoreth again, withered and dead members to their former strength, if they be nourished with the fomentation thereof' (Coles, 1657).

That both these concepts were nonsense does not need saying here. Other 16th and 17th century uses of willow sap or extracts were for removing the pterygium that grows across the cornea, and as an addition to sedative medicines.

Salix was in the College's pharmacopoeias in 1618 through to the 6th edition of 1650 but it disappears from pharmacopoeias for most of the 18th century. An exception is the *Pharmacopoeia Universalis* (James, 1752), where the leaves are recommended for nosebleeds, footbaths, to procure sleep, and to 'cool the heat of fevers', and the ashes of the bark to remove warts and corns. More typical is Brookes' *Natural History of Vegetables* (1772) which notes previous uses and decoctions were 'now out of date'. *Salix* bark is mentioned in the first *American Dispensatory* (Coxe, 1806) as having been recommended for fevers but having little effect.

Modern history

Considering the lack of pharmaceutical interest in *Salix* in the 17th century, honours for its revival must go to a country vicar. The Rev. Edward Stone (1702-1768), from Chipping Norton, Oxfordshire, is credited with providing the first scientific description of the benefits of an aqueous extract of powdered willow bark for treating 'the agues' (fever caused by malaria) in a letter he wrote to The Royal Society of London in 1763: 'There is a bark of an English tree, which I have found by experience ... to be very efficacious in curing aguish [feverish] and intermitting disorders [*i.e.* fevers which show cycles of fever peaks]. About six years ago, I accidentally tasted it, and was surprised at its extraordinary bitterness; which immediately raised me a suspicion of its having the properties of Peruvian bark [quinine – which is very bitter and used for malarial fevers]'. For this reason, thinking that if it tasted like quinine it might work like quinine, he tried it on 50 of his patients.

Stone's choice of willow bark was influenced at least in part by the Doctrine of Signatures, for he added: 'As this tree delights in a moist or wet soil, where agues chiefly abound, the general maxim that many natural maladies carry their cures along with them or that their remedies lie not far from their causes, was so very apposite to this particular case that I could not help applying it; and that this might be the intention of Providence here, I must own, had some little weight with me'.

What is frequently omitted in the story is that in those cases which did not get better with willow bark, he 'added one fifth part of Peruvian bark [cinchona, the source of quinine] to it and with this small auxiliary it totally routed its adversary [the fever]' (Stone, 1763). That many fevers remit spontaneously, and that quinine clearly was the effective agent, has not prevented the Rev. Stone from being heralded as the man who led to the discovery of salicin, the aspirin precursor. The distinguished Swedish physician Peter Bergius (1782) tried using willow bark in intermittent fevers [malaria] and found it of no effect.

Attempts were made to make a medicine from willow which would be more effective, not taste so bitter, and be better tolerated. Salicin was extracted from the bark by the German pharmacologist Johann Buchner in 1828 and entered the London, Edinburgh and Dublin pharmacopoeias 'as a febrifuge like the Sulphate of Quinine' (Royle, 1847). The chemical structure was determined and in 1852 salicylic acid, the active component, was synthesised by Wilhelm Gerland, a PhD student of Hermann Kölbe at Marburg University in Germany (Gerland, 1853). When Kölbe published the synthesis of salicylic acid in 1859 he omitted to include his student's discovery so is frequently quoted as the discoverer.

In 1876, two German physicians, Strickler and Reiss, working independently, reported on the anti-inflammatory and pain-relieving properties of salicin in acute rheumatic fever. Thomas MacLagan, a Scottish physician, performed a clinical trial in acute rheumatic fever in which he also demonstrated that salicin controlled both the fever and the inflamed joints. MacLagan stated that salicin was 'the most effective means yet for the cure of acute articular rheumatism and it may even show itself to be a specific for the disease'. The Doctrine of Signatures also played a part in MacLagan's choice of salicin, extracted from willow bark. He wrote: 'A low-lying damp locality, with a cold rather than warm climate, gives the conditions under which rheumatic fever is most readily produced. On reflection, it seemed to me that plants whose haunts best corresponded to such a description were those belonging to the natural order Salicaceae, the various forms of willow' (MacLagan, 1876; Jack, 1997; Montinari *et al*, 2019).

Salicin was better tolerated than the active component, salicylic acid, which was most unpleasant to take, irritated the stomach causing nausea and abdominal pain, and induced continuous ringing in the ears (tinnitus). Something better was needed.

Finally in 1897, Felix Hoffmann, working in the laboratory of Bayer in Germany, synthesised a less toxic, better-tasting, and medicinally-effective derivative, acetylsalicylic acid (aspirin) in which the irritant acidity was neutralised by adding the acetyl molecule (Jack, 1997; Montinari *et al*, 2019). One of the factors that encouraged Hoffman was that his father was taking salicin for rheumatism and found it extremely unpleasant, and Hoffman wished to make something effective and palatable.

Medicinal uses

Salicylic acid

Salicylic acid is a topical medicament that encourages the uppermost keratin layer of the skin (the horny layer) to shed. Synthetic salicylates are also added to sunscreens (they have a weak ability to block ultraviolet light B (UVB)) and are added as preservatives to many foods and cosmetics.

Chemistry

Salicylic acid, mode of action

When applied to the skin, salicylic acid acts as a keratolytic meaning that it both softens keratin, the protein that forms part of the outer horny layer, and loosens the attachments between the cells in this outermost layer. These properties assist desquamation (shedding of the top layer of the skin) (Lebwohl, 1999).

Acetylsalicylic acid, mode of action

The effect of aspirin and all other NSAIDs depends upon preventing the action of the cyclo-oxygenase (COX) enzymes that are needed to make prostaglandins, a group of chemicals that contribute to the fever, pain and swelling seen in acute inflammation (Montinari *et al*, 2019).

A cyclo-oxygenase-enzyme, COX-1, is present in platelets. Blocking COX-1 with a low daily dose of aspirin (75-100mg) prevents the platelets from becoming sticky so that clots are less likely to form. Activated platelets contribute to the growth and spread of tumours, and it has also been suggested that low-dose aspirin reduces the risk of developing some cancers by altering the behaviour of platelets.

Creams, gels, paints or plasters containing salicylic acid in concentrations of between 10% and 50% are available for purchase 'over the counter' and may be used to soften and reduce the thickness of corns and callosities. Salicylic acid preparations are also used to treat viral warts commonly found on the hands or feet (verrucae). Such treatment may possibly stimulate an immune response to the virus in addition to promoting shedding of skin cells. Most of these viral warts eventually disappear without any treatment, and salicylic acid has only a modest beneficial effect, but treatment does at least occupy the patient until the warts go away (nature cures the disease) (Gibbs & Harvey, 2006; Sterling *et al*, 2014).

Lotions or creams containing salicylic acid combined with a corticosteroid may be helpful in localised inflammatory skin conditions with a lot of scale such as psoriasis affecting the scalp (Lebwohl, 1999).

Acetylsalicylic acid (aspirin)

Aspirin was the first of the class of medicines known as NSAIDs. Aspirin proved so remarkably effective in relieving pain and reducing fever that its use spread rapidly throughout the world. The drug entered the Guinness World Records in 1950 for being the most frequently sold painkiller. But it has many other uses.

Aspirin reduces inflammation

Aspirin in intermediate to high doses (650mg to 4g/day) was widely used to control the pain and swelling in acute inflammatory diseases such as rheumatoid arthritis and pericarditis (inflammation of the membrane covering the heart muscle), but nowadays NSAIDs with fewer side effects are often preferred. Aspirin may be used to treat mild to moderate pain in conditions such as headache, period pain (dysmenorrhoea) or strained muscles, but stomach irritation can be troublesome (see toxicity), and alternatives such as paracetamol are often more appropriate. For severe pain, aspirin is sometimes prescribed in combination with codeine phosphate, a very strong pain reliever related to morphine.

Aspirin reduces temperature in mild fevers, for example those caused by viral infections (Montinari *et al*, 2019).

Aspirin prevents blood clots

Blood clots (thrombi) are formed when tiny particles in the blood called platelets become activated and stick together. A low daily dose of aspirin (75-100mg) prevents the platelets from becoming sticky so that clots are less likely to form.

Blood clots in vessels supplying the heart can lead to angina (chest pain caused by reduced blood flow to the heart muscles) or a heart attack (blockage of an artery supplying the heart muscle). Clots in blood vessels in the brain may cause temporary loss of consciousness (transient ischaemic attacks) or a stroke. People with conditions such as high blood pressure and diabetes, or those with fatty deposits in the arteries (atherosclerosis or 'hardening of the arteries') may be at risk of these problems. Low doses of aspirin are also used to prevent clots in people having cardiac procedures and if given shortly after a heart attack reduce the risk of death.

Low-dose aspirin can be used to prevent obstetric complications in women with the antiphospholipid antibody syndrome, an autoimmune disease associated with increased likelihood of blood clots and recurrent miscarriages. Aspirin may also have a role in preventing clots in Kawasaki disease, a rare disease of childhood in which clots may develop in the arteries supplying the heart (Montinari *et al*, 2019; Rife and Gedalia, 2020).

Aspirin in pre-eclampsia

Pre-eclampsia, a rare life-threatening condition which can occur after about 20 weeks of pregnancy, causes very high blood pressure, fluid retention and damage to organs such as the kidney and liver. Low doses of aspirin may be indicated to prevent pre-eclampsia in selected high-risk patients (Rolnik *et al*, 2017).

Aspirin reduces the risk of bowel cancer

The action of aspirin on platelets plays a central role in another remarkable finding. Activated platelets contribute to the growth and spread of tumours. Some people have underlying conditions that carry a high risk of developing bowel (colorectal) cancer. Studies have shown that a small dose of aspirin taken every day may reduce this risk. There is also some evidence that aspirin may reduce the risk of cancer of the oesophagus (gullet) but much weaker evidence suggesting that aspirin reduces the risk of other cancers (Song *et al*, 2020; Patrignani & Patrono, 2018).

Aspirin and infection

Studies in the laboratory have demonstrated that aspirin prevents replication of many viruses, including chickenpox virus, cytomegalovirus and hepatitis C virus. Clinical studies in critically ill patients with conditions such as severe sepsis (a life-threatening reaction to an infection) showed that aspirin improved survival, possibly by reducing clotting and inflammation. Aspirin may improve survival in adults with COVID-19 infection in which clotting is a well-recognised complication, but clinical studies are required to study this further (Bianconi *et al*, 2020).

Toxicity

Overuse or very concentrated topical preparations of salicylic acid may irritate the skin. Salicylic acid applied to the skin is absorbed, particularly if applied to broken skin. Toxicity and (rarely) death have been reported from overuse (Madan & Levitt, 2014).

Even a low dose of aspirin may cause dangerous internal bleeding from the lining of the stomach. Bleeding is caused by a combination of gastric irritation and blood that is less likely to clot. The risk is much higher in people with peptic ulcers or inflammation in the wall of the stomach and in association with alcohol. Aspirin also increases the risk of bleeding within the brain although this is a much less common complication. The potential benefits of treatment must always be balanced against potentially life-threatening complications (Arif & Aggarwal, 2021).

Aspirin may trigger Reye syndrome, a rare problem with a fatality rate of between 30% and 40%, that occurs most often in children or young adults 3-5 days after the onset of a viral infection such as influenza, a cold or chickenpox. The syndrome causes a fall in blood-sugar levels, vomiting, and swelling in the liver and brain with personality changes, confusion, seizures, and loss of consciousness. Some brain damage may be permanent. Aspirin should be avoided in children or teenagers with viral infections or flu-like symptoms. Reye syndrome also occurs in the absence of aspirin and the mechanism by which aspirin is associated with this condition is not completely clear.

Salicylate poisoning (salicylism) due to accidental overdose or intentional ingestion of excessive aspirin is a medical emergency. Poisoning causes tinnitus, nausea, abdominal pain and rapid breathing. High doses may be associated with potentially fatal complications such as swelling of the brain, fluid in the lungs, seizures and cardiac arrest.

Aspirin-exacerbated respiratory disease presents in some people with an underlying tendency to inflammation in the nose and sinuses and/or asthma. Aspirin and other NSAIDs make these symptoms worse. This reaction is not an allergy but is classed as a form of hypersensitivity. In this condition, aspirin activates inflammatory cells leading to a persistently runny nose, asthma that is difficult to control and nasal polyps.

Rarely, people develop a true allergy following repeated exposure to any form of salicylate. This may cause problems such as wheezing, nasal congestion, hives, itching and diarrhoea. Severe allergy can lead to a life-threatening anaphylaxis with swelling of the face, lips, tongue or throat and difficulty breathing (Arif & Aggarwal, 2021).

Summary

Whether the powdered bark of *Salix* as prepared by the Rev. Stone had sufficient salicin in it to act as an antipyretic may need to be revisited to see if his results were a placebo effect or secondary to the addition of quinine from *Cinchona* bark. Certainly, it was not an antimalarial. Nevertheless, it contributed to the extraction, investigation and pharmacological modification of salicin to produce aspirin, one of the world's great medicines. The use of willow bark, now replaced by salicylic acid, for treating warts, corns and other skin conditions is a medicine which has truly been 'used for thousands of years'. The range of conditions which respond to aspirin is a good example of the benefit of prolonged observation of a medicine's effects to detect previously unsuspected uses.

References

Arif H, Aggarwal S. *Salicylic Acid (Aspirin)*. Treasure Island, Florida, StatPearls Publishing, 2020

Bergius PJ. *Materia Medica E Regno Vegatibili*, 2nd edn, corrected, vol 2. Stockholm, Petri Hesselberg, p839, 1782

Bianconi V, Violi F, Fallarino F, *et al*. Is acetylsalicylic acid a safe and potentially useful choice for adult patients with COVID-19? *Drugs* 2020;80:1383-96

Coxe JR. *The American Dispensatory*. Philadelphia, A. Bartram for Thomas Dobson, 1806

Gerland W. New formation of salicylic acid. *Quarterly Journal of the Chemical Society of London* 1853;5(2):133-155 (in a letter from Dr Kölbe, about Gerland's work in achieving this, to the *Journal*). Note: This is usually quoted as H. Gerland, and dated 1852, but vol 5 was published in 1853.

Gibbs S, Harvey I. Topical treatments for cutaneous warts. *The Cochrane database of systematic reviews* 2006;3: CD001781

Jack DB. One hundred years of aspirin. *The Lancet* 1979;350:437-9

James R. *Pharmacopoeia Universalis*, 2nd edn. London, J. Hodges, 1752

Lebwohl M. The role of salicylic acid in the treatment of psoriasis. *Int J Dermatol* 1999;38:16-24

Madan RK, Levitt JA. Review of toxicity from topical salicylic acid preparations. *J Am Acad Dermatol* 2014;70:788-92

MacLagan T. The Treatment of acute rheumatism by salicin. *The Lancet* 1876;343 and 383

Montinari MR, Minelli S, De Caterina R. The first 3500 years of aspirin history from its roots – A concise summary. *Vascul Pharmacol* 2019;113:1-8

Patrignani P, Patrono C. Aspirin, platelet inhibition and cancer prevention. *Platelets* 2018;29:779-785

Rife E, Gedalia A. Kawasaki Disease: an update. *Curr Rheumatol Rep* 2020;22:75

Rolnik DL, Wright D, Poon LC, *et al.* Aspirin versus placebo in pregnancies at high risk for preterm preeclampsia. *New Eng J Med* 2017;377:613-22

Song Y, Zhong X, Gao P, *et al.* Aspirin and its potential preventive role in cancer: an umbrella review. *Front Endocrinol* 2020;11:3

Sterling JC, Gibbs S, Haque Hussain SS, *et al.* British Association of Dermatologists' guidelines for the management of cutaneous warts. *Br J Dermatol* 2014;171:696-712

Stone E. An Account of the Success of the Bark of the Willow in the Cure of Agues. In a Letter to the Right Honourable George Earl of Macclesfield, President of R. S. from the Rev. Mr. Edmund Stone, of Chipping Norton in Oxfordshire. *Philosophical Transactions of the Royal Society* 1763;53:195-200

Salix alba, source of salicylic acid and salicin, growing near a river edge (photo by Henry Warriner)

CHAPTER 38 – GRAHAM FOSTER

Silybum marianum
The source of silymarin and Legalon-SIL

Introduction

Milk thistle, *Silybum marianum*, is widely marketed as a herbal aid to liver and gall bladder health with the active constituents being extracted from the seed heads. Its use for other conditions is supported by references to antiquity, including Pliny in the 1st century and Culpeper in the 17th, but the only original reference to its use for liver diseases comes from Pietro Mattioli's *Commentaries on Dioscorides* published in the middle of the 16th century. However, it is not entirely clear which thistles these august authors were referring to, and they would have had no concept of the function or pathophysiology of the human liver.

Silymarin is the name for a standardised extract from the seed heads of *Silybum*, and Legalon-SIL is a compound derived from silymarin which has potential for reversing liver fibrosis.

Chemistry

Assessing the value of extracts of *Silybum* is complicated by the multitude of different chemicals within the seed heads which include a complex mixture of major flavonolignans: silybin A, silybin B, isosilybin A, isosilybin B, silychristin, isosilychristin and silydianin, which are of variable solubility in water and differentially absorbed after oral ingestion. Rottapharm/Madaus (Cologne, Germany) have developed a soluble extract of silymarin by converting silybin A and B into succinate salts to create Legalon-SIL that is administered intravenously and is likely to be considered for clinical use by the appropriate authorities in the near future for the treatment of liver damage due to hepatitis C.

Plant profile

Silybum marianum, a member of the Asteraceae family, naturally occurs widely throughout the Mediterranean and Southwest Europe to Afghanistan and Ethiopia.

The shiny, marbled, spiny leaves emerge from a basal rosette and are very distinctive, with their pale veins standing out against a jade green background. This plant is a biennial and so the solitary thistle flower, its purple centre ringed with spiny bracts, appears on a tall and sturdy stem in the second year of growing. Milk thistle's natural habitat is scrub and wasteland so it will grow in poor soil and full sun, but the size of the plant varies enormously depending on its situation.

CARDO DI SANTA MARIA.

Silybum marianum – woodcut from Mattioli's *Discorsi*, 1568

History

A thistle-like plant called *Silybum* appears in the *Materia Medica* of Dioscorides [70 CE] and in Pliny [70 CE], noting that the white juice of its roots prepared with honey and vinegar, causes vomiting – and with no mention of any effect on the liver. The plants from Dioscorides were illustrated in the *Juliana Anicia Codex* (514 CE) but the illustration of *Silybum* here resembles the carline thistle, *Carlina vulgaris*. Dioscorides also describes two other plants, *Leucacantha* and *Spina Alba*. The *Juliana Anicia Codex* illustration of the former, although identified as *Cirsium vulgaris*, is much closer to *S. marianum*. *Leucacantha* was used for sciatica, pleurisy, hernias, toothache and spasms, according to Dioscorides.

Silybum does not appear to have been known to Galen [200 CE] or the physicians of the Golden Age of Islam and disappears from the herbals and pharmaceutical literature. The Italian Mattioli (1564), commenting on Dioscorides, says it is unknown to him but describes a plant which he calls *Carduus lacteus* (Latin for milk thistle) and *Divae mariae* (Latin for the divine [Virgin] Mary) with an excellent woodcut of our *S. marianum*. The origin of the English names, milk thistle and Our Lady's thistle, can be traced back to this point as well as its use in liver disease, for Mattioli recommends it for the liver, unblocking veins, curing urinary obstruction, for dropsy, liver and kidney disease, as an emmenagogue and for pain in the side. Our Lady's thistle in the translation of the 12th century *Physica* by the French Abbess Hildegard von Bingen [1200] was used for a 'stitch' in the side.

The first Early Modern reference to the uses of the plant with the name *Silybum* appears in Parkinson (1640) with its English names of Our Lady's thistle and milk thistle. Here he follows Mattioli's account of *Carduus lacteus* although he does not mention it by name, saying it is good for fevers, the plague, 'obstruction of the liver and spleen' and to cure jaundice as well as treating renal stones, dropsy, and as a diuretic. He adds that the seeds are distilled, and the distillate drunk or applied with cloths over the liver 'to cool the distemperature thereof'. Culpeper does not give hepatic uses of Our Lady's thistle until 1651, when he copies Parkinson *verbatim*.

Carduus mariae was a pharmaceutical drug in the 18th century (James, 1752; Linnaeus, 1749, 1789) for pleurisy, dropsy and hydrophobia and was renamed by Linnaeus to *S. marianum* (Linnaeus, in Gaertner, 1791).

Silybum and all the variants of milk thistle and *Carduus lacteus* are otherwise mostly absent from pharmacopoeias of the 18th to 20th century. Barton and Castle (1877) quote Mattioli's *Commentaries on Dioscorides* regarding *Carduus marianus* (sic) that a 'decoction of the leaves is useful in dropsy, jaundice and nephritis'. Mrs Grieve (1931) quotes Culpeper on the benefit of milk thistle to the liver.

The history of *S. marianum* being used for liver disorders reappears in the 21st century thanks to modern chemistry and pharmacy.

Liver damage and its treatment

The liver can be damaged by a variety of different means, including regular administration of toxins such as alcohol or herbal medicines, but globally the commonest cause is persistent ('chronic') infection with either the hepatitis B or hepatitis C virus, which affects many hundreds of millions of people. Whatever the cause of the insult to the liver, the response is the same with inflammation leading to scar formation (fibrosis) that eventually takes over the whole liver to form cirrhosis that progresses to liver failure and death. Reversing liver fibrosis is a longstanding goal of many pharmaceutical companies but given that the collagen in the scarred liver is similar to the collagen that, literally, keeps the body together, a high degree of specificity is required for any 'anti-liver collagen' preparation. The heart for example is joined to its blood vessels by collagen and any reduction in their strength would inevitably lead to the heart disintegrating, with catastrophic consequences. Given the difficulties in developing a safe medication for reversing liver fibrosis, most research into new medicines for liver disease has focussed on identifying compounds to treat the underlying cause, and great progress has been made in developing drugs that cure or control chronic viral hepatitis.

However, the hunt for agents that will improve liver function and reverse scarring is continuing, but research has been hampered by the lack of an animal model for human liver diseases and the challenge of clinical trials in a slowly progressing disease with a large placebo effect. Any trial to evaluate new drugs for liver damage must involve a large group of people randomly assigned to either the active drug or a suitable placebo. Many of those motivated to take part in these clinical trials modify their lifestyle, chiefly by reducing their alcohol intake, and thereby derive benefit whatever drug they have been taking. In some of the recent trials in metabolic liver disease, up to 20% of participants in the placebo arm derived benefit, highlighting the challenges of identifying beneficial products. Given these difficulties, it is not surprising that the few, relatively small-scale trials conducted with silymarin derivatives have yielded conflicting results, and the evidence of benefit does not reach the standard required by most drug licensing authorities. This does not of course preclude selective reporting of the more successful studies in internet advertisements.

Legalon-SIL

An Austrian trial of the impact of the well-characterised Legalon-SIL preparation in patients with liver scarring included a number of patients with liver damage due to the hepatitis C virus. The investigator, Peter Ferenci, monitored the level of virus in the blood and was rather surprised to find a dramatic fall in the viral load (Polyak *et al*, 2013). Subsequent work in laboratory models of infection have shown that this silymarin preparation inhibits the replication of the hepatitis C virus, and the preparation has now completed a number of trials demonstrating clinical efficacy. During the development of the intravenous preparations of silymarin derivatives as antiviral drugs for hepatitis C infection, a massive research effort led to the development of a portfolio of medications from other sources that are extraordinarily effective. These well-tolerated drugs cure over 95% of treated people and are being deployed in an international drive to eliminate hepatitis C infection around the globe. At the present time, further development of silymarin derivatives is on hold as there is no apparent market for these difficult to administer preparations. However, as viral variants that are resistant to the current drugs emerge, it is probable that silymarin will be one of the next generation of drugs to be licensed as a rescue therapy for those whose virus has become resistant to the first-line drugs.

References

Gaertner J, Linnaeus C. *De fructibus et seminibus plantarum* vol 2. Stuttgart, Academiae Carolinae, p378, t162, 1788

Grieve M. *A Modern Herbal*. London, Jonathan Cape, 1931

Polyak SJ, Oberlies NH, Pécheur EI, *et al*. Silymarin for hepatitis C virus infection. *Antivir Ther* 2013;18(2):141-7 doi: 10.3851/IMP2402

Silybum marianum growing by the roadside in Italy

CHAPTER 39 – ANTHONY DAYAN

Tanacetum cinerariifolium
The source of pyrethrins for pesticides

Introduction

Tanacetum cinerariifolium, previously called *Pyrethrum cinerariifolium* and *Chrysanthemum cinerariifolium*, is known by the common name of pyrethrum, and less frequently as Dalmatian chrysanthemum and Dalmatian pellitory. It, and the closely-related *T. coccineum*, became the source of a very effective pesticide, called pyrethrum or pyrethrin. From the early 1900s to the late 1950s this was the basis of research that led to the synthetic pyrethroids, the most widely-used pesticides in the world today, and which have helped to save at least as many lives as any other plant in the College Garden.

Plant profile

Tanacetum cinerariifolium is a short-lived perennial native to Albania and the former Yugoslavia where it can be found on grassy sites, rocky ground, by roadsides and by the sea.

It forms an attractive mound of aromatic silver green dissected foliage above which the daisy-like white-petaled flowers are held on stems 50cm tall. These flowers have a central yellow disc and are pollinated by bees and flies.

Pyrethrum is mainly concentrated in oil glands on the surface of the seed with the other parts of the plant containing much lower concentrations. The growing plant releases an aphid alarm pheromone which attracts ladybirds and repels aphids and for this reason is used in 'companion planting'.

The closely-related species *Tanacetum coccineum* has pink, red or white flowers (occasionally yellow at the apex) which also contain insecticidal pyrethrum substances, but it is a poor source

Tanacetum coccineum

compared to *T. cinerariifolium*. Its native range is wider and spans Eastern Europe through to Central Asia and Iran where it is found in sunny, dry, mountainous habitats. This species has fragrant foliage and attracts butterflies.

Nomenclature

Tanacetum comes from the Greek for immortality. *Cinerariifolium*, in Greek, means *ash-coloured foliage*.

Called *Anthemis pyrethrum* by Linnaeus and *Anacyclus pyrethrum* by AP de Candole, *Tanacetum* became its modern botanical name in 1844 after being classified in the genus *Pyrethrum* and *Chrysanthemum* in the years between. Confusingly, the name *Chrysanthemum*, which we all understand to mean the horticulturally important florists' plants, now called *Dendranthema*, is still used in popular articles as the source of the pesticide, pyrethrum. It was not involved in the development of pyrethrum-based pesticides.

The use of *T. cinerariifolium* (or *T. coccineum*) as a pesticide before the 19th century does not appear to be documented.

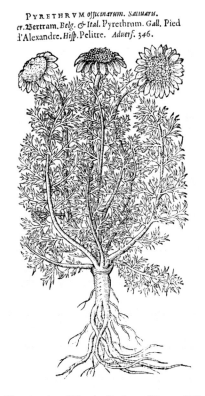

Tanacetum cinerariifolium (as *Pyrethrum officinarum*, Pellitory) from l'Obel's *Plantarum seu Stirpium Historia*, 1576

History

Anointing with the oil from a plant called *Pyrethrum* was recommended by Dioscorides [70 CE] for inducing sweating and treating paralysed limbs. Chewing the leaves, which had a very 'hot' taste, induced salivation. Galen [200 CE], calling it *Pyrethron*, repeats this with the addition of it treating toothache. Ibn al-Bayṭār (d. 1248) wrote that it came from Constantine in Algeria and was called *sandasab*, its Arabic name being *Akulkara* or *Aáquaqarhú*, and was exported to India via Egypt (Flückiger & Hanbury, 1874). In Avicenna's *Canon of Medicine* (1490, 1522), as translated from Arabic by Gerard de Cremona, it has the Italian name of *piretro*. It is in the College *Pharmacopoea Londinensis* (1618) as pyrethrum; Culpeper (1650) calls it *Pyrethri*, *Salivaris* and pellitory of Spain as does Gerard (1633) who also calls it *Pyrethrum officinale*. Gerard's woodcut resembles *T. cinerariifolium*.

Parkinson (1640) has the same woodcut, calling it *Pyrethrum vulgare officinare* and pellitory of Spain. All of them leave the uses unchanged, with the root being chewed to induce salivation in Quincy's *Dispensatory* (1718) and through to the 20th century (Remington, 1894; Martindale, 1967).

The dried powdered flowers of *Pyrethrum*, widely used as a commercial insecticide in the 19th century, were not in the *British Pharmacopoeia* (1864) but are included in *Squires Companion to the British Pharmacopoeia* (1899).

The need for pesticides

In the 18th-19th centuries the growing industrialisation of agriculture everywhere created increasing demand for safe, effective and easy-to-use pesticides to protect food crops, fruits and especially cotton, as well as other fabrics, animals and people. The chemicals then available, such as arsenical compounds and copper-sulphur mixtures, were reasonably active but they were toxic to many animals and humans and to the environment.

Industrial production of pyrethrum powder

In the mid-19th century *T. cinerariifolium* had been recognised as having pesticidal properties in the Caucasus and, as the climate and local circumstances were favourable, its commercial cultivation was developed into a major industry to supply European needs. Initially this was in the Caucasus (the area between Turkey and Russia and the Black and Caspian Seas). An enterprising German merchant, Johann Zacheri (1814-1888) from Munich, was responsible for initiating its mass production and pan-European distribution with the flowers being grown and ground up in the Caucasus and vigorously marketed as 'Zacheri's Insect Killing Tincture' – trade name 'Zacherlin'. It was a huge success. In 1880, his son took over the business, changing the source of *Tanacetum* to the Balkans in the more convenient (closer) Dalmatia, on the Adriatic coast of what is now Croatia, and building a factory in Vienna (with an elaborate mock-Moorish façade) to grind the powder. Caucasian powder became known as Dalmatian powder.

In 1885 the seeds of *T. cinerariifolium* and *T. coccineum* were introduced to Japan. Shortly afterwards, Eiichiro Ueyama set up an industry to produce the plants and pyrethrum powder and became a major supplier to Asia and the USA, producing 13,000 tons of dried flowers at the peak in 1935. He invented pyrethrum sticks, which burnt like joss-sticks, to kill mosquitoes and, later, pyrethrum coils that would burn all night, so providing good protection during sleep.

The industry in the Balkans collapsed during the First World War, and planters in Kenya took it over and developed an extensive trade in pyrethrum powder, produced from *T. cinerariifolium* flowers, exporting to the USA and Europe for use on humans, plants, animals and textiles of many sorts.

Nowadays Kenya and Tasmania are the principal world suppliers of pyrethrum powder and the pyrethrin oleoresin extract which is one of the major 'organic' pesticides.

One of the consequences of the Second World War was a greatly increased demand for better and safer pesticides for agricultural and medical purposes. Developments in chemistry led first to the unrelated organochlorines, such as DDT, which were very successful for a while until their persistence and environmental toxicity became apparent, Similarly, the organophosphorus pesticides, such as Malathion, also became hugely important until their harmful properties were also realised, and their use became increasingly restricted. Additionally, resistance to those pesticides became a serious problem in the late 20th century.

From pyrethrum to pyrethrins

Pyrethrum is a reasonably effective and rapidly-acting pesticide but it is not stable on exposure to light or moisture in the environment. Its popularity has been due in part to its very low toxicity to humans and other warm-blooded mammals and birds though valuable insects, such as bees and butterflies, are killed if exposed to the powder or spray. It is also very dangerous to fish. The active constituents, the pyrethrins, work by disrupting conduction along nerve fibres resulting in death of certain arthropods and fish. Mammals are much less susceptible because they are naturally able to detoxify pyrethrins quickly. Direct application of concentrated pyrethrins to the skin of humans can produce blistering and local tingling and other abnormal sensations for a short while.

The pyrethrins are the six closely-related naturally-occurring insecticidal compounds in pyrethrum and are present in relatively high concentrations in the seed capsules – up to 1.1% in *T. cinerariifolium* and 0.5% in *T. coccineum* – and only at much lower concentrations in other species. Pyrethrum from the dried and powdered flowers was replaced in the 1920s by extracting the pyrethrins from the seed capsules with kerosene to make liquid oleoresins, resulting in much more convenient sprays. At that time world consumption of pyrethrum had risen to more than 20,000 tons a year.

Until the chemical nature of the pyrethrins was discovered by the Swiss chemists Staudinger and Ruzicka in 1916 (not published until 1924) it was impossible to develop analogues with better properties (Roth & Vaupel, 2018). This work was later described as one of the great classics in chemical discovery of the era. Staudinger and Ruzicka later gained Nobel Prizes for other work.

In the late 1950s research by Drs Tattersfield and Elliott at the Rothamsted Experimental Station in Hertfordshire identified the key insecticidal molecular features of the pyrethrins. They and others in the agrochemical industry worldwide were then able to synthesise new chemical analogues with greatly increased potency and much better stability.

Pyrethrins in horticulture and agriculture

Since then, the world production of synthetic pyrethroids, such as deltamethrin, permethrin, and alpha-cypermethrin, has risen to several hundred thousand tons per year to protect fruit, grain and fibre crops in the field, and the ornamentals in our gardens. They are used to kill insects that eat the plants and those that transmit diseases and to defend many fabrics, such as carpets, curtains and woolly garments, from insect attack in the home.

Prevention of insect-transmitted human and animal disease

Amongst their valuable medical applications have been the treatment of lice and other infestations of pets, farm animals and human beings. A very important public-health use has been spraying wide areas, mostly in the tropics, to kill disease-carrying insects that bite humans and livestock, transmitting such diseases as malaria, the Zika virus, Lyme disease, typhus, the plague, trypanosomiasis, viral fevers, blowfly strike of sheep, and many others. Sadly, these pesticides too are being shown to carry long-term toxicity problems to wildlife, but their benefits outweigh their risks until environmentally-safer compounds can be found.

One of the most striking examples of their value has been in preventing malaria in tropical countries. In the 1970s-80s in large parts of Africa and Southeast Asia there were probably up to 20 million deaths yearly due to malaria, with a particularly high incidence amongst babies and young children. Groups most at risk lived in areas where medical treatment with drugs was either not available, or resistance had developed to convenient drugs. Spraying large areas to kill mosquitoes was not feasible because of the unaffordable cost of repeated application of pesticides over many thousands of square kilometres – as well as causing the extermination of thousands of other insect species and fish with other environmental disasters.

Research, particularly in Africa, showed that most bites by malaria-carrying mosquitoes occurred at night, so partial prevention could be obtained by supplying bed nets to protect people. Much greater effect was obtained by treating the bed nets with potent synthetic pyrethroids that would kill insects on contact with the net. In Africa until such medicated bed nets became available there were more than 500,000 deaths per year of children under five. With better drug treatments, too, there are now fewer than 90,000 deaths per year of these youngsters, and the number is still falling rapidly (Roth & Vaupel, 2018).

Conclusion

It has been a lengthy and worldwide journey from 'Zacheri's Insect Killing Tincture' to today's synthetic pesticides against insect vectors of horticultural, agricultural, human and animal diseases, but this small, white-flowered plant and its pink-flowered cousin were the beginning of successful progress along a remarkable chemical, biological and industrial pathway.

References and Reading Suggestions

BBSRC *The History of the Pyrethroid Insecticides*. https://webarchive.nationalarchives.gov.uk/ukgwa/20200302111045/https://bbsrc.ukri.org/documents/pyrethroid-timeline-pdf/ accessed 7 January 2021

British Pharmacopoeia. London, Spottiswode & Co., 1864

Casida JE. Pyrethrum Flowers and Pyrethroid Pesticides. *Environ Hlth Perspect* 1980;34:189-202

Glynne-Jones A. Pyrethrum. *Pesticide Outlook* 2001;5:195-8

Martindale. *Extra Pharmacopoeia*, 25th edn. London & Bradford, Percy Lund, Humphries & Co. Ltd., 1967

Matsuo N, Mori T (eds). *Pyrethroids. From Chrysanthemum to Modern Industrial Pesticide*. Heidelberg, Germany, Springer Verlag, 2012

Quincy J. *Pharmacopoeia Universalis Extemporanea or a Complete English Dispensatory*. London, A. Bell, 1718

Remington JP. *The Practice of Pharmacy*, 3rd edn. Philadelphia, J.R. Lippincott & Co., 1894

Roth K, Vaupel E. Pyrethrum: History of a Bio-Insecticide – Parts 1-6. *Chem Views Magazine* 2018 doi: 10.1002/chemv.201800105

Schleier JJ, Peterson RKD. Pyrethrins and Pyrethroids. In: Lopez O, Fernandez JG (eds) *Green Trends in Insect Control*, No. 11. London, UK, RSC, pp94-114, 2011

Squire P. *Companion to the British Pharmacopoeia*, 17th edn. London, J & A Churchill, 1899

Staudinger H, Ruzicka L. Insektentötende Stoff. *Helv Chim Acta* 1924;7:177-201

Ueyama E. Introduction of pyrethrum flowers (130 years in Japan). *Acta Hortic* 2017;1169:1-6

Boisduval scale is a persistent pest on *Phalaenopsis* orchids, and can be controlled by pyrethroids, although it develops resistance to them

CHAPTER 40 – ARJUN DEVANESAN

Taxus baccata and *Taxus brevifolia*
The source of paclitaxel, docetaxel and cabazitaxel

Introduction

The anti-cancer drug paclitaxel (brand name Taxol) was first discovered in the bark of the Pacific yew (*T. brevifolia*) and subsequently a precursor was found in the needles of the European yew (*T. baccata*). Paclitaxel is still used today as a treatment for breast and ovarian cancer, though it is no longer sourced directly from trees. Other useful semi-synthetic analogues of paclitaxel, like docetaxel and cabazitaxel used in prostate cancer as well as lung and other cancers, also owe their existence to the yew. It is a historical lesson that poisons and medicines are often two sides of the same coin and an amazing example of how a compound used by a plant for self-preservation, and is produced both by the tree and its symbiotic endophytic fungus to extend its life enormously, has been used to extend the lives of people with cancer.

Plant profiles

Taxus baccata is often called the English yew although its native distribution covers much of Eurasia. The ancient specimens found still growing in holy places were clearly planted intentionally but plants can be found growing naturally in woods and scrubs throughout the British Isles, especially on calcareous soils. One of the most familiar of trees, it is a densely branching, hardy evergreen that can grow for centuries or even millennia. It can reach up to 20m tall and develop an enormous trunk with a potential girth of up to 8m. It has wonderful powers of regeneration, re-sprouting if damaged or cut back, which is why it is so useful for hedging or topiary. The bark is scaly and brown,

Taxus brevifolia, Pacific yew

and the deep green needle-like leaves are arranged spirally on the stem (though they look as though they are arranged in two rows). The species is dioecious (male and female flowers grow on separate plants).

Female plants develop insignificant flowers on the undersides of one-year-old branches followed by fleshy red fruits, known as arils, containing a single seed. Pollination is by wind and the seed dispersal is primarily done by birds feeding on the fruits. They are able to do this because the aril, or fleshy seed covering, is the only non-toxic part of the plant and the seeds are not digested.

Taxus baccata is the backbone of many English gardens: it benefits wildlife and will grow in any well-drained soil on sites in full sun or shade. It tolerates exposure, dry soils and urban pollution.

Taxus brevifolia is very similar in appearance but is smaller (growing to 15m) and has shorter leaves (hence the species name). It is commonly known as the Californian, Oregon, or Pacific yew and has a much narrower distribution, being native to South Alaska and Western USA. Here it is found in lowland to mountain forests growing as a tree beneath a closed forest canopy. In drier, open forests it adopts a shrub-like habit similar to *Juniperus communis*. Due to the felling of large trees in the wild for use as a wood and in medicine, the species is rated as Near Threatened on the IUCN Red List.

Because many other conifers produce their seed in cones, some botanists did not consider *Taxus* to be a true conifer. However, due to studies of its evolutionary relationships, the yew family (Taxaceae) has now been placed firmly within the conifers.

History and mythology

It is no trivial matter to determine the age of an ancient yew. Many of these hoary hulks which brood over the southern or western corners of churchyards all over England, Scotland and much of Western Europe are some of the oldest trees we have. A conservative estimate puts the birth of some of these ancients long before the Romans first landed in Britain. In Scotland, it is sometimes said that Pontius Pilate was born under a Scottish yew, while his father was on a diplomatic mission to a Pictish King. 'Beneath those rugged elms, that yew-tree's shade, Where heaves the turf in many a mouldering heap,' so goes Thomas Gray's *Elegy Written in a Country Churchyard*. Perhaps it is their age, the hardness of their wood and the general spookiness of their appearance, or the fact that they seem to live forever and can regenerate even from their centre, that yews have often been used as a symbol of death and rebirth. As the tree most associated with Hecate, the Greek goddess of death and necromancy, the yew was revered also by the Celts and druids and found its way into early Christian superstitions. Most churchyard yews predate their churches, and it is likely that their mystique inspired the location of so many old burial sites. Yew has also

been long associated with war and weapons. Its
wood is remarkably strong and has been used for
the making of tools since the Neolithic. Virgil
(70-90 BCE) praises the yew tree bows of the
Syrians – and the bows on the Assyrian carvings
of the 8th century BCE in the British Museum
show them well.

Taxus baccata from Mattioli's *Discorsi*, 1568

The English bows at Agincourt were famously
made of yew, but the earliest known yew
longbow dates approximately 3300 BCE and
was discovered in the Ötztal Alps with a natural
mummy known as Ötzi. For a tree associated with
war and death, it should come as no surprise that
all parts of the yew (and especially its needles and
seeds – but not the scarlet surrounding aril) are
well known for being poisonous. Pliny [70 CE],
in his *Natural History*, noted that 'even wine flasks
for travelers made of [yew] wood in Gaul are known to have caused death'. In Shakespeare's
Richard II, it is called the 'doubly fatal yew', both for its poison and as the material from which
the English longbow was made. Indeed, Kings Harold, William II, and Richard the Lionheart
were killed by arrows from yew bows.

Its fatal nature was clearly known to Dioscorides [70 CE] who placed *Taxus*, the yew tree, called
Smilax in Greek, among other very poisonous plants, *Aconitum* (aconite), *Conium* (hemlock), *Nerium*
(oleander), *Mandragora* (mandrake) and mushrooms. He gives no medicinal uses, just a warning
that little birds that eat the berries are turned black and people suffer diarrhoea, adding that
sleeping under it carries the risk of death. Pliny [70 CE] similarly, writes 'The fume and smoke
of any Yeugh tree killeth mice and rats', our knowledge of its toxicity being extended when a
retired soldier who smoked the leaves died within 12 hours (Tree Poison Death Shock, 2017).
Pliny attributes the etymology of the word 'toxic' as being derived from *Taxus*. Mattioli (1569),
commenting on Dioscorides, adds that cows that eat the leaves are killed. Galen [200 CE] only
has '*Smilax* or *Cactus* [the latter meaning any unpleasant plant] is a tree with poisonous properties'.
It never entered the mainstream pharmacopoeias of the Golden Age of Islam or of Europe,
but the 13th century nun Hildegard von Bingen [c. 1200] recommends inhaling the smoke from
the burning wood for nose and chest disorders – likely to be fatal if the leaves were included,
as the soldier discovered. It does not appear in herbals until l'Obel (l'Obel & Pena, 1571) has

the details from Dioscorides without comment and Parkinson (1640) notes that the berries are eaten by men and children without problems (but omits that it is the seed inside the aril that is poisonous, so these still kill people). In the 19th century a few physicians were recommending yew for rheumatism, cough, epilepsy and hysteria or even saying it was not poisonous, to the horror of their contemporaries (Scotti, 1872).

Modern history and the discovery of paclitaxel

It was warm and sunny in August 1962 when Arthur Barclay decided to sample a tall Pacific yew tree (*T. brevifolia*) in Washington State's Gifford Pinchot National Forest. He labelled it B-1645 as it was the 1645th plant sample he collected. He and his three assistants also collected two other samples: PR-4959, stem and fruit; and PR-4960, stem and bark. Barclay was collecting samples for the American Cancer Chemotherapy National Service Center (CCNSC). In 1955 the National Cancer Institute (NCI) started a monumental drive to screen substances for anti-cancer activity. Between 1955 and 1960 the screening programme mainly investigated known synthetic chemical substances – various chemical compounds sat on laboratory shelves at the time with as yet undiscovered medical uses. However, probably driven by a dearth of strong candidates, NCI expanded its screening programme to include natural plant and animal products, in partnership with the US Department of Agriculture. Over the next 20 years, over 30,000 compounds were screened.

The NCI plant programme was run by Jonathan Hartwell, an organic chemist who became Assistant Chief for Program Analysis Activities at the CCNSC in 1958. While the history of the yew never seems to figure in reports of the discovery of paclitaxel, it is worth noting that Hartwell put together a document detailing the traditional uses of plants to treat cancer in the Chinese, Greek, Egyptian and Roman traditions. This work led to investigations into *Juniperus sabina* which showed some anti-cancer activity but never amounted to much. However, the active interest in evergreen trees and conifers continued.

Barclay's samples, both the above and a larger sample taken in 1964 after yew bark showed some interesting cell toxic effects, eventually made their way into the hands of Dr Monroe Wall and Dr Mansukh Wani. At the time, both researchers were working on camptothecin, a natural extract from *Camptotheca acuminata* (*q.v.*) also with anti-cancer activity. The strong cytotoxicity of *T. brevifolia* led to scepticism in the NCI of its potential medicinal use but having worked with *Camptotheca*, Wall and Wani were more optimistic. Partly because they dedicated most of their time to the more promising *Camptotheca* (and because of the complex chemical structure of the active compound in *T. brevifolia*) it was not until 1971 that the chemical structure was discovered ($C_{47}H_{51}NO_{14}$). Even before the full chemical structure had been painstakingly worked out, it was clear to researchers that it was an alcohol – hence the suffix 'ol' – and compounds were usually named after their source. Hence the brand name Taxol came from *Taxus* with the suffix 'ol' as it was an alcohol.

Another major stumbling block in the discovery and subsequent investigation into the anti-cancer activity of paclitaxel was that the yield of paclitaxel from natural samples was extremely low. Add this to the fact that *T. brevifolia* is famously slow-growing and we can anticipate a major problem. It took about 12kg of bark to extract 0.5g of paclitaxel, and each tree only produced about 2kg. Also, once the bark was stripped, the tree would die. This made the sourcing of paclitaxel deeply unattractive to the NCI administrators. As a result, paclitaxel was once again relegated to a back shelf until 1979, when it was shown once again to be particularly effective against breast cancer in a mouse model, and that its mode of action (microtubule stabilisation) was unique.

Paclitaxel

Also, just when it seemed that paclitaxel might be the next great breakthrough in cancer research, researchers realised that paclitaxel was insoluble in water! Without a means to administer paclitaxel intravenously, the product was next to useless. Luckily, at just that time, a new product was sweeping water-insoluble chemicals off the shelves and back into the testing laboratories – Cremophor EL. This was an ethoxylated compound of castor oil from *Ricinus communis* (*q.v.* Excipients for details). Paclitaxel could be dissolved in Cremophor EL and injected intravenously. With this final discovery, clinical trials could begin.

Showing remarkable activity in breast and ovarian cancer, paclitaxel passed through Phase I and II trials. However, supply remained a major issue. If paclitaxel were used in all ovarian and breast cancer patients in America alone, hundreds of thousands of slow-growing Pacific yews would have to be cut down every year. It was clear that such a strategy was not sustainable. And if that was not obvious to the NCI, a host of environmental groups were quick to point it out. The race was then on to synthetically manufacture it. In the meantime, however, a French scientist Pierre Potier, working for Rhône Poulenc Rorer (now Sanofi-Aventis), managed to extract a similar compound (10-deacetyl baccatin III or 10-DAB) from the quicker-growing and more common European yew, *T. baccata*. Most importantly, 10-DAB comes from yew needles, a renewable resource, rather than the bark, and by adding a small synthetic tail onto the 10-DAB molecule, Potier managed to produce docetaxel (brand name Taxotere), which works in a similar way to paclitaxel and was licensed for prostate cancer under the name docetaxel. With Andrew Greene, the same team then went on to find a way of producing paclitaxel from docetaxel and so gardeners of the time in England may remember donating their yew hedge clippings for extracting these drugs. Eventually Robert Holton at Florida State University developed a commercial semi-synthetic manufacturing process which was taken up by Bristol-Meyers Squibb in 1991 for the commercial production of paclitaxel. A year later, it was approved by the Food and Drug Administration for the treatment of ovarian cancer and in 1994 for breast cancer.

Other uses

Stents (a tubular scaffold), which are inserted into narrowed coronary arteries, keep them open and allow sufficient blood and oxygen to reach the heart muscle. The cells of the lining of the coronary artery tend to grow and block the stent and paclitaxel has been used to coat the stent to inhibit this new growth and reduce the need for re-operation.

Paclitaxel-producing fungi

With its growing popularity, the cost of production of paclitaxel (even a semi-synthetic version) remained prohibitively high. It was therefore a delight to plant biologists Gary Strobel and Andrea Stierle that in 1991 they discovered a fungus *Taxomyces andeanae* which grew deep in the bark of an ancient Pacific yew in Montana and seemed to produce paclitaxel all on its own. Following their discovery, paclitaxel is produced relatively cheaply by growing the fungus in large-scale fermentation tanks. It remains one of the most common treatments for advanced ovarian cancer.

This endophytic paclitaxel-producing fungus living in the tissues of a yew tree is a true symbiosis in nature. For the tree, paclitaxel is a defensive anti-fungal compound, produced by the tree which has activity against a range of wood-destroying fungi that could colonise the bark and destroy it. More than 20 different fungi have now been found in yew trees that produce paclitaxel, augmenting the paclitaxel produced by yew trees themselves. When a pathogenic fungus infects the bark of a yew, the chemicals produced by decaying wood stimulate the symbiotic fungus to produce paclitaxel inside hydrophobic bodies in their hyphae which then augments the tree's own paclitaxel synthesis as an anti-fungal agent. The presence of paclitaxel in the extracellular hydrophobic bodies means that the symbiotic fungus is not itself damaged by paclitaxel and that the rapidly-dividing cells of new branches are also preserved from harm. Only the invading pathogenic fungus is affected (killed) when it comes into contact with the hydrophobic bodies. Paclitaxel itself is also sequestered into the bark of the tree as a permanent defence. Similar paclitaxel-producing endophytic fungi exist in other long-lived trees including *Ginkgo* and the *Wollemia* pine.

That yew trees can live for perhaps 3000 years is an indication of the success of this symbiosis, which also has an evolutionary benefit to the fungus in that its host never dies (Talbot, 2015).

References

American Chemical Society *Discovery of Camptothecin and Taxol*. https://www.acs.org/content/acs/en/education/whatischemistry/landmarks/camptothecintaxol.html accessed 12 December 2020

Chast F. Pierre Potier (1934-2006). *Nature Obituaries* 2016;440:291

Scotti G. *Flora Medica della Provincia di Como*. Como, Carlo Franchi, 1872

Stone R. Surprise! A fungus factory for Taxol? *Science* 1993;260(5105):154-5

Talbot NJ. Plant immunity: A little help from fungal friends. *Current Biology* 2015;1074-6

Tree Poison Death Shock. *Oldham Evening Chronicle* 4 July 2017 https://www.oldham-chronicle.co.uk/news-features/8/news-headlines/104333/tree-poison-death-shock accessed 3 March 2022

Wall M, Wani M. Camptothecin and Taxol: Discovery to Clinic. Research Triangle Institute, Progress Report No. 21, June 26, 1967. *Cancer Research* 1995;55:757

Showing remarkable activity in breast and ovarian cancer, paclitaxel passed through Phase I and II trials. However, supply remained a major issue. If paclitaxel were used in all ovarian and breast cancer patients in America alone, hundreds of thousands of slow-growing Pacific yews would have to be cut down every year. It was clear that such a strategy was not sustainable.

And if that was not obvious to the NCI, a host of environmental groups were quick to point it out.

CHAPTER 41 - GRAHAM FOSTER

Valeriana officinalis
The source of sodium valproate

Introduction

Valerian, *Valeriana officinalis*, is a clump-forming perennial with fleshy branching stems arising from rhizomes. The plant has a long history of medicinal use. In more modern times the plant has a reputation for having sedative properties and is widely available as an over-the-counter medication to treat insomnia. It is claimed to be without side effects, but in liver clinics around the country it is commonly seen as a cause of a marked hepatitis (inflammation of the liver) which recovers on stopping the drug. The plant is the origin of valproic acid, which is converted into sodium valproate, a drug widely used in the treatment of epilepsy and some other neuroexcitatory disorders including mania.

Plant profile

Valeriana officinalis is native to Europe through to Northwest Iran where it is found growing on roadsides, wasteland and amongst scrub.

A herbaceous perennial, the scented, bright green, pinnate leaves form mounded clumps high above which rounded clusters of white or pink flowers are held on 1.5m tall stems. The flowers are sweetly scented and attract many insects, particularly hoverflies. It spreads by rhizomes, and it is the pungent aroma of the fleshy white roots which are its most distinguishing characteristic. It requires a moist soil in full sun or dappled shade to grow well and reach its full height and, in this case, staking may be required.

Fragrant roots of *Valeriana officinalis*

History and nomenclature

This herb is believed to be the one recorded by Dioscorides as Phu, which refers to the peculiar strong odour of the root that combines pleasant and unpleasant fragrances. The herb may have been later named valerian after the Roman Emperor Valerian (Publius Licinius Valerianus, 253-260 CE), or named for its quality (in Latin, *valere* means 'to be in health' and when used to describe a medicine it means 'efficacious'). Other popular names are *Nardus sylvestris* and setwal. However, the name valerian is not mentioned by early authors and the earliest references to indicate that it is synonymous with Phu are stated in the writings of the Egyptian physician Isaac Judeus (d. 932 CE) and Constantinus Africanus (d. 1087) from the School of Salerno in southern Italy (Flückiger & Hanbury, 1879).

Centranthus ruber is known as red valerian, but is no relation of *Valeriana* and has no medicinal properties

Valeriana officinalis in Egenolph's *Plantarum Arborum*, 1562

Confusion arises as the pink-flowered *Centranthus ruber* (and *C. ruber* 'Albus' the white form) are also called valerian. Culpepper, calling them 'Behen, Valerian white and red' noted that the Arabians used it to 'comfort the heart and promote lust' whilst the Graecians (sic) hold them to 'stop fluxes and provoke urine' (Culpepper, 1649). While it is in the same family, Caprifoliaceae, as *V. officinalis*, it is not the subject of this chapter. Numerous other plants have also been included as 'valerian', before Linnaeus established the genus in 1753, which further confuses the tracing of the history of its pharmaceutical effects.

Culpeper (1649), calling it Phu, *Valerinae maioris, minoris*, valerian or *setwal*, says the greater valerian is a diuretic; induces menstruation; treats cystitis (strangury), headaches, testicular swellings and gastric wind; heals wounds and ulcers; and draws out thorns. He makes no mention of sedation. He is simply quoting Dioscorides' properties of Phu written around 70 CE. While popular herbalism regards valerian as being used as a sedative for millennia, this property is not to be found in the literature before the 18th century. Linnaeus's *Materia Medica* (1749) gives one of its uses as 'narcotica' and this may be the earliest reference, but by the beginning of the 19th century it was common for dispensatories to note that it was used as a sedative (Duncan, 1819; Coxe, 1806; Thomson, 1811). In Martindale's *Extra Pharmacopoeia* (1936) it appears as: 'Given in hysterical and neurotic conditions as a sedative. Its action has been attributed to its unpleasant smell'.

The use of *V. officinalis* in epilepsy stems from Fabius Columna (1567-1640) who wrote in his *Phytobasanus* (1592) that he had cured himself of epilepsy by using it (but relapsed). It was advocated widely for epilepsy through the 18th century but both Cullen (1789) and Woodville (1792) noted that it did not work when they used it.

Valerian contains small amounts of the unpleasant-smelling valeric acid that is used to generate volatile esters widely used to manufacture perfumes as they differ from their parent in having pleasant odours. The carboxylic derivative of valeric acid, valproic acid, is an inert liquid that is widely used to dissolve compounds of interest that are insoluble in water. Its value as an anticonvulsant was discovered by Eymard in 1962 during his search for anticonvulsant drugs when he used valproic acid as the solvent for such compounds. Sodium valproate is the salt of valproic acid and remains in use for the treatment of epilepsy and mania, and as protection against the development of migraines.

Chemistry

Higher brain functions are a delicate balance between excitatory and inhibitory neural pathways that are linked together by chemical transmitters. Disrupting the balance of activity leads to either excitation or lassitude. Gamma-amino butyric acid (GABA) is a major inhibitory neurotransmitter and its activity is increased by sodium valproate leading to an overall reduction in neuronal excitation, thereby helping to prevent epileptic seizures and reduce the over-excited state of mania. In addition to its effects on GABA, sodium valproate also affects nerve impulse transmission by interfering with the chemical channels in nerves that allow electrical pulses to be propagated. Hence valproate is a general central nervous system depressant, and it is of proven clinical value in the treatment of epilepsy, mania and migraines where it is widely used.

Sodium valproate is toxic to the liver and a significant proportion of people develop minor abnormalities of the liver whilst taking this medication but, happily, most recover whilst still taking the drug. Of concern are a small number of patients who develop very severe liver dysfunction that may lead to liver failure and death.

Chemistry

Sodium valproate causes liver dysfunction by damaging the the mitochondria, the energy-producing sub-cellular organelles. Mitochondrial toxins lead to a characteristic form of liver failure associated with deposits of small fatty droplets in the liver (microvesicular fatty change) and this is the hallmark of severe sodium valproate toxicity. It is unclear why some individuals are more prone to this than others, but children are particularly sensitive.

In pregnant women the drug can lead to foetal abnormalities leading to difficult choices about exposing the foetus to harm or destabilising maternal epilepsy. The incidence of congenital malformations consequent on valproate use in pregnancy is given as 10% (cleft lip and palate; craniostenosis; cardiac, renal and limb defects, and many other anomalies); developmental disorders in 30-40% including delayed milestones; a lowering of IQ by 10 points; poor language skills; poor memory, and attention deficit disorder. Autism is five times more likely and autism spectrum disorder three times more likely.

These complications have led to a reduction in the use of valproate in young women, particularly in countries where alternatives are available. However, since the patent on sodium valproate is time-expired the drug is relatively inexpensive and therefore remains widely used in poorer nations.

Toxicity of valerian as a herbal medicine

An extract of the dried root of *V. officinalis* is licensed by the UK's Medicines and Healthcare Products Regulatory Agency (MHRA) as a traditional herbal medicinal product for the temporary relief of sleep disturbances and mild anxiety. The word temporary is used deliberately as it is addictive and there are withdrawal symptoms on stopping it – the European Medicines Agency (EMA) licences it 'for up to four weeks'. It appears to be as effective as benzodiazepine sedatives. However, the well-documented severe effects of valproate on the liver are also seen with herbal valerian. Minor abnormalities of the liver are common in people taking it and, occasionally, they are associated with more severe liver damage. The impact of the herbal product is poorly documented or understood but it is highly probable that it will have an adverse effect on pregnancy: it is not licensed by the MHRA for use in pregnancy or while breastfeeding.

Conclusion

Sodium valproate is derived from *V. officinalis* and is an inexpensive, widely-available anticonvulsant that has been of great value in the treatment of epilepsy for many decades. Its major side effects, liver toxicity and foetal abnormalities in pregnant women, have led to its replacement by less toxic agents but it remains in widespread use, particularly in resource-poor countries. Its plant of origin remains in widespread use as a sedative, despite its well-documented association with liver damage.

References

Bethesda MD. LiverTox: Clinical and Research Information on Drug-Induced Liver Injury. National Institute of Diabetes and Digestive and Kidney Diseases 2012. https://www.ncbi.nlm.nih.gov/books/NBK547852/ accessed 16 July 2022

Coxe JR. *The American Dispensatory*. Philadelphia, A. Bartram for Thomas Dobson, 1806

Cullen W. *A treatise of the Materia Medica*. Edinburgh, Charles Elliot, 1789

Duncan A. *The Edinburgh New Dispensatory*. Edinburgh, Bell & Bradfute, 1819

Flückiger F, Hanbury D. *Pharmacographia. A History of the Principal Drugs of vegetable origin met with in Great Britain and British India*. London, Macmillan and Co., 1874

Johannessen CU. Mechanisms of action of valproate: a commentatory. *Neurochem Int* 2000;37:103-10

Martindale. *The Extra Pharmacopoeia*. London, The Pharmaceutical Press, 1936

Thompson T. *The London Dispensatory*. London, Longman, Hurst, Rees, Orme, and Brown, 1811

Woodville W. *Medical Botany* vol 1-3. London, James Phillips, 1790-3

CHAPTER 42 – MICHAEL DE SWIET

Veratrum album
Veratrum californicum
Veratrum nigrum
The sources of protoveratrine, cyclopamine and sonidegib

Veratrum album and *V. nigrum* were the source of medicines like protoveratrine that were used to treat eclampsia and pre-eclampsia. *Veratrum californicum* contains a chemical which can be converted to a medicine, sonidegib, which disables the genetics of the growth of basal cell carcinomas.

Veratrum californicum

Veratrum nigrum

Veratrum album

Introduction

Veratrum album has white flowers, tinged green, *V. nigrum* has very dark red flowers, and *V. californicum* also has white flowers, tinged green; all are members of the family Melanthiaceae. *Veratrum album* and *V. nigrum* are widely distributed in Eurasia from Spain to Korea, and from the Himalayas to the Arctic islands of northern Russia. *Veratrum californicum* is a native of North America. All these species contain a range of powerful alkaloids which affect the autonomic nervous system and the mechanism of cell division in embryos and certain cancers.

Plant profiles

All three of these species are extremely striking herbaceous perennials with gorgeous, pleated folds of bright green foliage and tall flower spikes.

Veratrum album, commonly known as the white false hellebore, is native to Eurasia where it is found in moist, grassy, sub-alpine meadows and open woods. The star-shaped white flowers have green centres and are held on statuesque flowering spikes up to 1.5m high above the large, bright green, pleated leaves typical of the genus. The plant needs moist soil and shade for its foliage to look at its best, but it will also grow in full sun as long as lots of well-rotted organic matter is incorporated into the soil.

Veratrum californicum, the Californian false hellebore, is a native of North America where it occurs in western USA, down to New Mexico and Mexico. It thrives in the extremely damp conditions of swamps, creek bottoms, moist woodlands and meadows from lowland to sub-alpine zones. The plant can be distinguished from *V. album* by its height (it can reach up to 2.5m), by the upper leaves which are more elongated, and by the blunted tepals of its white and green flowers. It is in flower from August to September.

Veratrum nigrum, the black false hellebore, has a large native range, distributed from Central Europe to Korea in dry glades and water meadows, mountain slopes and scrub. Similar in height and presence to *V. album*, it has deep reddish-brown, star-shaped flowers and generous, broad, pleated leaves.

Veratrum species are pollinated by bees, flies, moths and butterflies. They should be sheltered from cold drying winds and can be propagated by division of their rhizomes or from root cuttings.

Nomenclature

Plants called *Helleborus* and *Veratrum* are confused, and their Latin names – *niger, nigrum* meaning black, and *album* meaning white – and their English vernacular names contribute to this.

White (false) hellebore is *V. album* and has yellow-green flowers. Carl Linnaeus in his *Materia Medica* (1749) calls it *Veratrum caule ramosa* (*Veratrum* with branched stem) and *Helleborus albus flore viridi* (white hellebore with green flowers) with the pharmaceutical name of *Helleborus albi*. This then became *V. album*.

Black (false) hellebore is *V. nigrum* and has dark red flowers. Linnaeus regarded it as a colour variant, *Helleborus albus flore, atro rubente* (the white hellebore with dark red flowers) also known as *Veratrum pedunculis corolla patentissima longioribus* (a Latin descriptive name) which became *V. nigrum*.

The Christmas rose is *Helleborus niger* (Latin for black hellebore) and has white flowers, but black roots. It is in the family Ranunculaceae.

The chemicals in *V. album* and *V. nigrum* are similar and it easier to lump them together when considering their historical medicinal uses.

History

Dioscorides [70 CE] writing about the white hellebore, *V. album* was well aware of it being poisonous, saying the roots cause choking, with multi-coloured vomiting (even when given as a suppository), was an abortifacient when inserted as a pessary and kills mice. However, given in food he says it purges without doing too much harm.

Veratrum nigrum in Parkinson's *Theatrum Botanicum*, 1640   *Veratrum album* from Egenolph's *Plantarum Arborum*, 1562

Pliny [70 CE] says that the white hellebore is called *Veratrum* in Latin, and was far more terrible than black hellebore, so much so that if the tiny threads of root that were used medicinally caused too much vomiting, another emetic or purgative would be used to get rid of it from the body. Pliny is somewhat scornful and says that it is much safer to give a big dose of white hellebore *(V. album)* and then the root is ejected from the body so much quicker. He writes that cattle confidently feed upon the white hellebore *(V. album)*. There does seem doubt as to the identity of this plant as the cattle on the Swiss alpine meadows avoid eating *V. album* (which has green leaves and both Pliny and Dioscorides describe it as having reddish leaves).

From the 1st to the 17th century, in the writings of the Arabian and Persian physicians, and through the European Renaissance, the uses of *Veratrum* did not change substantially, with the historic uses of what are *H. niger* and *V. album* being combined. Paulus Aeginetus [625-690 CE] explains that because they were emetics and purgatives, they were used for all the conditions for which they were indicated in Humoral medicine – gout, epilepsy, dropsy, melancholia, elephantiasis, and others. So violent was the action of 'hellebore' that elaborate methods of administration to avoid overdosage were employed. In one, thin strings of its root were inserted into radishes and allowed to remain overnight, then removed, and the chopped radishes taken by the patient. In another, wool containing macerated hellebore root is inserted as a suppository, and when 'sufficient' vomiting has been achieved, the suppository can be pulled out (Adams, 1847; Aegineta, 1542).

Linnaeus (1749) says of *V. album* that it is a violent purgative and a sternutatory (a medicine to induce sneezing), as well as a treatment for mania, epilepsy, quartan fevers, scabies, lice and cophosis (total deafness).

Veratrum

By the mid-19th century, Bentley (1861) wrote that the rhizomes of white false hellebore, *V. album*, contain the alkaloids Veratria and Jervin amongst many others, and that it was a narcotico-acrid poison, used externally as an errhine (a medicine to be snuffed up the nostrils), and for destroying vermin, and internally as a purgative, and an analgesic in gout. In the 1850s, extracts of *V. album* were found to reduce the heart's action and slow the pulse. Bentley called it an 'arterial sedative'. This was later shown to be a reflex action, the Bezold Jarisch reflex (Bezold & Hirt, 1867; Jarisch & Richter, 1939) whereby receptors in the great vessels are stimulated, increasing vagal output from the medulla causing bradycardia, vasodilatation and also vomiting. Vomiting is a reflex effect originating at the nodose ganglion (Swiss, 1952).

In 1859 *Veratrum*, mistakenly believed to be an anticonvulsant, was used orally in a woman who was having convulsions (eclampsia) due to pre-eclampsia, a dangerous condition in which high blood pressure occurs in the later stages of pregnancy or during or after delivery. Dr Paul Baker in Alabama treated her with drops of a tincture of *V. viride* and she recovered (DeLacy Baker 1859). By 1947 death rates at the Boston Lying-in Hospital from eclampsia had been reduced from 30% to 5% by the use of Veratrum (Irving 1947). It was subsequently used as the of treatment of first choice for eclampsia. When blood pressure measurement became routine practice, it was realised that *Veratrum* worked by reducing the high blood pressure that occurs in eclampsia rather than as an anticonvulsant.

Veratrum as initially used therapeutically is a mixture of alkaloids, one of which is protoveratrine. This was discovered to be itself a mixture of protoveratrine A and protoveratrine B which differ by a single hydroxyl group (Nash & Brooker, 1942). Protoveratrine B when given intravenously caused less vomiting than protoveratrine A and was therefore preferred for use in hypertensive emergencies (Winer 1960).

Veratrum album, the white hellebore, was described as narcotico-acrid poison in the mid-19th century

Another preparation was alkavervir, a mixture of alkaloids from *Veratrum viride* that was given orally, intravenously or intramuscularly. This and the protoveratrines were the staple treatment of eclampsia into the 1970s. Overdose of any of them caused severe hypotension, epigastric burning, vomiting, vertigo, cardiac arrhythmias, bradycardia, bronchiolar constriction and respiratory depression. Death occurring from respiratory arrest was rare with oral treatment as it was vomited out before being fatal. The *Veratrum* alkaloids were also used for treating hypertension not associated with pregnancy. (Martindale, 1967).

Toxicity of *Veratrum*

It is classed as a Class 1a poison, *i.e.* extremely hazardous. It is a skin irritant, and its alkaloids activate Na$^+$ channels, are teratogenic and hallucinogenic. It is toxic to the heart and nervous system and causes death by respiratory and cardiac arrest (Wink, 2009).

The severe side effects and very narrow therapeutic window were always a problem (Schep *et al*, 2006). Therefore, several other anticonvulsants and antihypertensive drugs have been tried in eclampsia. *Veratrum* and its derivatives are no longer used as antihypertensives. The agent of choice for both treatment (Eclampsia Trial Collaborative Group 1995) and prophylaxis (The Magpie Trial Collaborative Group 2002) of eclampsia is now intravenous magnesium sulphate. Additional drugs are used to control blood pressure.

Veratrine

Veratrine and veratridine were other medicines containing *Veratrum* alkaloids. Their action was 'like aconite', neurotoxins, and violently irritative especially on mucous membranes and 'should be handled with great care' and never taken internally. They were used topically for the treatment of headlice and as an analgesic with the warning that they were absorbed through the skin and systemic poisoning (vomiting, burning in the mouth, purgation, hypotension, bradycardia, respiratory depression and muscle weakness) could occur.

Veratrum californicum

A further use of *Veratrum* species was developed when it was noted that *V. californicum* and other species if eaten by sheep in California resulted in foetal malformations, in particular having only one eye. The agent in the plant that was responsible for the cyclops defect was therefore called cyclopamine. It was found to act on certain genetic pathways responsible for stem-cell division in the regulation of the development of bilateral symmetry in the embryo. Synthetic analogues have been developed which act on what have come to be called the 'hedgehog signalling pathways' in stem-cell division, and these 'hedgehog inhibitors' are being considered for the treatment of various cancers such as advanced basal cell carcinoma which cannot be treated by surgery or radiotherapy. One analogue is sonidegib (trade name Odomzo) which has been approved for this by the European Medicines Agency (European Medicines Agency, 2021). By contrast, vismodegib, another analogue, has been rejected by NICE for use in similar circumstances (NICE 2017).

Editor's Note: The hedgehog gene was so named as fruit flies that lacked the gene developed spiky projections on their embryos, like a hedgehog. Three hedgehog genes in vertebrates were named after hedgehogs, Indian hedgehog, desert hedgehog and Sonic hedgehog – the latter after a character in a video game. Two other genes were named Tiggywinkle hedgehog, from a character in Beatrix Potter's books, and echidna hedgehog after the spiny anteater of that name.

As a herbal medicine *Veratrum* is 'Prescription Only', via a registered dentist or medical practitioner (UK Medicines and Healthcare Products Regulatory Agency).

References

Adams F. *The seven Books of Paulus Aegineta, translated from the Greek* 3 vols. London, Sydenham Society, vol iii, pp107, 506, 1847

Bentley R. *A Manual of Botany*. London, John Churchill, 1861 von Bezold A, Hirt L. Uber die physiologischen Wirkungen des essigsauren Veratrins. *Untersuch a d physiol Lab Wurzburg* 1867;1:73

DeLacy Baker P. Veratrum viride in chorea and other convulsive diseases. *South Med Surg J* 1859;15:579-591

European Medicines Agency. Odomzo (sonidegib). https://www.ema.europa.eu/en/medicines/human/EPAR/odomzo accessed 14 May 2021

Irving FC. The treatment of eclampsia and preeclampsia with Veratrum viride and magnesium sulphate. *Am J Obst Gyn* 1947;54:731-7

Jarisch A, Richter H. Die afferenten Bahnen des Veratrineffektes in den Herznerven. *Arch f exper Pat. u Pharmakol* 1939;193:355

Nash HA, Brooker RM. Hypotensive alkaloids from Veratrum album, protoveratrine A, protoveratrine B and germitetrine B. *J Am Chem Soc* 1953;75:1942-48

National Institute for Health and Care Excellence *Vismodegib for treating basal cell carcinoma. Technology appraisal guidance [TA489]* 22 November 2017. https://www.nice.org.uk/guidance/ta489 accessed 14 May 2021

Schep LJ, Schmierer DM, Fountain JS. Veratrum poisoning. *Toxicol Rev* 2006;25(2):73-8

Swiss ED. The emetic properties of Veratrum derivatives. *J Pharmacol & Exper Therap* 1952;104:76

The Eclampsia Trial Collaborative Group 1995. Which anticonvulsant for women with eclampsia? Evidence from the Collaborative Eclampsia Trial. *The Lancet* 1995;1455-63

The Magpie Trial Collaborative Group. Do women with pre-eclampsia, and their babies, benefit from magnesium sulphate? The Magpie Trial: a randomised placebo-controlled trial. *The Lancet* 2002;359:1877

Winer BM. Comparative Studies of Protoveratrine A and Protoveratrine B Intravenously in Hypertensive Man. *Circulation* 1960;22:1074-82

Wink M. Mode of Action and toxicology of plant toxins and poisonous plants. *Mitt Julius Kuhn-Inst* 2009;421:93-111

CHAPTER 43 – NOEL SNELL

Visnaga daucoides
The source of nifedipine, amiodarone, sodium cromoglicate

Introduction

Visnaga daucoides (formerly *Ammi visnaga*) is known locally as khella in North Africa and the Middle East, where it has been employed for two millennia, particularly as an antispasmodic in the management of renal colic. It has several vernacular names in modern day Egypt, all being derivatives from the Arabic word *khallala* meaning 'to pick the teeth'. The dry rays of the umbels become so hard that they are useful toothpicks and give it its European vernacular name of 'toothpick plant'.

In recent times it has been the source of effective medicines for heart disease (nifedipine and amiodarone) and for asthma (sodium cromoglicate and nedocromil sodium).

Plant profile

Visnaga daucoides is a member of the Apiaceae family (related to *Ammi majus*, *q.v.*) and is a half-hardy annual distributed throughout the Mediterranean and Western Asia where it can be found growing on sandy soils.

An extremely attractive plant, the flowers turn from lime-green to white and are held in large, dome-shaped flowerheads on sturdy stems above a mass of delicate feathery leaves. They tend to come into flower slightly after the mid-summer peak which makes their display doubly welcome. Best grown from seed started under glass and planted out into moist but well-drained garden soil; plants can grow in semi-shade or sun.

History

The seeds from the tiny fruits of khella have been used from biblical times in the treatment of renal colic, caused by kidney stones (which may have been particularly common in the Nile delta due to a common parasitic infestation of the renal tract, schistosomiasis).

Clay tablets in the British Museum, from the library of Ashurbanipal (ruler of the Assyrians 668-626 BCE) include the following prescription: 'If a man is affected in the lungs, thou shalt bray [crush] *Ammi*, spread it over a thorn-fire, let it (the smoke) enter…his nostrils, it shall make him cough [and he will recover]' (Thompson, 1934). This usage does not identify the plant with any of the uses of *Ammi* recorded in the past two millennia.

The historic identity of *Visnaga daucoides*

Ammi

Historically, *V. daucoides* was known as *Ammi* about which Dioscorides says that the seed is used for the griping colic and dysuria (pain on passing urine). These are the symptoms of renal colic. It was also used for snake bite. The painting entitled *Ammi* in the *Juliana Anicia Codex* of 514 CE (the earliest extant illustrations of the plants named by Dioscorides) is clearly *V. daucoides*. Its use continued unchanged in the Arabic literature of the Golden Age of Islam and in Egypt through into the modern era. For further discussion see the historical section in the chapter on *Ammi majus*.

Visnaga daucoides – herbarium specimen from the Pharmaceutical Society's herbarium at the Royal College of Physicians

Gingidium

Tracing the history of a plant and its medicinal use is not always easy. There is a plant, called *Gingidium* by Dioscorides, which he says is a good potherb, raw and cooked, and drunk with wine is good for the bladder. The painting of it in the *Juliana Anicia Codex* has been identified as *Thapsia garganica* (*Der Weiner Dioskurides*, 1998, folio 140), in the Apiaceae family, but it is neither *Ammi majus* nor *Visnaga daucoides*. The 7th century Byzantine Greek, Paulus Aegineta, repeats Dioscorides' use of *Gingidium* as a food and condiment. It is otherwise absent from medieval Arabic literature. As *Thapsia garganica* is poisonous and has the common name of death carrot, it is not going to be the *Gingidium* – 'the good potherb' – of Dioscorides.

Visnaga

However, the names, descriptions, and woodcuts of *Gingidium*, *Visnaga*, *Ammi*, in the herbals of the 16th and 17th centuries appear interchangeable and confused. Dodoens' *Crüÿdeboeck* (1554) says

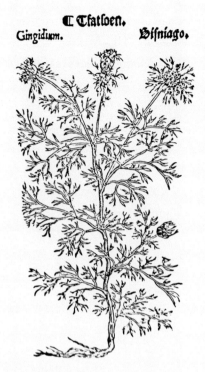

Visnaga daucoides as *Gingidium visniago* in Dodoens'
Crüÿdeboeck, 1554

Visnaga daucoides as *Gingidium alterum* in Dodoens'
Pemptades, 1583. This is Linnaeus' (1753) lectotype
for *Daucus visnaga* (now *Visnaga daucoides*)

A variety of *Gingidium* called *Visnaga* by Mattioli in
his *Discorsi*, 1568. Possibly *Visnaga daucoides*

Visnaga daucoides entitled *Ammi parvum foliis foeniculi*,
small bishop's weed, from Parkinson's *Theatrum
Botanicum*, 1640

that *Gingidium* was called *Visnagia/Visnagio* (sic) in Italy and Spain, used for toothpicks and for treating gravel and stone (renal calculi). He called it *Gingidium alterum* in his *Pemptades* (1583) and Linnaeus (1753) based his naming of *Daucus visnaga* on the latter description and woodcut. The woodcuts have finely dissected leaves, but not as hair-thin as the plant we now called *V. daucoides*. Mattioli (1568), commenting on the *Gingidium* of Dioscorides, wrote of a variety called *Visnaga* whose dried flower stalks (the petioles) were used as toothpicks. These attributes are correct for *V. daucoides* but the identification of the woodcuts is questionable. The proliferation of woodcuts of unidentifiable Apiaceae and new names, led Parkinson (1640) to write that the identity of the true *Ammi* was unknown, but his woodcut of '*Ammi parvum foliis foeniculi*, small bishop's weed' (Parkinson, 1640) has the hair-thin leaves of *V. daucoides* and is probably the plant we know today. Bishop's weed is the English name for *Ammi majus*.

The taxonomic similarity of the Apiaceae was, and is, confusing.

Visnaga daucoides is not in the College's *Pharmacopoea Londinensis* of 1618 or 1650 under any name and has a brief mention only as *Ammi alterum semine apii* [the other *Ammi* with seeds of parsley] in Alleyne's *Dispensatory* (1733) as being imported from Egypt via Alexandria, Venice and Germany and little used. Cullen (1789) says about *Ammi*, referring to several species, that it is in the *London Dispensatory* but justly neglected as it is difficult to obtain. The knowledge that the seed of *V. daucoides* was an effective treatment for renal colic and stones then disappeared from European texts until the 20th century. Linnaeus gives no medicinal uses for it in his *Materia Medica* (1749, 1782).

Modern nomenclature

Finding that it did not fit into Linnaeus' genus *Daucus*, Joseph Gaertner renamed it as *V. daucoides* in 1788 using a common botanical option of reversing genus and species names (*daucoides*, meaning resembling *Daucus*). Aylmer Bourke Lambert relocated it from Linnaeus's *Daucus* into the genus *Ammi* as *Ammi visnaga* in 1799. Recently, modern botanical authorities have established that it fits better into the genus *Visnaga* so Gaertner's name for it has been restored.

Active ingredients and pharmacology

The active principle in the seeds is a chromone, khellin. This was first isolated in an impure form in 1879, and its chemical structure was determined in 1932; it was shown to relax smooth muscle, including that in the ureters, explaining its benefits in renal colic. Its detailed pharmacology was determined in 1945-1947 by the Russian-born Gleb von Anrep, then Professor of Pharmacology in Cairo. It is said that his laboratory technician, who suffered from severe cardiac angina, took a preparation of *khella* for renal colic and noted that his angina improved; this stimulated Anrep to investigate the effect of khellin on the heart. He studied it initially in animals, first in a heart-lung preparation, then in live dogs, showing it to be an effective coronary vasodilator; he then treated human subjects, showing an improvement in 35 of 38 patients with angina. A subsequent paper (Anrep, 1947) reported encouraging results in over 150 patients, and also gave details of the treatment of 45 asthmatics, 41 of whom were said to have experienced 'complete and prolonged relief' of their symptoms (backed up by lung function measurements). Anrep noted that the onset of action of khellin in asthma was slower than that of adrenaline or ephedrine, but more prolonged, and it was very safe to use – the only side effects were a feeling of warmth and occasional mild nausea (though others have reported more severe nausea, vomiting and anorexia). Subsequently, several preparations of khellin were marketed, including (in the UK) Benecardin from Benger Laboratories and Eskel from Smith, Kline & French, both formulated as tablets and used as smooth muscle relaxants and coronary artery dilators; and in Germany, Khelfren, a 0.5% solution of khellin given for asthma as an aerosol.

Uses and development as modern medicines: preparations and commercialisation

Cardiac medicines – amiodarone, nifedipine

As a result of Anrep's research, several pharmaceutical companies attempted to make derivatives of khellin which were better tolerated, more effective, and soluble enough to be given by injection as well as orally. In 1957 Belgian researchers looking at the benzofurane portion of khellin synthesized benzodiarone, which was marketed by the Labaz company as a coronary vasodilator for angina. Four years later a derivative of benzodiarone which it was hoped would also have anti-arrhythmic effects was synthesized. Named amiodarone, this drug was again marketed as a coronary vasodilator, but found its niche in the management of resistant cardiac arrhythmias, for which it was licensed in the UK in 1980. Unfortunately, amiodarone can cause a wide range of adverse effects: it can cause an unusual cardiac arrhythmia, 'torsade de pointes', and can also interfere with thyroid hormone metabolism and release, leading to hypothyroidism or hyperthyroidism (usually subclinical). It can produce side effects in the eye, may cause liver damage, and can lead to accumulation of phospholipids in the lung and subsequent pulmonary inflammation. It may lead to persistent slate-grey pigmentation on sun-exposed skin. Its use is now restricted to the treatment of cardiac arrhythmias, usually as second-line therapy.

Both the Bayer company in Germany and Smith, Kline & French in the USA took khellin as their starting point for synthesising and studying a series of dihdropyridines in the search for a novel coronary vasodilator. Both groups independently discovered nifedipine, but Bayer patented it first. Pharmacologically, this agent turned out to act as a calcium antagonist, causing vasodilatation and hence being valuable in the treatment of angina, hypertension, Raynaud phenomenon and peripheral arterial disease. Some patients experience an excessive fall in blood pressure with dizziness and faintness, and also headache and ankle oedema; these usually respond to a reduction in dose.

Medicines for asthma – sodium cromoglicate and nedocromil

In 1952 a very detailed study of the effects of khellin in patients with obstructive airways disease (both asthma and chronic bronchitis) was published (Kennedy & Stock, 1952). Although this was an open study, careful objective measures of several lung function parameters were made. The investigators showed that khellin was an effective bronchodilator in about half the patients, with a slow onset of action but sustained for up to 48 hours. Its effects were more marked in asthma than chronic bronchitis. Nausea was a common side effect but was less of a problem when given by injection than as tablets. The khellin for this study was provided by Benger Laboratories, and it was probably these findings that persuaded Benger to initiate a research programme in their laboratories at Holmes Chapel, Cheshire, UK, to find better-tolerated analogues of khellin.

A number of chromone analogues based on the structure of khellin were produced and were shown to prevent fatal bronchoconstriction induced in guineapigs by inhalation of histamine or methacholine, and to inhibit the formation of mediators of the allergic response in the lung. However, the Benger's researchers were looking for a novel bronchodilator, not an anti-allergic drug, and their findings were not fully followed up until the appointment of a new medical liaison officer in 1956, Roger Altounyan.

Roger Altounyan and sodium cromoglicate

Altounyan was born in Syria in 1922 to an English-educated medical family of Armenian extraction. As a child on holiday in the Lake District his family and their boat 'Swallow' are said to have been the inspiration for Arthur Ransome's classic children's book, *Swallows and Amazons* (although Ransome later denied this). During the Second World War he became a pilot instructor in the RAF, gaining the Air Force Cross. After the war he trained in medicine at Cambridge before joining Benger's Laboratories; he suffered from severe chronic asthma and was determined to discover novel anti-asthmatic agents. Despite official disapproval he tested a range of new chromone compounds on himself; eventually he discovered sodium cromoglicate, which had good anti-allergic properties when inhaled. He also designed a delivery device, the 'Spinhaler'.

The product was successfully marketed as Intal by Fison's Laboratories (who had taken over Benger's) and, because it proved to be very safe, was particularly popular for treating asthmatic children. In addition to reducing the number of allergic asthma attacks when taken regularly, it was also effective in preventing exercise-induced asthma (Edwards & Howell, 2000). Preparations for treating other allergic disorders, including allergic rhinitis and conjunctivitis (hay fever) were also marketed.

Side effects of treatment with sodium cromoglicate are uncommon and include skin reactions, muscle pains, headache and gastroenteritis; nasal and throat irritation from the powder formulation; cough and bronchospasm; and very rarely anaphylaxis due to an allergic reaction (Ibanez

1996). Further chromones were investigated in an attempt to find one that was orally active, unfortunately orally absorbed compounds seemed to cause liver toxicity.

Nedocromil sodium

Eventually another inhaled chromone was marketed, nedocromil sodium (trade name Tilade). This was as effective in asthma as cromoglicate but no better and had an unpleasant taste on inhalation which was masked by including mint flavouring in the formulation.

The anti-allergic mechanism of action of chromones is still not fully understood, and may include stabilisation of mast cells, and inhibition of chloride channels and of immunoglobulin-E production.

Summary

Khellin, the seed of *Visnaga daucoides* used to treat the smooth muscle spasm of renal colic, was described by Dioscorides and continues to this day. Medicines derived from it relax the smooth muscle of the coronary arteries, so treating angina, and have other actions. Few plants have such a long history of effective therapeutic use.

References

Alleyne J. *A New English Dispensatory*. London, Thos. Astley and S. Austen, 1733

Anrep GV, *et al*. Therapeutic uses of khellin. *The Lancet* 1947;i:557-8

Cullen W. *A treatise of the Materia Medica*. Edinburgh. Charles Elliot, 1789

Edwards A, Howell J. The chromones: history, chemistry and clinical development. A tribute to the work of Dr R E C Altounyan. *Clinical & Experimental Allergy* 2000;30:756-774

Gaertner J, Linnaeus C. *De fructibus et seminibus plantarum* vol 2. Stuttgart, Academiae Carolinae, p378, t162, 1788

Ibanez M, Laso M, Martinez-San Irineo M, *et al*. Anaphylaxis to disodium cromoglycate. *Ann Allergy Asthma Immunol* 1996;77:185-6

Kennedy M, Stock J. The bronchodilator action of khellin. *Thorax* 1952;7:43-65

Lamarck J-BPAdeMde. [*Ammi visnaga* (L)Lam.]. *Fl. Franç. (Lamarck)* 1779;3:462

Thompson RC. Assyrian prescriptions for diseases of the chest and lungs. *Revue d'Assyriologie et d'archéologie orientale (Paris)*. 1934;31(1):1-19

CHAPTER 44 – ANTHONY DAYAN

Excipients and solvents
The plant sources of excipients and solvents

Excipients are substances added to a medicine to allow it to be usable, storable or palatable; they include solvents, flavouring agents, preservatives and materials required to make ointments, creams and lotions.

Formulation of a medicine with excipients and solvents turns a therapeutically active substance, a drug, into an acceptable medicine, the practical dosage form that a manufacturer can consistently produce and deliver in bulk, that pharmacists can store and dispense and that can be taken by or administered to patients to produce a consistent action. Formulation is a scientifically and technically challenging process for pharmacists and chemists, but without it many of the synthetic and herbal medicines on which we rely could never have left the laboratory and become useful in day-to-day healthcare.

Excipients and solvents are used by pharmacists in formulation to provide:
- inert bulk that permits practical handling or inhaling of doses by patients
- stabilising agents to prevent decomposition due to light, moisture, age or bacterial and fungal contamination
- colouring agents to aid identification
- flavouring agents to disguise unpleasant tastes
- substances that protect sensitive drugs from stomach acid that can hasten their decomposition
- solvents to make injectable or spreadable solutions, mixtures and dispersions and sometimes agents to delay release of the active substance so that its activity in the body can be prolonged.

For technical and manufacturing reasons, excipients are needed in making tablets with high-speed presses, as aids to controlled and reliable disintegration once swallowed, and to stabilise suppositories and pessaries before insertion. Without one or more excipients, solid and liquid medicines, whether to be swallowed, inhaled, injected, dropped into the eye or ear, rubbed on the skin or used as pessaries and suppositories, would be almost impossible to manufacture or use in an acceptable way.

A complete list of excipients and solvents would be impossibly long but there are some derived from plants that are used so often as to deserve a brief account here. There are others that are inorganic in nature, such as talc, silica or buffers to control pH, and an increasing number that are synthetic, such as antioxidants for protection against the air, many colouring agents, sweeteners and so on of great practical importance that are not discussed here. There is a recent encyclopaedic account of excipients in Sheskey *et al*, (2020).

The principal excipients and solvents derived from plants include:

EXCIPIENT and plant of origin

Carnauba wax *Copernicia prunifera*

Castor oil and chemical derivatives *Ricinus communis*

Cellulose and many derivatives Partly-hydrolysed plant fibres and wood pulp. Chemical reactions to make special types

Cyclodextrins Specific bacterial fermentation of starches

Dextrins Partly hydrolysed starch, usually corn (maize, *Zea mays*) starch

GUMS

Acacia Exudate from bark of *Acacia* trees

Guar Endosperm of *Cyamopsis tetragonoloba*

Tragacanth Dried sap of *Astragalus* spp.

OILS

Coconut Fruit of *Cocos nucifera*

Corn oil *Zea mays*

Cotton seed *Gossypium* spp.

Olive *Olea europaea*

Palm *Elaeis guineensis*

Rape (canola) *Brasssica napensis*

Starches and many derivatives Maize, *Zea mays*; potato, *Solanum tuberosum*; wheat, *Triticum aestivum*; and rice, *Oryza sativa*

Carnauba wax

Extracted from leaves and leaf buds of a palm tree, *Copernicia prunifera*, formerly *C. cerifera*, native to northeastern Brazil. Sometimes called the 'wax palm' or 'tree of life' because it has so many uses: the fruit is eaten by man and animals; fibres from the leaves are made into textiles, hats, mats and bags; and the wood is valued in building because of its resistance to boring insects.

To obtain the wax sustainably, leaves are cut from the palm every three years and dried in the sun. Their natural wax coating forms a fine white powder that is collected by shaking the leaves, and cleaned mainly by sieving for use in pharmaceuticals, cosmetics, some foods and polishes (Johnson & Nair, 1985).

Carnauba wax is a hard wax with a high melting point because it contains several high molecular weight hydrocarbons. This makes it valuable in pharmaceuticals as a polish to protect the sugar coating used for many tablets and to help to make slowly-dissolving preparations that release an active drug over a period to ensure a sustained action.

Ricinus communis

Castor oil is obtained by cold expression from the seeds of *Ricinus communis*, the castor oil plant, known also as *Palma Christi*, from the palmate (hand) shape of the leaves, and from the reputation of the oil to help to heal wounds

Seeds of *Ricinus communis*

and cure various disorders. The name castor comes from its use in perfumes as a base replacing castoreum, the dried secretions of the perineal glands of the beaver, called *Castor* in Latin. *Ricinus* is the Latin for *tick* as the seed, which has the common but incorrect name of 'bean', has some resemblance to the shape of a tick. It is a member of the spurge family, Euphorbiaceae.

Castor oil is an old-fashioned laxative used in humans and animals, and of Cremophor EL, a solvent used in injections and in ocular, capsule and tablet formulations of insoluble drugs, topical creams and ointments, and in cosmetics. Its most important use in medical practice is as a solvent for paclitaxel in cancer chemotherapy. The stability of the oil under high temperature and pressure conditions has made it an important lubricant for engines and gear boxes.

Plant profile

Ricinus communis in its natural habitat in Northeast Tropical Africa is an evergreen tree-like shrub quickly reaching 10m in height. In more temperate climates it is cultivated as an annual and is much smaller, often just 1m tall. It is frost tender and so it can only be grown outdoors during the warmer months, but it is highly ornamental. Its exotic appearance with large palmate leaves, unusual flowers and spiky-looking seed pods repays the extra effort required to grow it. Seed is sown under glass in late spring and planted out once all danger of frost is passed, preferably into an open sunny site with well-drained, moisture-retentive soil. As it grows quickly and looks so dramatic it has long been extremely popular as a 'dot plant' in summer bedding schemes. The species has spikes of white flowers and purplish, palmately divided leaves supported by hollow stems but this use in cultivation has led to the development of many named varieties whose leaf and flower colours range from red to green, bronze and purple. *Ricinus communis* is monoecious (both male and female flowers are on the same plant) and is wind pollinated. It may need staking in exposed areas.

The whole plant is very poisonous.

Pure castor oil consists of triglycerides of certain long-chain fatty acids. The seeds contain the extremely toxic, water-soluble protein, ricin, which must be completely removed before its use as a pharmaceutical grade oil or in compounds derived from castor oil. The seeds, after processing to remove the ricin, are a high calorie, protein and fat-rich animal feed. By mouth, ricin is much less toxic than by injection, and the amount in 5-20 seeds can be fatal, but ingestion is commonly followed by vomiting and survival. By injection, it is famous (infamous) for the 'umbrella murder' of Georgi Markov on Waterloo Bridge in 1978. The KGB are alleged to have killed Georgi Markov, a dissident Bulgarian journalist, with a pellet containing 0.28mg of ricin fired into his leg using a specially adapted air gun in an umbrella. While his symptoms were those of ricin poisoning, no ricin was ever found in the pellet that was extracted from his leg.

Chemistry

Removal of ricin is easily done by a mixture of heating to 100-102°C or immersing in hot water, chemical treatments (with alkalis such as ammonia or sodium hydroxide) or by using bacteria that contain enzymes that remove ricin. Genetic engineering to remove the gene that produces ricin can be done, but there are considerable legislative hurdles to be overcome before this would be allowed commercially.

Polyoxyethoxylene castor oil derivatives, the important compounds used in medicine and cosmetics, are manufactured by reacting ethylene oxide with castor oil under conditions that control the number of added molecules. The products range from waxes and semi-solids to viscous liquids mostly known as macrogols and cremophors. These are important solvents of otherwise insoluble drugs for injections, tablets, creams and ointments. An example is a solvent called Cremophor EL for administering paclitaxel (Taxol) intravenously, which is described in the account of *Taxus baccata* (*q.v.*).

Cellulose

Cellulose is the group of long-chain carbohydrate polymers characteristic of many structures in all plants. Cellulose itself is not directly used in medicines. It is the diverse properties of its many derivatives that have made it so valuable and so very widely employed in a vast range of products. Cotton wool and cotton fibres, both from the seed pods of one or more species of the cotton plant (*Gossypium* spp.) are used to make bandages and dressings.

Mechanical treatment of cellulose, and sometimes the addition of fine particles of silicon dioxide, is done to produce microcrystalline cellulose powders suitable for controlling the viscosity of solutions and for providing inert bulk and as an aid to regular disintegration of tablets.

Cellulose acetate is used to coat tablets and granules as a mechanical protection and to control the rate of release of their drug content, to hide an unpleasant taste in a chewable tablet or to prolong a desired action. It can also be made into special membranes which have been used in artificial kidneys and analytical machines.

Hydroxymethyl, -ethyl and related chemical derivatives of cellulose are manufactured by controlled reaction with ethylene or propylene oxide and other reagents. They are used pharmaceutically to control the viscosity of liquids to be drunk or applied to the eye; to film coat tablets; as tablet bulking, binding and stabilising agents; in giving the appropriate mechanical and heat-sensitive properties to suppositories, pessaries and topical ointments and lotions. They are used in much the same way in cosmetics and toiletries and in the food industry, where some have particular value in providing the correct mouthiness feeling in low- and fat-free items as well as adding a moist feel to some baked goods.

Cyclodextrins

The type in use today is β-cyclodextrin produced by a carefully managed enzymatic action on starch to give a product with a molecular weight between about 950 and 1200.

Chemistry

Cyclodextrins are ring-shaped molecules made up of repeating smaller sugar molecules, sometimes with subsequent chemical modifications, like cellulose derivatives, to provide particular physical or chemical properties. Their most valuable feature is that the ring has a hollow core into which insoluble molecules can be inserted during manufacture, and from which they will be released in the body at a predictable rate. It is this inclusion function that makes cyclodextrins so valuable pharmaceutically – they can suspend and carry drug molecules into the body to produce a predictable dose- and time-related effect, and stabilise certain drugs inside the ring. They can also be used to control viscosity.

Medicines based on them may be administered either topically or more often by injection and sometimes orally. They have also been used to make cosmetics.

Dextrins

These compounds are polymers of varying size of glucose (once called dextrose after the way it rotated polarised light) ranging from chains of fifteen to more than a hundred glucose molecules linked together.

They are manufactured by the controlled hydrolysis of pure maize or potato starch.

Depending on the length of the chain they range from syrupy liquids to powders and adhesives, so are used in tablet-making as binders, disintegrants and coatings, and in other applications as adhesives and thickeners for suspensions of medicines.

They are given intravenously as a source of energy and carbohydrate to patients who cannot absorb food by the usual route.

Gums

It is simplest to regard all gums as sticky, highly-viscous materials derived from plant saps whose specific properties and hence their pharmaceutical uses depend on what combination of viscosity, stability, perhaps flavour and digestibility are required. As dried, solid forms they are used as binders and disintegrants in making tablets. In liquid medicines to be swallowed or applied to the skin they serve as thickening, suspending and stabilising agents and aids to localised adherence.

They are manufactured by purification of dried sap, *e.g* gum arabic from *Senegalia senegal* and gum tragacanth from *Astragalus* spp., or from ground endosperm of seeds of *Cyamopsis tetragonoloba* for guar gum. They are all complex carbohydrate polymers comprising long chains of linked molecules of several different sugars. They are readily digested in the mouth and intestines, and are usually odourless, tasteless or bland which makes them valuable in the manufacture of foods and cosmetics.

Oils

There is considerable overlap between pharmaceutical, cosmetic and kitchen uses of all the plant oils because of their physical properties, relative stability, taste (or lack of it) and the overlap of their chemical compositions.

They all consist of triglycerides of variable mixtures of shorter- and longer-chain, saturated and unsaturated fatty acids making it possible to match particular features to suit specific needs for stability, dissolving or suspending capacity, viscosity, taste, smell and appearance.

In making medicines their principal uses are as solvents for drugs to be taken by mouth or injected, and sometimes as slow-release agents to obtain a prolonged drug action. Quite often they are used to make emulsions of active agents. They are also used in topical preparations applied to the skin as bland liquids, lotions or ointments; oil may speed and enhance the absorption of drugs through the skin as well as increasing the flexibility and improving the appearance of the treated area.

Starches

Starch is used as a binder, bulking agent and tablet disintegrant. Starch powder is a good absorbent and is employed in dusting powders and the mucilage has been applied as protective dressing to skin wounds and to cover areas treated with ointments. Starch-based dusting powders for rubber gloves exploit its absorptive property and lack of irritancy.

Starch is manufactured by milling grain followed by successive sievings and washings to obtain powder with the appropriate composition. Like cellulose, its composition and properties can be modified by chemical reactions to produce, for example, hydroxyethyl starch with modified physical characteristics, molecular weight, etc. to enhance biological suitability for particular uses, usually in preparations to be injected.

Chemistry

All starches consist of carbohydrate polymers whose molecules are linked to produce high molecular weight substances that serve as energy reserves in all plants. The individual units are glucose-based amylopectin and amylose. The former molecules have a branched chain structure, and the latter has both helical and linear forms so varying proportions of the two, and polymers of different lengths, result in starches with diverse properties.

Saponins

A further family of complex natural compounds of considerable and growing pharmaceutical, therapeutic, agricultural and industrial importance is the saponins. The name reflects their original recognition as surface active agents with soap-like properties, such as producing foam when agitated in water. Many plants produce them; the original source was the soap plant in the genus *Saponaria*; the Latin for soap is *sapon*. They occur widely in many genera including *Ranunculus*, *Gentiana* and *Primula*. As most are bitter and may be toxic if eaten in large amounts they probably serve as defence molecules deterring insects and herbivores from eating the plants (Vinken *et al*, 2007; Abdelrahman & Jogaiah, 2020).

Chemistry

Chemically, saponins are glycosides comprising carbohydrate chains attached to various triterpenoid structures which may be steroidal, a steroidal alkaloid or a more conventional triterpene. These together make saponins able to dissolve both in watery and oily solutions and so to be surfactants, and to react readily with components of cell walls in all sorts of organisms (Bar *et al*, 1998; Abdelrahman & Jogaiah, 2020).

If eaten they may be bitter or sweet depending on their composition and they may either interfere with or enhance the absorption of fat-soluble vitamins and other substances (Cheok *et al*, 2014; Abdelrahman & Jogaiah, 2020).

Medical interest in saponins extracted from the soap bark plant *Quillaja saponaria* is in their potency as adjuvants to enhance vaccines.

Adjuvants are substances added to a vaccine to enhance and steer the magnitude and nature of the immune response in a particular direction, possibly to enhance cell-mediated rather than humoral immunity or to raise the level of an immunoglobulin of a particular type produced in response to immunisation (Magedans *et al*, 2019). A saponin called 'Quill A' from the root of the soap bark plant is a popular example, but other types are being examined because they are often particularly powerful in increasing immunity against cancer cells and parasites, as in vaccines against malaria and viruses, including at least one candidate vaccine against COVID-19. They tend to provoke a strong inflammatory and cellular reaction at the site of injection which has limited their application so far, but manipulation of the molecules may lead to effective and less inflammatory saponin-based adjuvants.

References

Bar IG, Sjölander A, Cox JC. ISCOMs and other saponin based adjuvants. *Adv Drug Delivery Rev* 1998;32:247-271

Cheok CY, Karim HA, Sulaiman R. Extraction and quantification of saponins: A review. *Food Res Internat* 2014;59:16-40

Franke H, Scholl R, Aigner A. Ricin and Ricinus communis in pharmacology and toxicology-from ancient use and 'Papyrus Ebers' to modern perspectives and 'poisonous plant of the year 2018'. *Naunyn-Schmiedeberg's Archives of Pharmacology* 2019;392:1181-1208

Jogaiah S, Abdelrahman M (eds). *Bioactive Molecules in Plant Defense: Saponins.* Springer Nature Switzerland, 2020

Johnson DV, Nair PKR. Perennial crop-based agroforestry systems in Northeast Brazil. *Agroforestry Systems* 1985;2:281-92

Magedans YVS, Yeendo ACA, de Costa F, *et al*. Foamy matters: an update on Quillaja saponins and their use as immunoadjuvants. *Future Med Chem* 2019;11:1485-99

Sheskey PJ, Hancock BC, Moss GP, Goldfarb DJ (eds). *Handbook of Pharmaceutical Excipients*, 9th edn. London, Pharmaceutical Press, 2020

CHAPTER 45 – ANTHONY DAYAN

Vitamins
The plant sources of vitamins

Vitamins are small organic molecules required by humans for life and well-being. Many, especially the B group and C, are essential co-factors in the metabolism of cells and body systems; vitamins A, D and E have multiple vital actions on the integrity and normal functioning of particular cells in important organs such as bone, blood and gonads, as well as more general roles; vitamin K has a central place in normal blood clotting.

The B vitamins and vitamin C are known as the water-soluble vitamins; vitamins A, D, E and K make up the fat-soluble group. While not all of them are derived from plants, and not usually regarded as 'medicines' they are included in this book because of their importance to health.

Vitamins are usually absorbed from a normal mixed diet, and each was discovered initially by studying the nature of human and animal diseases caused by a deficiency of it, followed by detailed research into the nature of the missing factor and its roles in the body. The histories of the discovery of vitamins and deciding the amounts we need are epics of fundamental human and veterinary medical research (Semba, 2012a). Their irregular naming and numbering scheme arose from claims made for the many independent discoveries of them at different times in various laboratories.

Very brief accounts of our knowledge of the nature, actions and sources of vitamins are given here and readers wishing to know more about the daily amounts required by adult men and women, children and before and during pregnancy, and the potentially harmful effects of excessive doses should consult general sources such as the website of the British Nutrition Foundation and NHS/UK.

Daucus carota

Daucus carota – a carrot

Vitamin A

Plant sources in the College Garden:

> *Daucus carota* – carrot

The principal sources in the diet are cheese, eggs, oily fish, milk, liver and fortified breakfast cereals but useful amounts occur in certain vegetables and fruits, such as carrots, mango, papaya and apricots.

Vitamin A is a group of unsaturated compounds known as provitamin A, comprising retinol and certain carotenes, which the body converts into active forms. These are essential for normal development, reproduction and immune functions; as retinoic acid for the growth and integrity of skin cells; and as retinol for the light-sensitive pigment rhodopsin in the eye required for normal vision. The beguiling story that British night fighter pilots in the Second World War were fed on carrots to improve their night vision was allegedly devised to conceal our secret new device of airborne radar as the source of their successes in combat. Unfortunately, there is no evidence to support it and eating carrots will not improve night vision in someone on a normal mixed diet.

History

As one of the earliest vitamins discovered, its history forms part of the origin of the concept of a 'vitamin' (Semba, 2012a, b). In 1816 François Magendie in Paris put dogs on a deficient diet and found they developed corneal ulcers and eventually died. Hopkins in Cambridge in 1912 showed that rats on a milk diet only thrived if given an accessory factor beyond its protein, fat and carbohydrate content and later showed that the factor was present in high concentration in cod liver oil. This was labelled as a 'fat-soluble vitamin' and subsequently identified as vitamin A, retinol. In the 1940s the molecular structure of vitamin A was determined, and practical methods of synthesising it and analogues were found.

Vitamin A deficiency causes night blindness and dryness of the eyes, followed by corneal ulceration, infertility and general ill-health. It is a major cause of blindness and poor health in less-developed countries where there is little access to oily fish, eggs, milk and eggs.

An excess of vitamin A can cause acute brain damage, birth defects and deafness. It has caused the death of unwary polar explorers, who ate the livers of polar bears and husky sledge dogs, which contain a high level of vitamin A.

The retinoids, derivatives of vitamin A, are prescribed either systemically or topically for many conditions. For example, acute promyelocytic leukaemia is treated with all-trans retinoic acid (tretinoin); another derivative, acitretin may be used to reduce the risk of developing skin cancers in immunosuppressed patients; the treatment of severe acne was revolutionised by the advent of isotretinoin (13-cis retinoic acid); and bexarotene has proved helpful in cutaneous T-cell lymphoma. These effective treatments for serious diseases can be accompanied by major side effects and all are teratogenic, risking birth defects if given in pregnancy.

The B vitamins
B1, B2, B3, B5, B6, B7, B9, B12

The B vitamins are all derived from plants, except for vitamin B12 which is therefore considered no further.

These eight chemically unrelated compounds are grouped because they all have vital though disparate roles in the metabolism of all cells in animals, plants and microorganisms. Their discovery, biochemical and chemical characterisation, and understanding their roles as coenzymes essential for many synthetic and energy-producing steps and some regulatory functions in all cells form part of the initial recognition of vitamins as a class of essential compounds mostly absorbed by higher species from their diets and the equally fundamental roles of the cellular metabolic processes they support. Deficiencies of them in higher species, such as humans, result in general ill-health and failure to thrive as well as disorders and pathological lesions more or less characteristic of each individual vitamin.

The medical and biochemical work on many of them has been so important that the scientists and clinicians involved have been awarded several Nobel Prizes (Carpenter, 2004).

The B vitamins occur in fruits, green vegetables and intact, unpolished cereals, except vitamin B12. As dietary deficiencies are common even today, and usually involve several B and other vitamins, mixtures of all or most B vitamins and vitamin E are usually given, often with vitamin D, too. Deliberate fortification of bread is a common public health measure. Similar higher-strength vitamin mixtures are found in fortified breakfast cereals and most proprietary supplements.

Despite our understanding of these substances, deficiencies still occur in groups on limited diets, such as prisoners and the poor. The production of vitamins for use in human and veterinary medicine is now mainly by industrial synthesis, sometimes using an enzyme extracted from a plant, yeast or bacterium.

Vitamin B1 thiamine

Plant sources in the College Garden:

> *Triticum aestivum* – wholegrain wheat and pasta
>
> *Oryza sativa* – rice, including the husk
>
> *Cucurbita maxima, C. pepo* – squash
>
> *Phaseolus vulgaris* – French bean (also other legumes)

Many foods, like bread, breakfast cereals, yoghurt and white rice are often fortified with added thiamine. Pork, fish and seafood are good sources of thiamine, as well as eggs, nuts, and other vegetables.

Thiamine, vitamin B1, is an essential vitamin which we can only obtain from our diet. A disease called beriberi and several neurological syndromes are caused by thiamine deficiency.

It is a sulphur-containing thiazole-pyrimidine molecule required by about 150 enzymes in bodily function. Deficiency of it and other members of the vitamin B and C groups have always been common in association with inadequate diets.

Cucurbita pepo – squash

Equisetum praealtum – horsetail

History

Beriberi was described in the Chinese textbook of medicine, the *Nei Ching* in 350 BCE. The first modern description of the loss of power and sensation in hands and feet, and the trembling was by the Dutch physician, Jacob Bontius, in the Dutch East Indies (now Indonesia) in *De Paralyseos quadam specie, quam Indigenae Beriberii vocant* [The type of paralysis which the inhabitants call beriberi]. He said it was so called because it was the noise made by sheep (*quod Ovem sonat*) (Bontius, 1642). The Sinhalese word for weakness is *beri* so the name implies a double weakness, but it seems that Bontius mistakenly used the Malay word for sheep, *beri*, as the etymology, saying that those affected walked like sheep 'because the knees first give way, and the patient raises his legs'. However, the Malay word *biribi* is used to describe several sorts of gait disorder, which seems another more likely origin. Bontius suffered from it and also lost the power of speech. The Professor of Anatomy who was the demonstrator in Rembrandt's *The Anatomical Lesson*, Nicholaes Tulp, gave a detailed account a few years later (Tulp, 1652) but it was the blind botanist Rumphius in Ambon who first reported an effective cure in 1705 – mango beans, *Phaseolus minimus* or *P. radiatus* (Bruyn & Poser, 2003). In the 19th century, sailors in the Japanese navy developed a severe form of weakness and neuritis when their diet was changed from brown to white polished rice (the outer bran coat is removed).

Editor's note: It was not just Japanese sailors who experienced this. Prisoners of war in Singapore during the Second World War developed beriberi when fed on polished rice but recovered when their captors were persuaded to supply them with the husks rather than let them be thrown away.

Dutch, British and American research showed that the disorder could be reproduced in birds and animals on deficient diets, and by the 1930s the missing compound had been completely identified, and synthesised (Williams *et al*, 1936).

Beriberi in humans, produced by thiamine deficiency, has two forms – the wet form dominated by breathlessness and leg swelling due to heart failure; and the dry form marked principally by nervous system damage with numbness and weakness of the limbs (polyneuritis). Certain areas in the brain are also particularly susceptible to thiamine deficiency and result in such features as loss of balance, disturbance of recent memory and confusion, and Wernicke and Korsakoff syndromes. A poor diet is the principal cause in sub-Saharan Africa, but rare now in more affluent countries. Currently, in the latter, it is more commonly associated with chronic alcoholism due to excessive drinking and poor eating habits and particularly to sudden alcohol withdrawal, which may precipitate it.

Treatment of thiamine deficiency due to alcoholism or during alcohol withdrawal causing delirium tremens, Wernicke's encephalopathy and Korsakoff syndrome, is available only on prescription as 200mg intravenously or orally three times a day until symptoms disappear, or for less severe cases, 25-100mg orally once a day. Similar doses are advised for treating beriberi. Smaller doses of thiamine (usually around 5mg) are present in compound vitamin B tablets that can be purchased at pharmacists. In the absence of coexisting medical problems, the daily requirement for an adult is 1.5mg.

Thiamine deficiency producing severe neurological damage (called cerebrocortical necrosis or polioencephalomalacia) has been reported in horses that eat hay containing 20% or more of *Pteridium aquilinum* (bracken fern) or *Equisetum* spp. (horsetail) for three weeks (or smaller amounts for longer), because they contain a thiaminase, even when dry. Injections of thiamine early in the onset of the illness can reverse it.

Vitamin B2 riboflavin

Plant sources in the College Garden:

Phaseolus vulgaris – French bean (also other legumes)

Eruca vesicaria spp. *sativa* – rocket (also other leafy vegetables)

Malus domestica – apple

Asparagus officinalis – asparagus

A helping of any of these contains about 8% of one's daily requirement of vitamin B2. The best dietary sources are milk, cheese, eggs, liver, kidneys, lean meats, mushrooms and almonds but it is also synthesised by bacteria in the large bowel.

Malus domestica 'Court Pendu Plat' – apple

Eruca sativa – rocket

Chemistry

Riboflavin, vitamin B2, is an is essential component of two major coenzymes, flavin mononucleotide and flavin adenine dinucleotide. These coenzymes play major roles in energy production; cellular function, growth, and development; and metabolism of fats, drugs, and steroids. These are needed for the conversion of the amino acid tryptophan to niacin, and vitamin B6 to the coenzyme pyridoxal 5'-phosphate. Riboflavin helps maintain normal levels of homocysteine, an amino acid in the blood (see US National Institute of Health, Riboflavin – Health Professional Fact Sheet).

Riboflavin is a combination of a sugar and a flavin compound that gives it a striking yellow colour. It was recognised first as a factor that could restore growth in animals on a deficient diet. Subsequent studies were aided by its bright fluorescence under UV light, a property exploited in tracing leaks in complex industrial plants.

The structure of riboflavin was identified in 1934 (Ellinger & Koscharer, 1934), and deficiency, as an illness called ariboflavinosis, was reported four years later (Sebrell & Butler, 1938) when the typical clinical findings were caused by a B2 deficient diet fed to 18 adult women for over four months. Because of its similarities to pellagra with angular stomatitis but without the skin lesions, it had previously been called *pellagra sine* [without] *pellagra* (*q.v. pellagra* due to vitamin B3 deficiency).

The signs and symptoms of riboflavin deficiency in humans are weakness, inflammatory lesions in the mouth, moist scaly skin, anaemia, cataracts, ocular keratitis, thyroid disorders, and teratogenic deformities in babies born to B2-deficient women. As only minimal amounts are stored in the body, people on poor diets for any reason are liable to develop at least subclinical deficiency. Sometimes this is due to being put on a special diet because of another illness. Its full value in maintaining health is still being explored.

Vitamin B3 niacin

Principal plant sources of vitamin B3 in the College Garden:

> *Oryza sativa* – brown rice
>
> *Phaseolus vulgaris* – French bean (also other legumes)
>
> *Capsicum annuum* – bell peppers
>
> *Triticum aestivum* – wholegrain wheat
>
> *Solanum tuberosum* – potatoes

Solanum tuberosum, potato

The main sources are in meat, fish, nuts, yoghurt and mushrooms. Normal daily requirements are 1.3mg for men, 1.1mg for women. Tryptophan, present in some foods, is also converted to niacin in the body and is an important source (Niacin – Health Professional Fact Sheet, online).

Vitamin B3, niacin, is the generic name for nicotinic acid, and nicotinamide and related derivatives, found in many foods.

Chemistry

Nicotinic acid and nicotinamide are converted in the body to the metabolically vital coenzymes nicotinamide adenine dinucleotide and its reduced form nicotinamide adenine dinucleotide phosphate, vital coenzymes for the functioning of the principal enzymes that enable cells to produce energy (adenosine triphosphate, ATP) from foods and many other essential activities.

Zea mays – maize

An inadequate diet causes the serious illness pellagra, characterised by 'the three 'D's' – dermatitis (the only sign in 3% of patients), diarrhoea, dementia and even death (which makes four). The dermatitis consists of erythema and superficial scaling on skin exposed to the sun, heat, friction or pressure, which fades to leave a reddish-brown pigmentation. Additionally, there is a 'butterfly rash' (malar rash) on the cheeks and a well-demarcated erythema on the front of the neck called 'Casal's necklace' after the first person to describe it. Depression, behavioural changes and general ill-health are also common.

It was formerly a particular problem amongst the poor in the southern states of the USA at times when maize, *Zea mays*, formed most of the diet. It is seen in anorexia, alcoholism, and in patients being treated with anti-tuberculous medication (isoniazid and/or pyrazinamide) which blocks niacin production from tryptophan.

History

Niacin's importance in pellagra was first discovered and isolated by scientists investigating the 'black tongue disease' in dogs and found it to be due to a dietary deficiency, pellagra-preventing factor

(Elvehjem *et al*, 1938), following previous work on foods, such as beef and butter, which prevented pellagra and 'black tongue disease' (Goldberger *et al*, 1926). Nicotinic acid had been synthesised and used in photography in the 19th century and, unsuccessfully, for the treatment of beriberi (Funk, 1912).

Pellagra had become a widespread disease in peasants in Spain after the introduction of maize from South America, and was first called *mal de la rosa*, referring to the rose-colour of the skin when the scabs from the dermatitis fall off (Thiérry, 1755), based on the patients observed by Gaspar Casal (Casal, 1762 – written 1735, published posthumously). The name, pellagra, was first used when it had achieved epidemic proportions in Italy in the late 18th century due to the introduction of maize there (Frapolli, 1771). It later became common in France and Central Europe, and by 1900 in the southern states of the USA.

Niacin also shows great promise in preventing the development of sun-induced non-melanoma skin cancers, and in people taking drugs that suppress the immune system, for example those with organ transplants (Forbat *et al*, 2017; Snaidr *et al*, 2019).

It has been given to patients with arterial diseases such as heart attacks and claudication, as it increases local blood flow by producing vasodilatation, but the therapeutic action is so brief as to be of limited clinical value.

Niacin is one of the few vitamins which can be toxic in overdosage causing flushing due to vasodilatation and even liver damage.

Vitamin B5
pantothenic acid

Principal plant sources of vitamin B5 in the College Garden:

Oryza sativa – brown rice

Triticum aestivum – wholegrain wheat

Solanum tuberosum – potato

Solanum lycopersicum – tomato

Daucus carota – carrot

Malus domestica – apple

Pantothenic acid is an essential nutrient, present in almost all foods, especially meat, eggs and dairy foods including yoghurt with beef liver having the greatest amount. The daily requirement is about 5mg.

Solanum lycopersicum – tomato

Chemistry

This seemingly simple molecule was discovered in 1879 and its structure discovered in the 1930s. It is essential for the synthesis of coenzyme A (CoA) and acyl carrier protein. CoA is essential for fatty-acid synthesis and degradation, and a multitude of other anabolic and catabolic processes. Dietary precursors are converted to pantothenic acid in the intestine and the intestinal wall and this is absorbed directly into the bloodstream (Pantothenic Acid – Health Professional Fact Sheet).

Deficiency states in humans and farm animals are very rare and have not been well studied. The most prominent symptoms in general undernutrition, for example in prisoners on a limited diet and volunteer studies, have included apathy, loss of energy, abdominal and muscle cramps, burning feet, and peripheral numbness and weakness. Animals may show poor survival and limited weight gain; affected birds produce fewer and smaller eggs.

Allium cepa – onion

Vitamin B6 pyridoxine

Principal plant sources of vitamin B6 in the College Garden:

> *Cucurbita maxima* – squash
> *Solanum tuberosum* – potato
> *Allium cepa* – onion
> *Oryza sativa* – brown rice and bran
> *Triticum aestivum* – wholegrain wheat

Vitamin B6 is the generic name for six different compounds (vitamers) of which pyridoxine is the most important. They are normally absorbed from dietary fruits, starchy vegetables, grains and substantial amounts from liver. Pyridoxine was discovered by Paul Györky (Györky, 1934), isolated and crystalised from rice bran by Samuel Lepkovsky in 1938, and its structure determined a year later by Stanton Harris and Karl Folkers (Rosenberg, 2012).

Chemistry

Chemically, it is a simple substituted pyridine and is the essential co-factor for key enzymes used by cells to make carbohydrates, certain amino acids and fats, as well as enabling muscle cells to store and produce energy from locally-stored carbohydrates. The vitamers pyridoxal and pyridoxamine are equally effective because they are all converted in the body to the active form pyridoxine phosphate, an essential co-factor in enzymes that make nucleic acids as well as performing the other metabolic functions noted above.

Deficiency states do sometimes occur in people and animals on very restricted diets, in rare genetic disorders, and after treatment with isoniazid, a principal antituberculosis drug. Patients with renal failure, autoimmune disease, malabsorption and alcoholism, and those taking some anti-epileptic medications and theophylline are also prone to develop deficiency.

Prevention and treatment with the vitamin (now produced synthetically) or more often with a standard multivitamin mixture is simple and effective. High doses are used in relieving rare forms of metabolic diseases and pyridoxine-dependent epilepsy and in treating those poisoned by eating false morels (*Gyromitra* mushrooms). The latter contain hydrazine chemicals that may cause fits and damage to red blood cells by destroying pyridoxine in the body and so blocking various enzymes.

B6 deficiency is rare, and the consequences include microcytic anaemia (so called because red cells become abnormally small), electroencephalographic abnormalities, dermatitis with cheilosis (scaling on the lips and cracks at the corners of the mouth) and glossitis (swollen tongue), depression and confusion, weakened immune function, fatigue, weakness, depression and peripheral nerve damage. Eating gingko seeds to excess also depletes vitamin B6 and causes vomiting and convulsions.

Chronic overdosage of oral pyridoxine causes sensory neuropathy, ataxia, photosensitivity and skin disorders (Vitamin B6 – Health Professional Fact Sheet).

Vitamin B7 biotin (sometimes incorrectly called vitamin H)

Principal plant sources of vitamin B7 in the College Garden:

Triticum aestivum – wholegrain wheat

Malus domestica – apple

Helianthus annuus – sunflower (seeds)

Helianthus annuus – sunflower, whose seeds are rich in vitamin B7

Biotin is common in our diet, a major source is meat (especially viscera like liver), fish and eggs, also bananas, nuts, green vegetables and milk products. Like most vitamins, it is available as an oral supplement. It was first synthesised in 1943 (Harris *et al*, 1943).

The clinical features of biotin deficiency, which is very rare, include conjunctivitis, scaly red rashes on the face and genital area, depression, hallucinations and peripheral neuropathy with numbness and tingling in the hands and feet. Brittle nails and hair loss also occur. The latter may be the source of the unfounded belied that anyone, even those eating a normal diet, might benefit from taking extra biotin supplements.

As it is also produced by normal bacteria in the gut, deficiency is very uncommon. It may occur in those eating very poor diets, with severe disorders of the intestines, during treatment with certain anti-epileptic drugs and in chronic alcoholism. An unusual but notable cause of biotin deficiency is its tight binding by avidin, a specialised protein found in egg white that only affects people who eat large amounts of raw egg white, as in types of mayonnaise and eggnog.

Chemistry

Biotin has at least two major functions in the body – as a co-factor for key enzymes involved in the synthesis and metabolism of carbohydrates and it binds to special proteins in the nuclei of cells involved in gene expression and chromatin stability.

Biotin has become very important in laboratories and in the biotechnology industry because of the strong binding of biotin to avidin and its bright fluorescence under UV light. Once biotin has been bonded to proteins, even in mixtures, they can be separated by allowing the protein to bind to avidin in columns and using UV fluorescence to show each protein as it is separated from the avidin under conditions that discriminate between molecules with different properties.

Biotin supplements (even a single dose of 10mg) can interfere with laboratory results, including thyroid function tests, causing false readings. Some anti-epileptic drugs significantly lower biotin levels. People taking biotin supplements should inform their doctors.

Vitamin B9, folate, better known as folic acid

Principal plant sources of vitamin B9 in the College Garden:

> *Triticum aestivum* – wholegrain wheat
>
> *Phaseolus vulgaris* – French bean (also other legumes)
>
> *Eruca vesicaria* subsp. *sativa* – rocket (also other leafy vegetables)
>
> *Malus domestica* – apple
>
> *Asparagus officinalis* – asparagus

The principal sources of folate in the diet are dark green leafy vegetables, such as kale, broccoli and spinach, beans and peas, fresh fruit juices and wholegrains. There are also useful quantities in liver and seafoods.

Folate, vitamin B9, is the generic name for a group of food folates including folic acid. It is a vital co-factor in several fundamental biochemical functions in cells.

Chemistry

Folic acid is required for the synthesis of nucleic acids, for activation of another vitamin, B12 (cobalamin), and indirectly it influences other bodily pathways that safely metabolise sulphur-containing amino acids that could produce arterial damage if not closely controlled. The first two pathways make it very important in all dividing cells, so a deficiency can result in a special form of anaemia and may affect both fertility and normal embryonic development.

Folate deficiency is a risk in those on poor diets, those with intestinal diseases, and after treatment with certain anti-epileptic drugs. It can manifest as a sore mouth, depression, megaloblastic anaemia (so called because red cells become abnormally large), impaired immunity, early greying of hair, and infertility in both sexes.

In pregnancy in a folate-deficient mother the embryo is more likely to become malformed with a particular risk of spinal canal (spina bifida) and cardiac deformities. Folic acid supplements before and during pregnancy reduce these risks. It also reduces the incidence of autism spectrum disorder by 40% in children. The role of folate in preventing anaemia of pregnancy was first shown by Lucy Wills who used brewer's yeast from which folate was later isolated (Wills, 1931).

Because the activity of folate in nucleic-acid synthesis and cell division is impaired by the anti-folate drugs used to inhibit cell division in treating forms of cancer and leukaemia, as well as malaria and some other parasitic diseases, induced folate deficiency resulting in serious anaemia may complicate such therapies unless the folate status of the patient is controlled. Folic acid supplements can increase the incidence of, and mortality from, cancer in the absence of anticancer treatment (see Folate – Health Professional Fact Sheet online).

Asparagus officinalis – asparagus

Semi-synthetic folic acid is given to prevent the induced deficiencies. Women who are or who wish to become pregnant are advised to take it as an oral supplement. Folate supplementation itself is not known to be toxic but its administration can mask vitamin B12 deficiency by correcting the anaemia without preventing its other harmful effects. Folate fortification of bread is a common public health practice in Britain and several other countries.

Vitamin B12 cobalamin

This is included here for completeness, but there are no plants that contain sources of B12 so vegans should rely on B12-fortified food or supplements. Deficiency can result in megaloblastic anaemia (anaemia with abnormally large red blood cells), a sore inflamed tongue, and becoming pale and short of breath because of the anaemia. Neurological symptoms include loss of sensation and tingling in hands and feet, difficulty in walking, memory problems and confusion (see Vitamin B12 – Health Professional Fact Sheet online).

Vitamin C – see *Citrus* x *limon*

Vitamin D calcitriol

This is included here for completeness, but there are no plants which contain sources of vitamin D so vegans should rely on vitamin D-fortified foods or supplements. A low vitamin D level is due to a deficient diet or relatively limited exposure to sunlight, but is commoner in people with dark skins. Vitamin D deficiency in childhood causes the bony deformities called rickets.

Olea europaea – the olive tree, source of olive oil

Vitamin E

Plant sources in the College Garden:

Triticum aestivum – wholegrain wheat, cereals and pasta

Olea europaea – olive (oil)

Glycine max – soya (bean)

Helianthus annuus – sunflower (seeds)

Capsicum annuum – bell peppers

Vitamin E comprises a group of eight closely-related, fat-soluble compounds – tocopherols and tocotrienols. In the body they act as antioxidants, but there is still some uncertainty about their overall importance.

The principal dietary sources of vitamin E include wheat germ (the central part of a grain of wheat), most seeds and nuts, the oils expressed from them and such dark, leafy vegetables as spinach. Some foods, such as margarines, breakfast cereals and brown bread, have added tocopherols mainly as an antioxidant to prevent rancidity and as a dietary supplement.

Deficiency of vitamin E is usually associated with inadequate absorption of fats due to a gastrointestinal disorder. Symptoms include failure to thrive of the young, infertility, disorders of nerves and the brain, and deposition of a dark brown pigment in cells regarded as debris of cell membranes damaged by the relative excess of free radicals of oxygen. There are unproven links between inadequate vitamin E intake and retinal macular degeneration and cataract, cancer and cardiovascular diseases.

Toxicity due to an excessive dose is not known in humans, but a very high dose in animals has interfered with blood clotting.

Vitamin K

Plant sources in the College Garden:

Glycine max – soya (bean)

Glycine max – the soya bean

This fat-soluble vitamin consists of two vitamers, K1 phylloquinones and K2 menaquinones. They act similarly in the body.

The principal dietary sources of vitamin K1 are leafy green vegetables, *e.g* spinach, kale and broccoli, and certain fermented soya bean products. There are low levels in animal and dairy products including egg yolk, meat and some fish.

The principal role of both is as a co-factor to make several blood-clotting factors effective in activating another protein involved in bone formation. The anti-blood clotting drug warfarin acts by preventing the action of vitamin K on precursors of those clotting factors (*q.v.* melilot, *Melilotus officinalis*). Vitamin K also has uncertain links with arterial diseases, brain function and ageing.

Deficiency of vitamin K is unusual except in people who suffer from malabsorption disorders, or who have alcoholic liver disease as the liver is its main store in the body. The commonest cause of bleeding and bone disorders due to a lack of the effect of vitamin K is long-term administration of warfarin and similar anticoagulants.

Treatment of a deficiency is usually by administration of a water-soluble oral formulation or by injection. It is common to give new babies an injection of vitamin K, as they are usually born with a low level and there is little in human milk.

References

General

Semba RD. The Discovery of the Vitamins. *Int J Vitamin Nutr Res* 2012a;82:310-315

Semba RD. On the 'Discovery' of Vitamin A. *Ann Nutr Metabol* 2012b;61:192-198

Carpenter KJ. The Nobel Prize and the discovery of vitamins. NobelPrize.org. Nobel Prize Outreach AB 2022. https://www.nobelprize.org/prizes/themes/the-nobel-prize-and-the-discovery-of-vitaminsC accessed 4 August 2022

Vitamin A

Semba RD. On the 'Discovery' of Vitamin A. *Ann Nutr Metabol* 2012b;61:192-198

B1 thiamine

Bruyn GW, Poser CM. *The History of Tropical Nutritional Disorders*. USA, Science History Publications, 2003

Williams RR, Cline JK. Synthesis of Vitamin B1. *J Amer Chem Soc* 1936;58:1504-5

B2 riboflavin

Sebrell WH, Butler RE. Riboflavin deficiency in man. A preliminary note. *Public Health Reports* 1938;53(52):2282-4

https://ods.od.nih.gov/factsheets/Riboflavin-HealthProfessional/ accessed 4 August 2022

B3 niacin

Casal G. *Historia Natural Y Medica de el Principado de Asturias*. Madrid, M. Martin, 1762

Elvehjem CA, Madden RJ, Strong FM, *et al*. The Isolation and Identification of the Anti-Black Tongue Factor. 1938. *J Biol Chem* 123:137-149

Forbat E, Al-Niaimi F, Ali FR. Use of nicotinamide in dermatology. *Clin Exp Dermatol* 2017;42:137-144

Frapolli F. *Animadversiones in morbum, vulgo pelagram* [Mediolani, apud J. Galeatium]. 1771

Funk C. The etiology of the deficiency diseases. Beri-beri, polyneuritis in birds, epidemic dropsy, scurvy, experimental scurvy in animals, infantile scurvy, ship beri-beri, pellagra. *Journal of State Medicine* 1912;20:341-68

Goldberger J, Wheeler GA, Lillie RD, *et al*. A Further Study of Butter, Fresh Beef, and Yeast as Pellagra Preventives, with Consideration of the Relation of Factor P-P of Pellagra (and Black Tongue of Dogs) to Vitamin B. *Public Health Reports* 1926;41(8)

Snaidr VA, Damian DL, Halliday GM. Nicotinamide for photoprotection and skin cancer chemoprevention: A review of efficacy and safety. *Expmntl Dermatol* 2019;28:15-22

Thiérry F. Description d'une maladie appelée *mal de la rosa*. *Jour Méd Chir Pharm* 1755;2:337-346

https://ods.od.nih.gov/factsheets/Niacin-HealthProfessional/ accessed 4 August 2022

B5 pantothenic acid

https://ods.od.nih.gov/factsheets/Pantothenic Acid-HealthProfessional/ accessed 4 August 2022

B6 pyridoxine

György P. Vitamin B2 and the Pellagra-like Dermatitis in Rats. *Nature* 1934;133(3361):498-9

Rosenberg IH. A History of the Isolation and Identification of Vitamin B6. *Ann Nutr Metab* 2012;61:236-238

https://ods.od.nih.gov/factsheets/vitaminB6-HealthProfessional/ accessed 4 August 2022

B7 biotin

Harris SA, Wolf DE, Mozingo R, *et al* Synthetic biotin. *Science* 1943;97:447-8

B9 folate, better known as folic acid

https://ods.od.nih.gov/factsheets/Folate-HealthProfessional/ accessed 4 August 2022

Wills L. Treatment of 'pernicious anaemia of pregnancy' and 'tropical anaemia' with special reference to yeast as a curative agent. *Brit Med J* 1931;1:1059-64

B12 cobalamin

https://ods.od.nih.gov/factsheets/vitaminB12-HealthProfessional/ accessed 4 August 2022

Olea europaea in the College Garden. Its fruits are a source of Vitamin E

APPENDIX 1 – HENRY OAKELEY

Historical references consulted
Source literature
before 1700
and for Linnaeus

Aegineta, Paulus [625-690 CE]. *Pauli Aeginetae Opus de re Medica*. Translated by Joannes Guinterium Adernacum. Venice, Andream Arrivabenum, 1542

Alexander of Tralles [525-c. 605 CE]. *De lumbricis*. Venice, Hieronymous Mercurialis, 1570 [not seen]

Alexander of Tralles [525-c. 605 CE]. *Alexandri Tralliani Medici Libri Duodecim*. Translated and edited by Joanne Guinterio Andernaco. Lyons, Antonium de Harsey, 1575

Alpini, Prospero. *De Plantis Aegypti*. Venice, Franciscum de Franciscis, 1592

Anicia Juliana Codex [514 CE]. *Der Weiner Dioskurides. Codex medicus graecus 1 der Österreichischen Nationalbibliothek*. Facsimile in two volumes of the paintings of Dioscorides' plants which had been given to Juliana Anicia, with commentary by von Otto Mazal. Graz, Akademische Druck- u. Verlaganstalt, 1998, 1999

Apuleius Barbarus, [Pseudo-Apuleius, 6th century]. *Liber Apulei Platonici de Medicaminibus Herbarum*. Edited by Gabriel Humelberger. Zurich, C. Froschouer, 1537

Averrhoes [Averrois/Ibn Rushd, 1126-1198]. *Colliget Averrois Totam Medicinam ingentibus voluminbus ab aliis traditam mira quadam brevitare & ordine sic ad amusim complectens*. Venice, Octavianum Scotus d. Amadei f., 1542

Avicenna [Ibn Sina, 980-1037]. *Liber canonis primus quem princeps aboali Avinsceni de Medicina edidit, Translatus a magistri Gerardo Cremonensis in Toleto ab Arabico in Latinus.Verba aboali avinsceni* Libri 1-V. [Canon of Medicine, books 1-5]. The medicinal plants are in book 2, *Incipit liber cannonis secundus Avicenne verba principis aboali*. Venice, Dionysus Bertochus, 1490

Avicenna [Ibn Sina, 980-1037]. *Liber canonis totius medicine ab Avicenna arabum doctissimo excussus, a Gerardo Cremone[n]si ab arabica lingua in Latina[m] reductus…* Lyon, Jacques Myt, 1522

Bado S. *Anastasis corticis Peruuiae, seu Chinæ Chinæ defensio*. Genuæ, Typis P. I. Calenzani, 1663

Bauhin C. *Pinax Theatri Botanici*. Basel, Ludovici Regis, 1623

Beck LY (translator). *Pedanius Dioscorides of Anarzarbus de Materia Medica*. New York, Hildesheim; Zurich, Olms-Weidmann, 2005

Bergius PJ. *Materia Medica E Regno Vegatibili*, 2nd edn, corrected, vol 2. Stockholm, Petri Hesselberg, p839, 1782

von Bingen H (c. 1200 CE – a French abbess). *Physica*. Translated by Priscilla Throop. Rochester, Healing Arts Press, 1998

Bodaeus J (ed). *Theophrasti Eresii. De Historia Plantarum libri decem*. Amsterdam, Henricum Laurentium (Commentary by JC Scaliger and R Constantin), 1644

Bontius, Jacob. *De Medicina Indorum*. Lugduni Batav. Franciscum Hackium [written 1631, published posthumously]. pp90-91, 1642

Bontius, Jacob. *Historia Naturalis et Medicae Indiae Orientalis. Libri sex*, p131, in Piso W, Bontius J, Marcgrave G. in *De Indiae Utriusque*, 1658

Clusius C. *Rariorum Plantarum Historia Liber Quartus*. Antwerp, Jan Moretus ex Officinia Plantiniana, 1601

Clusius C. *Curae Posteriorees seu plurimarum non ante cognitarum, aut descriptarum stirpium ...* . Leiden and Antwerp, Officina Plantiniana, 1611. Re correspondence from Gregorio da Reggio, 1611

Cobo B. *Inca Religion and Customs* [1653]. Translated and edited by Roland Hamilton. Austin, University of Texas Press, 1990

Coles W. *Adam in Eden or Nature's Paradise*. London, J. Streater for Nathaniel Brooke, 1657

Columna F. [Fabio Colonna] *Phytobasanos Sive Plantarus Aliquot Historia*. Naples, Horatio Salviani, Jacobus Carlino and Antonius Pace,1592

Culpeper N. *A Physical Directory or a Translation of the Dispensatory made by the Colledge of Physicians of London*, 1st and 2nd edns. London, Peter Cole, 1649, 1650

Culpeper N. *The English Physitian*. London, Peter Cole, 1652

Culpeper N. *The English Physitian Enlarged*. London, Peter Cole, 1653

[Culpeper N] Sibley E. *Culpeper's English Physician and Complete Herbal* … London, printed for the author and sold at the British Directory Office, Ave-Maria Lane; and by Champante and Whitrow, Jewry Street, Aldgate, 1799

Daléchamps J. *Histoire Générale des Plantes*. Lyon, Philip. Bord, Laur. Arnaud & Cl. Rigaud, 1653 (first edition in 1586)

Dioscorides [70 CE] *The Greek Herbal of Dioscorides Englished by John Goodyear 1655*, translated by John Goodyear, published by Robert Gunther. Hafner Publishing, New York, 1959

Dioscorides [70 CE] *Pedanius Dioscorides of Anazarbus De Materia Medica*, translated by Lily Beck. Hildesheim, Zurich, New York, Olms-Weidmann, 2005

Dodoens R. *Crǔÿdeboeck*. Antwerp, Van der Loe, 1554

Dodoens R. *Stirpium Historiae Pemptades Sex* (Dodoens' translation into Latin of his *Crǔÿdeboeck*). Antwerp, Christopher Plantin, 1583

Egenolph C. *Plantarum Arborum*. Frankfurt, Christian Egenolph, 1562

Fernellius Io [Jean Fernel]. *Therapeutices Universalis seu Medendi rationis, libri septem*. Frankfurt, Andreae Wecheli heredes, 1593

Fuchs L. *De historia stirpium commentarii insignes*. Basileae: In Officina Isingriniana, 1542

Gaertner J, Linnaeus C. *De fructibus et seminibus plantarum* vol 2. Stuttgart, Academiae Carolinae, p378, t162, 1788

Galen C. [200 CE]. *Cl. Galeni De Simplicium Medicamentorum facultatibus libri undecim*, translated by Theodorico Gerardo Gaudano. Paris, Jacobum Bogardum, 1543

Garcia ab Horto D. *Aromatum et Simplicium Aliquot medicamentorum apud indos nascentium historia*. (p 163,164 re *Lignum columbrinum*) Antwerp, Ex Officina Plantiniana, Viduam & Ionnem Moretum, 1593

Gerard(e) J. *The Herball or Generall Historie of Plantes* (a translation into English of Dodoens' *Stirpium Historiae Pemptades Sex*). London, John Norton, 1597

Gerard J. *The Herball or Generall Historie of Plantes*. Edited by Thomas Johnson. London, Adam Islip, Joice Norton and Richard Whitakers, 1633

de Gordon B. [1270-1330] *Opus Lilium Medicinae* ... Lugduni [Lyon], Guillaume Rouillé [printed by Philibert Rollet], 1550

Gunther RT. (ed) *The Greek Herbal of Dioscorides. Englished by John Goodyear 1655*. Edited and first printed 1933 Hafner Publishing Co, New York, 1959

Herbario volgare 1522 [written c. 1300], translated and printed by Alessandro de Bindoni, Venice. [An Italian translation of the *Tractatus de Virtutibus Herbarum* attributed to Arnold of Villanova (1240-1311)]. Facsimile, Italy, Edizione Polifilio, 1979

van der Heyden H. *Discours et avis sur les flus de ventre douloureux*. Belgium, Servais Manilius Ghent, Manilio, Sebastiano, 1643

von Hutten U. *De guaiaci medicina et morbo gallico liber unus*. Mainz, Ioannis Scheffer, April 1519

von Hutten U. D*e Morbo Gallico*, translated from the Latin by Thomas Paynell. London, Thomas Berthelet, 1533

(King James I) *A Counterblaste to Tobacco*. London, R.B., 1604 de L'Escluse C. *Histoire des Plantes … par Rembert Dodoens* (a translation into French of Dodoens' *Crüÿdeboeck*, 1554). Antwerp, Van der Loë, 1557

Linnaeus C. *Hortus Cliffortianus*. Amsterdam, 1738

Linnaeus C. *Materia Medica*. Holmiae [Stockholm], Laurentii Salvii, 1749

Linnaeus C. *Species Plantarum*. Holmiae [Stockholm], Laurentii Salvii, 1753

Linnaeus C. *Systema Naturae* Edit 10, Tomus II. Holmiae [Stockholm], Laurentii Salvii, 1759

Linnaeus C. *Materia Medica*. Lipsiae et Erlange, Wolfgang Walther, 1782

L'Obel M. *Plantarum seu Stirpium Historia*. Antwerp, ex Officina Christophori Plantini Architypographi Regii, 1576

L'Obel M, Pena P. *Stirpium Adversaria Nova*. London, Thomae Purfoetii, ad Lucretie symbolum, 1571

Lyte H. *A Niewe Herball or Historie of Plantes* (Rembert Dodoens' work in English, translated from L'Escluse's *Histoire des Plantes*). London, Gerard Dewes, 1578

Matthiolus (Mattioli) PA. *I Discorsi di M. Pietro Andrea Matthioli … nelli sei libri di Pedacio Dioscorides Anazarbeo della materia Medicinale*. Venice, Vincenzo Valgrisi, 1568

Matthiolus (Mattioli) PA. *Commentarii in sex libros Pedacii Dioscoridis Anazarbei de Medica Materia* [Commentary on the *Materia Medica* of Dioscorides]. Venice, Officina Valgrisiana, 1569

Matthiolus (Mattioli) PA. *De Plantis Epitome Utilissima*. Francofurti ad Moenum, Johann Feyerabend, 1586

Mentzel C. *Pinax. Nominum plantarum. Universalis … adjectus est Pugillus Rariorum Plantarum*. Berlin (Berolini), t11, 1682

Mesue Joannis [Pseudo-Mesue aka Mesuae 1100 CE]. *De Re Medica Libri Tres. Jacobo Sylvio Medico Interprete*. Translated by Jacques Dubois aka Jacobus Sylvius. Paris, 1553

de Molina C. *Account of the Fables and Rites of the Incas* [1573]. Austin, University of Texas Press, 2011

Monardes N. *Dos Libros. El uno trata de todas las cosas que traen de nuestras Indias Occidentales que sirven al uso de Medicina*. Seville, Sebastián Trugillo, 1565, (also another edition published in Seville, Fernando Diaz, 1569)

Monardes N. *Del Pepe lungo* [of the long pepper] part 2, book 2. In: *Delle cose che vengano portate dall'Indie Occidentali pertinenti all'uso della Medicina*, (translated from Spanish into Italian). Venice, Giordano Ziletti, pp101-2, 1575a

Monardes N. *Dell' Arboro contra il Flusso* [of the tree to treat diarrhoea] part 2, book 2. In: *Delle Cose che vengono portate dall'Indie Occidentalis pertinenti all' uso della medicina* (translated from Spanish into Italian). Venice, Giordano Ziletti, pp111-2, 1575b

Parkinson J. *Paradisi in sole. Paradisus in Terrestris*. London, Humfrey Lownes and Robert Young, 1629

Parkinson J. *Theatrum Botanicum*. London, Tho. Cotes, 1640

Pemel R. *Tractatus de Simplicium Medicamentorum facultatibus. A Treatise of the Nature and Qualities of such Simples as are most frequently used in Medicines…* London, M. Simmons for Philemon Simmons, 1652

Pharmacopoea Londinensis. London, Edward Griffin, May 1618 (facsimile, with epilogue by Dr Henry Oakeley). London, Royal College of Physicians, 2018

Piso W. *Mantissa aromatica* p198 in Piso W, Bontius J, Marcgrave G. in *De Indiae Utriusque*, 1658

Piso W, Bontius J, Marcgrave G. in *De Indiae Utriusque re Naturali et Medica*. Amsterdam, Ludovicum et Danielem Elzevirios, 1658

Pliny C [Plinius Secundus, 70 CE]. *The Historie of the World Commonly called the Natural Historie of C. Plinius Secundus*, translated by Philemon Holland. London, Adam Islip, 1634

Ray J. *Historia Plantarum* vol 1. London, Henricum Faithorne & Joannum Kersey, p680, 1686

Serapionis J. *Insignium Medicorum. Joan. Serapionis Arabis de Simplicibus Medicinis opus praeclarum & ingens*. Argentorati [Strasbourg], Georgius UIricher Andlanus, 1531

Sloane H. *Catalogus Plantarum quae in insula Jamaica sponte proveniunt*. London, D. Brown, 1696

Sylvaticus M. *Pandectarum Medicinae*. Florence, Luce Antonii, 1524 (written 1317)

Talbor R. *Pyretologia, a rational account of the cause and cure of agues*. London, R. Robertson, 1672

Theophrastus [c. 371–287 BCE]. *Theophrasti de Suffruticibus, Herbisque, ac Frugibus Libri Quattuor*, translated by Theodoro Gaza. Strasbourg, Henrico Sybold, 1528

Theophrastus – see Bodaeus, 1644

Theophrastus. *Enquiry into Plants* vol 2, English translation by Arthur Hort. Cambridge Massachusetts, Harvard University Press and London, William Heinemann Ltd., 1980

Treveris P (printer). *The Grete Herball*. London, Peter Treveris, 1529

Tulp N. *Observationes Medicae. Editio nova*. Amsterdam, Ludovic Elsevir, 1652

Turner W. *A New Herball* (in three parts). Collen [Cologne], Arnold Birkman, 1551-1568

Further reading re phytopharmacy

Earles MP. *Studies in the development of experimental pharmacology in the eighteenth and early nineteenth centuries*. PhD thesis, University College, University of London, 1961 https://discovery.ucl.ac.uk/id/eprint/1317887/1/296133.pdf accessed August 2021

APPENDIX 2 – HENRY OAKELEY

Dramatis Personae

The importance of primary source references for early beliefs and uses cannot be overstated, and the historic herbals in the Dorchester Library at the Royal College of Physicians are a valuable resource.

These are the more important *Dramatis Personae* whose writings are frequently referenced in this book.

Alexander of Tralles [525- c. 605 CE]. A Greek physician who worked in Rome, Spain, Gaul and Italy. Wrote *Medici Libri Duodecim* and a treatise on intestinal worms, *De lumbricis*.

Apuleius Barbarus, also known as Pseudo-Apuleius, author of an influential 6th century herbal, *Herbarium Apuleii Platonici*, based on Pliny and Dioscorides. Contains 130 plants with uses and synonyms.

Avicenna (Ibn Sina, 980-1037), famous Persian physician and polymath whose *Canon of Medicine*, c. 1020, was based on Greek texts, translated into Syriac in the 9th century by Nestorian Christians in Persia, and later into Arabic, enshrining medical knowledge at the end of the first millennium until c. 1650.

Culpeper, Nicholas (1616-1654). English herbalist and physician who translated the second edition

Alexander of Tralles, *Medici Libri Duodecim*, 1575

In Effigiem Nicholai Culpeper Equitis.
The shaddow of that Body heer you find,
Which serves but as a case to hold his mind,
His Intellectuall part he pleas'd to looke
In lively lines described in the Booke.

Nicholas Culpeper from his *A Physical Directory*, 1650

Rembert Dodoens, from Lyte's *Herball*, 1578

of the *Pharmacopoeia Londinensis* (December 1618) into English as *A Physical Directory* (1649, 1650) adding in the uses of the plants and medicines sourced widely from Classical and contemporary authors. His *The English Physitian* (1652) stemmed from this and was the precursor of *Culpeper's Herbal*, produced after his death, which continued in print into recent times.

Dioscorides, Pedanius of Anazarbus (40-90 CE). Greek physician from Cecilia. All Dioscorides references are to his *Materia Medica* of c. 70 CE, containing the uses of about 600 plants. This was the basis for medicine for nearly 1700 years. English editions by Beck (2005) and Gunther (1959), 16th century Latin commentaries by Mattioli and Ruel with their numerous woodcuts.

Dodoens, Rembert (1517-1585). Flemish physician and botanist; he wrote his ground-breaking *Crüÿdeboeck* in 1554 on the medicinal uses of plants, in Flemish with 715 woodcuts. He translated it into Latin as *Stirpium Historiae Pemptades Sex* (1583). Other translations by L'Escluse (1577), Lyte (1578) and Gerard (1597, 1633).

Fernel (Fernelius), Jean (1497-1558) a French cosmologist, mathematician and physician to Catherine de' Medici. Author of *Therapeutices Universalis*, one of the three great pharmacopoeias of the 16th century, with Occo's *Augustana* and Cordus's *Dispensatorium*.

Fuchs, Leonard or Leonhart (1501-1566). German physician and botanist. His 1542 herbal, *De Historia Stirpium commentarii insignes*, set a new standard in woodcut illustrations. He quotes Galen, Dioscorides, Paulus Aegineta and Pliny for their uses.

Galen, Claudius of Pergamon (129-c. 200/216). Greek physician, surgeon in Rome, who was the dominant medical authority until the 17th century. His book *De Simplicium Medicamentorum facultatibus libri undecim* is the principal source of Humoral concepts of plants as medicines.

Gerard(e), John (1545-1612). Herbalist, surgeon and gardener, he published the monumental *Herball, or Generall Historie of Plantes* in 1597 using the English translation done by a Dr Priest from the College of Physicians of the Latin *Pemptades* (1583) of Rembert Dodoens. L'Obel helped sort out the many errors. Thomas Johnson's revised edition of 1633 has 2766 woodblock illustrations.

Hippocrates (468-377 BCE). The 'father of medicine' and of the Doctrine of the Humors.

de l'Escluse, Charles (1526-1609). A Flemish doctor who translated Dodoens' *Crüÿdeboeck* into French as *Histoire des Plantes* (1577).

Linnaeus, Carl (1707-1778). Swedish Professor of Medicine and Botany at the University of Uppsala, zoologist and taxonomist, who classified the world's plants and animals using a binomial system. His *Species Plantarum*, listing all the known plants in 1753, is the starting point for plant names used today.

de l'Obel, Matthias (1538-1616). Flemish botanist, physician and botanist to James I, author of two important herbals and also helped with the corrections to Gerard's *Herbal*.

Lyte, Henry (1529-1607). Translated l'Escluse's *Histoire des Plantes* (1577) *q.v.* into English as *Nievve herball or historie of plantes* (1578) in black letter font, with 870 woodcuts, and his own opinions added.

Mattioli (Matthiolus), Pietro Andrea (1501-1577). Born in Sienna, trained in Padua, physician to

Leonhart Fuchs in *De Historia Stirpium*, 1542

Matthias de l'Obel

John Parkinson from his *Theatrum Botanicum*, 1640

Archduke Ferdinand and Maximillian II. Prolific publisher: his commentaries on Dioscorides sold tens of thousands of copies using the Latin translation from the Greek, made by Jean Ruel, but also published in Italian, French, German and Czech. He illustrated his commentaries with nearly 1000 woodcuts, so enabling a linking of his nomenclature with today's.

Parkinson, John (1567-1650). Apothecary to James I and botanist to Charles I, he wrote the most inclusive English herbal of all, *Theatrum Botanicum*, in 1640. It contained 3800 plants with detailed reviews of the writings of earlier authors, with woodcuts.

Pliny Secundus, or Pliny the Elder (23-79 CE). Roman philosopher, naturalist. He compiled the encyclopaedic *Naturalis Historia* in c. 77 CE. Magnificent English translation by Philemon Holland in 1601 (edition of 1644 used for this book).

Ruel, Jean (1474-1537) French physician and botanist, author of *De Natura Stirpium* (a botanical treatise; 1536) and the definitive translation of Dioscorides' *Materia Medica* into Latin in 1516 which was used verbatim by Mattioli in his commentaries on Dioscorides.

Serapionis, Joannis [Serapion the Younger, 11th century, a pseudepigraph for Ibn Wafid, d. 1067]. Author of a lost book in Arabic on simple medicines only available in Latin, a compilation of the works of Dioscorides and Galen and many Arabic authors, and much quoted by 16th century European writers.

Sylvaticus, Matthaeus (c. 1280-1342). Italian physician and botanist from the School of Salerno, whose *Pandectarum Medicinae*, a compilation of the works of Galen, Dioscorides, Avicenna and Serapion (completed in 1317), is a major reference source.

Theophrastus (371-287 BCE). Greek philosopher, physician, 'father of botany', author of *Historia Plantarum*.

Treveris, Peter (active 1525-1532). Printed the first illustrated English herbal, *The Grete Herball*, in 1526 and 1529. It is a translation of the French *Le Grant Herbier* (1498) which itself is mainly derived from the *Circa Instans* from the School of Salerno in c. 1140.

Turner, William (1509/10-1568). Physician, naturalist, herbalist, Dean of Wells cathedral, 'father of English botany'. His *New Herbal* (in three parts: 1551 with 150 plants; 1562 with 250 plants; and 1568 with 70 plants) with woodcuts copied from Leonard Fuchs, and with uses (mainly translated verbatim) from Dioscorides and Galen, is the second illustrated English-language herbal.

Pietro Mattioli (Matthiolus), from his *Discorsi* of 1568

Glossary

This glossary does not contain the meaning of the chemical names or the diseases used in the text. Words explained in the text are not included.

adrenal medulla – the inner part of the adrenal gland

agonists – something which increases an action

alkaloids – organic molecules containing at least one nitrogen atom. Most are poisonous

allele – a specific part of a gene responsible for the production of one chemical or characteristic

AMP – adenosine monophosphate

anabolic steroids – steroids that increase the bulk of the body, especially muscle

anaerobic – without oxygen

analogue – a similar substance/chemical

anodyne – painkiller

antagonists – something which decreases an action

anti-helminthic – against worm-like parasites

anti-tussives – a medicine to stop coughs

apothecary – pharmacist

arrhythmia – abnormal heart rhythm

astringent – binding, constricting, drying

ATP – adenosine triphosphate

atrial – referring to the smaller collecting chambers of the heart

atrial fibrillation – an abnormal heart rhythm in which the heart beats rapidly and chaotically – the pulse becomes 'irregularly irregular' – due to failure of conduction of the synchronising cardiac impulse between the atria and the ventricles

atrium/atria – smaller collecting chamber(s) of the heart

autonomic nervous system – controls automatic activity like sweating

BCE – Before the Common Era in dates – previously BC, Before Christ

beta2 receptor subtype – receptors are proteins which bind to a specific molecule. Beta-2 adrenoceptors are activated by the catecholamines norepinephrine (noradrenaline) and epinephrine (adrenaline)

blast crisis – a phase in chronic myelogenous leukaemia where more than 30% of the cells in the blood or bone marrow are immature

bone marrow – substance inside the long bones, ribs and vertebrae (backbone)

bronchodilator – an agent that increases the diameter of the bronchi

buffers to control pH – the pH of a liquid is a measure of its acidity or alkalinity. Buffers are liquids containing a weak acid or base and their salts that keep the original liquid at approximately the same pH despite the addition of acids or alkalis

cell-mediated – action governed by cell activity

calcareous – chalky

carcinogenic – inducing cancer

cardiac action potential – the electrical impulse causing the heart muscle to contract

cardiac glycosides – chemicals containing sugar molecules that affect the beating of the heart

CE – Common Era, previously AD (*Anno Domini*), in indicating a year

cell-cycle arrest – cessation of normal cell reproductive processes

Cheyne-Stokes respiration – repetitive intermittent breathing pattern of increasing and then diminishing rate and volume followed by a period of not breathing

cholagogue – a chemical or medicine which promotes secretion of bile from the gall bladder

cilia – hairy projections on cell surfaces

clotting factors – natural proteins which enable our blood to clot

cormous – with corms, swollen underground stems, *e.g* bulbs and tubers

cutaneous – relating to skin

cyanogenic glycoside – substances containing sugar molecules from which cyanide can be released

cyclical – recurrent

cysts – sac of membranous tissue that contains fluid

cytokines – signalling molecules that mediate and regulate immunity and other bodily processes

cytological – relating to tissue cells

cytotoxicity – poisonous to cells

decoction – a liquid mixture obtained from heating a substance in water, *e.g* tea

degree – in Humoral medicine, referring to the strength of a substance's Humoral properties, first degree being mild, fourth being extreme. Humoral substances could be hot, cold, dry, wet; pepper would be hot in the third or fourth degree; deadly nightshade, cold in the fourth degree. Fruit juice might be cold in the first degree (*i.e.* slightly cooling)

dendritic cells, dendrites – the extensions of certain nerve cells

derivatives – substances derived from

dermatologist – physician specialising in skin disorders

dextrorotatory stereoisomer – a chemical which has a molecular shape that rotates light clockwise (to the right as one looks at the light) as it passes through it. A laevorotatory one is a molecule with the same composition but arranged in a different pattern

discalced (monk) – a monk who goes barefoot or in sandals as part of his religious Order

dispensatory – a pharmacy, or a book which gives the uses of medicines

diuresis/diuretic – increased urine flow/ inducing an increased urine flow

DNA – deoxyribonucleic acid – the molecule of two chains of nucleotides in a helix that carry genetic information. Present in the nucleus of all living cells

dose-response curves – the measured response to a medicine as a consequence of increasing dosage

dropsy – archaic name for heart failure referring to the common finding of swollen, oedematous legs

dysrhythmia – abnormal heart rhythm

EMA – European Medicines Agency

emmenagogue – substance which causes menstruation

emetic – a substance which induces vomiting

endophyte/endophytic – an organism, usually a bacteria or fungus, which exists inside a plant without harming it

enteric coating – a coating of a medicine which stops it being digested in the stomach

epidemiology – scientific study of certain aspects of a disease – such as frequency and prevalence

etymology – the origin of a name

expectoration – spitting

extracorporeal – outside the body

family – plants have many levels of groups. The family is the group which includes many related genera (singular – genus), and a genus is a group which includes many related species

FDA – Food and Drug Administration. The organisation in the USA responsible for drug and food safety and licensing

fistula – an abnormal connection or passageway between two hollow organs or vessels, or between one and the outside

foci – small locations, *e.g.* sources of infection

formulation – method of making up a medicine

galactagogue – a substance which stimulates lactation

genus – a level of classification of plants viz family – genus – species.

gestational – relating to pregnancy

glycoside – a molecule, often of biological importance, to which one or more sugar molecules is attached.

GMP – guanosine monophosphate

graft versus host disease – when immune system cells in a transplant react against the host cells and tissues by developing antibodies and/or sensitised T-cells that attack them

H1N1 swine flu – a type of the swine flu virus

haem – the molecule containing iron which is present in haemoglobin and other vital proteins in the body

half-life – the time which it takes for half of a chemical or medicine (etc.) to disappear (*e.g.* by excretion or metabolism from the body)

helminth – parasitic worm

hepatobiliary system – the functioning system of the liver, gall bladder and bile duct

histamine – a chemical produced by the specialised cells in immune and inflammatory responses

HLA-DR – a gene responsible for 'graft versus host' disease. Acronym for Human Leukocyte Antigen and 'DR' is a variety of this gene involved in disease susceptibility and resistance

hot/cold properties – see degrees

hull – dried outer covering, husk, of a seed

humoral immunity – aspect of immunity due to antibodies etc. in blood, plasma and other extracellular fluids

Humors – the four substances of which the body was composed (blood, yellow bile, black bile and phlegm) in the Graeco-Roman system of medicine. An imbalance was believed to cause illness.

hydrolysis – adding a water molecule to another

hydrophobic – unwettable

hyperpigmentation – excess pigment

hyperproliferation – overgrowth of tissue

hypertension – high blood pressure

hyperthermia – high body temperature

hyphae – projecting filaments from a fungus

hypopigmentation – deficiency of pigment

iliac passion – severe intestinal pain

immunomodulatory – regulating the immune response

immunosuppression – suppressing the immune response

in situ – Latin, in place

indigenous – native to (a place)

inflorescence – flower stalk, scape

interleukin-1 – member of a group of chemicals called cytokines which regulate immune and inflammatory responses

ionic liquid – salt-like materials that are liquid at unusually low temperatures

ionisation – the process whereby a molecule gains an electrical charge by gaining or losing electrons

Ipecacuanha – a purgative medicine

ischial passion – severe hip pain

isomers – molecules of identical composition but different shapes

keratinocytes – skin cells

leukaemia cell line – permanent culture of one or more types of leukaemia cell

lifecycle – *e.g* of a parasite from egg through larva to adult

lymph nodes – *aka* 'glands', the swellings on the lymphatic system, which become enlarged with infection. Neck, groin and armpits are the commonest places for them to be found

lymphatic system – part of the immune system with vessels connecting lymph nodes carrying white cells to fight infection and removing waste from the body into the blood stream

medulla – structure at the base of the brain which connects it to the spinal cord

meristem – a stem cell, a type of undifferentiated cell which can evolve into any other cell

meta-analysis – analysing a large number of similar research studies to see if there is a common, correct conclusion

metabolic syndrome – combination of obesity, diabetes, high blood pressure

metaphase – a stage in the division of the chromosomes in growing cells

metastatic disease – distant spread of cancer cells from their site of origin

MHRA – Medicines and Healthcare products Regulatory Agency

microtubules – intracellular tubular structures that maintain the architecture of cells. Some transport vital materials and organelles within cells and their processes

mitochondria – self-contained, DNA-containing structures in the cytoplasm of a cell that have specific functions including generating energy (mitochondrial DNA)

mitosis – cell division

mitotic spindle – refers to the spindle shape that chromosomes adopt during cell division

monoamine oxidase inhibitors – chemicals that inhibit monoamine oxidase which facilitates the breakdown of chemicals which have a single amine group in their molecule, *e.g.* serotonin, noradrenaline

monoclonal antibody – an antibody that binds to only one chemical antigen on a cell or other structure

mucous membranes – lining of nose, mouth, stomach, etc.

mycobacterial – like mycobacterium; a group of bacterial species that cause tuberculosis, leprosy and other conditions

mydriasis/mydriatic – dilating the pupil

myocardium – heart muscle

neuromuscular – interaction of muscles and their nerve supply

neuropathic pain – pain due to disorder of the pain fibres

neuropathy – damage to the motor and/or sensory nerves

neurotoxins – nervous system poisons

neutrophil – type of white blood cell

NICE – National Institute for Health and Care Excellence

nodose ganglion – the swollen area on the vagus nerve (the 10th cranial nerve) as it exits the skull where there are nerve cells communicating both with the brain and the periphery

nuclear DNA – the DNA in the cell nucleus, as opposed to mitochondrial DNA

nutrient agar – a gel manufactured from seaweed on which seedlings can be germinated and grown: it contains nutrient chemicals such as nitrogen, phosphate, trace elements and sucrose

oestrogenic and progestogenic – reproducing the actions of oestrogens and progesterone

oxidative stress – a process whereby peroxides and free radicals are produced in a cell, causing damage

oxidation – being combined with oxygen

parenteral – something administered intravenously, intramuscularly, subcutaneously or into the spinal fluid

pathogenic – disease-causing

PDE – phosphodiesterase a group of enzymes, *e.g.* PDE3, PDE4

pessaries – medication administered via the vagina

pharmacokinetics – the rates of absorption, distribution in the body, metabolism and excretion of a substance

phase 1 trials – testing a drug on a small group of healthy volunteers

phase 2 trials – testing a drug for effectiveness and safety on a small group of patients

phase 3 trials – testing a drug for effectiveness and safety on large groups of patients

phase 4 trials – follow-up survey in practice of an approved drug for effectiveness and side effects on large groups of patients

photochemotherapy – treatment using sunlight or ultraviolet light plus a topical or systemic chemical sensitiser

photopheresis – a treatment for lymphomas involving treating white-blood-cell-rich plasma (taken from the patient) with methoxsalen and irradiating it with ultraviolet light before returning it to the patient

phototoxicity – skin inflammation (*e.g* sunburn) caused by increased sensitivity to light induced by chemicals, many present in plants

physiological actions – normal actions within the workings of the body

precursors – chemicals which are turned into other, usually useful, chemicals and medicines

pro-drug – a medicine which is converted into the active medicine in the body

properties, drying, heating, etc. – hypothetical properties in the Hippocratic/ Galenic theory of disease to explain how a plant/medicine has its effect

prophylaxis – prevention

protozoa – single-cell organisms, free living or parasitic

purgative – laxatives, medicines which cause bowel emptying

quartan fevers – fevers with temperature peaks every four days, common in malaria

QT interval – the time taken for ventricular electrical systole, measured from the electrocardiogram. It is related to but different from the time taken for mechanical contraction of the heart muscle (ventricle)

radio-ablative surgery – a form of surgery using multiple beams of high-energy radio waves that produce high temperatures where they intersect to destroy cancerous tissue without opening the body or skull

radiosensitive – sensitivity to X- or gamma rays or subatomic particles

randomised control trials – trials that compare treatments in comparable groups

of patients against each other or against a placebo, where the allocation to the therapeutic or other group is random

RCP – Royal College of Physicians, London

recombinant – breaking and recombining pieces of DNA to produce alleles with different functions

renal – relating to kidneys

rhinitis – inflammation in the nose

rhizome – a modified subterranean plant stem that sends out roots and shoots from its nodes

schizont stage – the stage in the lifecycle of the malaria parasite when it multiplies inside a red cell, rupturing it and then spreading to infect other red cells

screening programme – (in the context of this book) looking at plants or chemicals for possible therapeutic applications

sebaceous gland – the cells in the hair follicle that secrete sebum which lubricates the hair and skin

semi-synthetic – a drug produced by chemical modification of a naturally-occurring precursor

sequestered – hidden within

sinus rhythm – normal heart rate with regular rhythm

species – lower-level classification of individual plants, within the groups of family and genus

spermatogenesis – the production of sperm

spp. – abbreviation for species (plural)

styptic – a substance which stops bleeding

subspp. – abbreviation for subspecies (plural)

suppositories – a preparation of a medicine to be administered via the anus

sympathomimetic – a chemical or medicine which imitates the actions of the body's sympathetic nervous system which in turn is responsible for dealing with stressful situations

synthetic – artificial, made in the laboratory

systemic – within the body

tachycardia – fast heart rate

T-cells – a type of white blood cell responsible for many immune functions

teratogenic – inducing abnormalities in the developing embryo

therapeutic dose – the effective dose of a medicine

therapeutic index – the ratio of the therapeutic dose to the toxic dose

therapeutic window – the dose range within which a medicine is effective and not toxic

thrombin – a clotting factor that converts fibrinogen to fibrin which, with platelets, allows blood to clot

thrombo-embolism – blood clots which travel and block the blood supply to other parts of the body

tincture – archaic term for a solution or suspension, often in alcohol, of a medicine

topical – on the surface of the skin

transgenic – a genetically-altered plant or other organism (*e.g.* yeast)

transient receptor potential cation channel subfamily V member 1 (TrpV1) – the capsaicin receptor (the site on the nerve where capsaicin acts)

transmitter – usually a neurotransmitter, a chemical released from a nerve ending that passes to a receptor on another nerve or organ (*e.g.* muscle fibre) to produce a stimulatory or inhibitory effect

trimester – normal human pregnancy is divided into three equal trimesters of three months each

tubulin – the protein that forms long chains that give rise to microtubules, the skeleton of the cell

umbel – an umbrella-shaped flower head

UVA – ultraviolet light type A (short wavelengths)

vagal – relating to the 10th cranial nerve, the vagus, which is part of the parasympathetic control system responsible for heart rate, respiration and digestion, among other things

vasculitis – inflammation of blood vessels

vasoconstriction – narrowing of blood vessels

vectors – usually insects that transmit agents that cause diseases, *e.g.* the *Anopheles aegypti* mosquito is the vector for the *Plasmodium* parasite that causes malaria

ventricle(s) – the main chamber(s) of the heart which pump blood to the lungs and body

vermifuge – a medicine which kills intestinal worms

vernacular – the spoken language of the country

viral shedding – the transmission of viral particles

volatile free base – a basic chemical such as nicotine, cocaine or morphine that can be vaporised to promote absorption by inhalation

WHO – World Health Organisation

xenograft – a tissue graft or organ transplant from a donor of a different species from the recipient

Index
CHEMICALS and DRUGS

Aatrilla	17	Benecardin	333
Absinthe	25, 215-216	Benzedrine	141
Acetylcholine	41, 89-90, 159, 161-163, 257-258	Benzodiarone	333
		Benzofurane	333
Acetylsalicylic acid	281-283, 286-287, 290	Bergapten	18
Adrenaline	140, 189, 191-192, 198, 333, 372	Bexarotene	347
Ajmalicine	79, 273-274	Bupivacaine	192
Ajmaline	273		
Alcohol	124, 202, 275, 289, 295, 311, 349, 378	Cabazitaxel	307
		Camptothecin	63-67, 85, 310, 312-313
Aldosterone	131, 133, 177	Canakinumab	118
Alkaloid	41, 63, 79, 83, 85, 89, 97, 163, 170, 190, 198, 202, 233, 235, 238, 250, 252, 263, 273-274, 276, 343	Cannabis	38, 59, 252
		Capsaicin	69-70, 73-75, 378
		Carbenoxolone	175, 177-178
Alkavervir	325	Cardiac glycosides	121, 124, 373
Aminophylline	56	Carnauba wax	337, 339
Amiodarone	329, 333	Castor oil	311, 337, 339-340
Amodiaquine	27	Catharanthine	79
Amphetamine	141	Cathinone	141
Anandamide	59	Cellulose	211, 337, 340-341, 343
Anisatin	209	Cephalotaxine	83-85
Arsenic	16, 38	Chloroquine	26, 98
Artemether	26-27	Chlorpromazine	277-278
Artemisinic acid	29, 234	Chromones	335
Artemisinin	25-29, 98, 234, 240	Cilostazol	58
Artesunate	26-27	Cobalamin	357-358, 361
Ascorbic acid	103, 105	Coca	34, 145-149, 188
Aspirin	223, 281-283, 285-291	Cocaine	34, 53, 145-149, 188-189, 192, 378
Asthmatol	140		
Atabrine	96	Codeine	40, 249-250, 253, 287
Atracurium	90	Colchicine	79, 113, 116-119, 267
Atropine	31, 34, 37, 39-42, 140, 163, 257-258, 260, 263	Conazoles	223
		Cortisone	131-134, 178
		Cremophor	311, 339-340

Cromoglicate 329, 334-335
Curare 87-91, 98, 257-258, 260-261
Cytisine 235, 238, 240

Dabigatran 219
Deoxy-harringtonine 84
Deoxynojirimycin 225
Dexamphetamine 141
Dexedrine 141
Dextrins 337, 341
10-deacetyl baccatin 311
Diamorphine 149
Dicoumarol 219, 222-223
Digitalis 57, 99, 121-127, 273
Digitoxin 121, 124-125
Digoxin 56, 121, 123-126
Dihydroartemisinin 26-27
Dihydroetorphine 245
Diosgenin 129, 132-133, 135
Diprophylline 56
Docetaxel 307, 311
Dopamine 236, 276, 278
Doxofylline 56

Ecstasy 137, 141
Ephedrine 56, 137, 140-142, 333
Epigallocatechin 58
Episalvan 48
Ergometrine 195, 198
Ergotamine 195, 198
Esbriet 203, 205
Eskel 333
Esters 84, 155, 189, 317
Ethylene diamine 56
Etoposide 63, 67, 265, 269-270
Etorphine 245
Excipients 311
Febrifugine 201-203, 205
Fentanyl 245-246
Flavin adenine dinucleotide 105, 351

Flavin mononucleotide 351
Flavonoids 177
Furanocoumarin 15

Galanthamine 89, 159, 163-164
Galegine 170, 173
Glucose 75, 78, 81, 104-105, 110, 131, 167, 170-171, 225-226, 341
Glycosides 121, 124, 177, 269, 343, 373
Glycyrrhizic acid 177
Glycyrrhizin 175, 177
Godnose 110
Gramine 187, 190
Guanidine 170-171, 173
Gulonolactone oxidase 105
Gum arabic 177, 342

Halocur 203
Halofuginone 201, 203-205
Happy Tree 63, 65-66
Harringtonine 63, 67, 83-85
Hecogenin 129, 133
Heroin 203, 253, 255
Hexuronic acid 110
Histamine 117, 334, 374
Homoharringtonine 83-85
Hydroxychloroquine 98
Hydroxyethyl starch 343

Ibuprofen 118
Ignose 110
Imatinib mesylate 84
Iminosugar 225
Indole 79, 190, 276
Ingenol mebutate 151-152, 154-156
Intal 334
Intercostrin 89-90
Inulin 213, 215-217
Ipecacuanha 252, 375
Irinotecan 63, 65, 67

Isoharringtonine	84
Isotretinoin	347
Jervin	324
Jesuit's powder	95
Khat	141
Khelfren	333
Khellin	332-335
Laudanum	115, 252
L-dehydroascorbic acid	105
Legalon-SIL	293, 296
Leurocristine	79
Levodopa	278
L-gulono-g-lactone	104
Lidocaine	148, 187, 189, 191-193
Lignocaine	187
Liquorice	175-179, 183, 215
Lobeline	235-236, 238, 240
Lysergic acid	197
Magnesium sulphate	326-327
Mandrax	202
Mefloquine	27
Menaquinones	359
Mepivacaine	192
Metformin	167, 170-173
Methacholine	334
Methaqualone	201-202, 205
Methedrine	141
Methoxsalen	15, 18, 22, 376
Methylamphetamine	141
3,4-methylenedioxymethamphetamine	141
Methylmydriatin	140
Methyl salicylate	281-282
Methylxanthines	52, 57-58
Migalastat	225-227
Miglitol	225-226
Miglustat	225-227

Milrinone	51, 58
Monoamine oxidase inhibitors	276, 376
Morphine	40, 115, 147-149, 245, 249-250, 252-255, 287, 378
Mydrin	140
Nardil	276
Nedocromil	329, 334-335
Niacin	351-353, 360
Nicotinamide	352, 360
Nicotine	229-230, 233-236, 238, 240-241, 258, 378
Nicotinic acid	352-353
Nifedipine	329, 333
Nintedanib	203
Noradrenaline	140, 276, 372, 376
Noscapine	249-250
Odomzo	326-327
Oestrogen	131-132, 135
Ofev	203
Olive oil	358
Opiates	243, 249, 252-254
Opium	34, 40, 243, 249-252, 255, 260
Oripavine	250
Oseltamivir	99-100, 207, 209-211
Oseltamivir epoxide	209
Oxycodone	245, 250, 253
Oxymorphone	245, 250
Oxytocin	198
Paclitaxel	63, 67, 307, 310-313, 339-340
Pantothenic acid	353-354, 361
Papaverine	40, 249-250, 254
Paraxanthine	56
Paregoric	252
Pentoxyfylline	51
Peplin	154-155
Peruvian bark	285
Phenol	283

Phenylindanes — 58
Phylloquinones — 359
Physostigmine — 41, 87, 89, 91, 257-258, 260-261
Picato — 155
Pilocarpine — 262-263
Pirfenidone — 203, 205
Podophyllotoxin — 63, 79, 265, 267-270
Pontefract cakes — 176
Prednisone — 118
Procaine — 149, 188-191
Progesterone — 131-133, 135, 376
Protopine — 245-246, 249-250
Protoveratrine — 13, 321, 325, 327
Prozac — 276
Pseudoephedrine — 137, 140, 142
Pyrethrum — 299-302, 304
Pyridoxine — 354-355, 361

Quaalude — 202
Quinic acid — 99, 209
Quinidine — 93, 99, 273
Quinine — 27, 55, 93, 95-100, 116, 202, 209, 285, 290
Quinoline — 63, 93, 98

Reserpine — 273-274, 276-279
Resiniferatoxin — 75
Retinol — 346
Rhoeadine — 243-245
Riboflavin — 350-351, 360
Ricin — 340, 344
Rivaroxaban — 219
Ro 64-0792 — 209
Roflumilast — 51, 58
Rohypnol — 202
Ropivacaine — 192
Salbutamol — 141, 233
Salicin — 281-283, 285-286, 290-291
Salicylic acid — 268, 281-283, 285-287, 289-291

Serpasil — 277
Shikimic acid — 207, 209-211
Sildenafil — 51, 57-58
Silymarin — 293, 296
β-sitosterol — 45, 47-49
Sodium cromoglicate — 329, 334
Sodium valproate — 315, 317-319
Sonidegib — 321, 326-327
Starch — 177, 337, 341, 343
Statins — 117, 223
Steroids — 129, 131-135, 351, 372
Stigmasterol — 129, 133
Strychnine — 89, 258, 260
Sucrose — 110, 216, 376
Synthalin — 170-171
Synthetic pyrethroids — 299, 303
Syntocinon — 198

Tamiflu — 99, 207, 209, 211
Taxol — 66, 307, 311-313, 340
Taxotere — 311
Teniposide — 265, 269
Terpene — 79
Terpenoids — 154, 216
Testosterone — 131-132, 134-135
Thebaine — 243-245, 250
Theobromine — 52, 55-56, 59
Theophylline — 51-52, 55-57, 59, 141, 355
Thiamine — 348-350, 360
Thiazide — 277
Thiocolchicoside — 118
Tilade — 335
Topotecan — 63, 65
Toxiferine — 88
Tretinoin — 347
Tubocurarine — 87-91
Tyrosine kinase inhibitors — 84

Valproic acid — 315, 317
Varenicline — 238, 240
Verapamil — 249-250, 254-255

Veratridine	326
Veratrine	326
Vermouth	25, 215
Viagra	51
Vinblastine	77-80
Vincaleukoblastine	78
Vincristine	77, 79-80
Vinculin	78
Vindesine	79-80
Vindoline	79-81
Vinflunine	79
Vin Mariani	147
Vinorelbine	79-80
Vismodegib	326-327
Vitamin A	346-347, 360
Vitamin B	347, 348, 350-358
Vitamin C	69, 103-106, 109-111, 345, 358
Vitamin C	105, 184
Vitamin D	347, 358
Vitamin E	347, 358-359, 361
Vitamin K	129, 219, 222, 223, 345, 359
Vitamins	109, 129, 344-347, 353, 356, 360
Warfarin	219, 221-223, 359
Xylocaine	191

Index
PLANT NAMES

Acerola cherries 106
Aconite 39, 309, 326
Aconitifolia humile bifolium 266
Agave sisalana 129-130
Agrobacterium tumefaciens 234
Allium cepa 354, **354**
Allium ursinum 113, 163
American mandrake 12
Ammi majus **14**, 15-19, **17**, 22, 329-330, 332
Ammi visnaga 16, 329, 332, 335
Anacyclus pyrethrum 251, 300
Anapodophyllum canadense 266
Angelica gigas 163, **164**
Angels' trumpets 31
Aniseed 38, 183, 209
Anisum Indicum stellata 208
Anisum Philippinarum insularum 208
Anthemis pyrethrum 300
Apium graveolens 21
Argemone mexicana 245, **246**
Aristolochia serpentaria 274
Artemisia absinthium 25
Artemisia annua **24**, 25, 28, 98, 234
Arundo donax 187-188, **188**, 190, **190**
Asparagus 350, 356-357
Asparagus officinalis 350, 356, **357**
Astragalus spp 337, 342
Atropa belladonna **30**, 31, **31**, **37**, 38, 42, 70, 260
Autumn crocus 113

Barbasco 132
Beans 52-55, 60, 131, 221, 223, 259, 261, 349, 357
Belladona 37
Belladonna 31, 37-39, 42-43, 70, 260
Bell peppers 69-70, 351, 358
Betula **46**

Betula pendula **44**, 45, 47, 49
Bishop's weed 16, 331-332
Bitter orange 103
Black currants 106
Black false hellebore 322
Black hellebore 323-324
Bracken fern 350
Brasssica napensis 337
Britannica 107
Broccoli 106, 357, 359
Brown rice 351, 353-354
Brugmansia suaveolens 31, 33, **33**,
Bullwort 15-16

Cabeza de negro 132, 135
Cacao beans 55
Calabar bean 41, 87, 89, 257, 259-261
Californian false hellebore 322
Camellia sinensis **50**, 51
Camptotheca acuminata 63, **62**, 66, 310
Cancer tree 63
Cancer weed 151
Candlemas bells 161
Capsicum annuum **68**, 69, 71-72, **71-73**, 110, 351, 358
Capsicum baccatum 70
Capsicum chinense 70
Capsicum frutescens 70
Capsicum pubescens 70
Carduus lacteus 294-295
Carlina vulgaris 294
Carline thistle 294
Castor oil plant 339
Catha edulis 141
Catharanthus pusillus 77, 78
Catharanthus roseus 77, **76**, 79, 81, 273
Cauda equina 139
Cayenne jasmine 77

Celery	21
Centranthus ruber	**316**, 317
Cephalotaxus acuminata	67
Cephalotaxus drupacea	83
Cephalotaxus fortunei var *alpina*	85
Cephalotaxus fortunei	83
Cephalotaxus griffithii	85
Cephalotaxus hainanensis	85
Cephalotaxus harringtonia	63, 67, 82-85, **82**
Cephalotaxus lanceolata	85
Chelidonium majus	245
Chicory	216, 220
Chilli pepper	69-70
Chinese star anise	207, 209
Chocolate	51, 53-56, 58-59
Chondrodendron tomentosum	**86**, 87-89, **88**
Chrysanthemum cinerariifolium	299
Cichorium intybus	216
Cinchona condaminea	94, **94**
Cinchona ledgeriana	96, **96**
Cinchona officinalis	**92**, 93, 95, **101**
Cinchona species	95
Cirsium vulgaris	294
Citron	103-104
Citrus medica	103-104
Citrus x *aurantium*	103
Citrus x *limon*	**102**, 103-104, **104**, 358
Claviceps purpurea	195, 199
Coca bush	145
Cochlearia officinalis	107
Cocoa	51-52, 54, 56, 58-59
Coconut	337
Cocos nucifera	337
Coffea arabica	51-54, **53-54**, **60**
Coffea robusta	51
Coffee	51-56, 58, 60
Cola acuminata	52-53
Colchicum autumnale	**112**, 113-114, **114**, 117, 119
Coltsfoot	38
Commelina communis	226, **227**
Common barley	190, 195
Common fumitory	245
Common spurge	151
Copernicia cerifera	337
Copernicia prunifera	337
Corchorus olitorius	129
Corkwood trees	31
Corn poppy	251
Corn-roses	244
Corona regia	220
Corydalis species	245
Cotton seed	337
Cow parsley	15
Cowtail pine	83
Crocus sativus	113
Cucurbita maxima	348, 354
Cucurbita pepo	348, **348**
Cuminum cyminum	16
Cummin royal	16
Curare	87-91, 98, 257-258, 260-261
Curare vine	87
Cyamopsis tetragonoloba	337, 342
Daffodils	159-160
Dalmatian chrysanthemum	299
Dalmatian pellitory	299
Datura ferox	30, 42
Datura innoxia	30
Datura stramonium	31-32, **32**, 36, **36**, 42-43
Daucus carota	16, 346, **346**, 353
Daucus visnaga	331-332
Deadly nightshade	31, 35, 37, 163, 373
Devil pepper	273
Dichroa febrifuga	201
Digitalis grandiflora	122
Digitalis lanata	121-122, **121-122**
Digitalis lutea	122
Digitalis purpurea	**120**, 121-122, **122**, 124, **126**, 127
Dioscorea batatas	129
Dioscorea communis	129
Dioscorea composita	129
Dioscorea elephantipes	**135**
Dioscorea mexicana	129
Dioscorea polystachya	**128**, 129
Dipteryx odorata	221, 223
Divae mariae	294
Dorycnion	37
Drimys winteri	103, 107
Duboisia leichhardtii	31

Duboisia myoporoides	41
Dulcis radix	175-176
Dwale	37, 42
Elaeis guineensis	337
Elecampane	213-216
Enula	213
Ephedra antisyphilitica	140
Ephedra californica	139
Ephedra distachya	138
Ephedra foemina	138
Ephedra gerardiana	**136**, 137, **137**, **139**
Ephedra intermedia	138
Ephedra nevadensis	139
Ephedra sinica	137-138, 140
Equisetum	137, 138, 143, **348**
Equisetum giganteum	**143**
Ergot	195-199
Erratick Poppies	244
Eruca sativa	**350**
Erysimum	160
Erythroxylum coca	**144**, 145, 147, 188
Erythroxylum coca var. *ipadu*	145
Erythroxylum novagranatense	146
Euphorbia peplus	**150**, 151-155, **152**, **156**, 157
Euphorbia resinifera	75, 153
European yew	307, 311
Fair maids of February	161
False Queen Anne's lace	16
Filipendula ulmaria	281-282, **32**
Flanders poppy	243
Foxtail barley	187
Fumaria officinalis	245
Galanthus nivalis	**158**, 159-161, **160-161**, 164
Galanthus woronowii	161, **161**
Galega officinalis	**166**, **169**, 167-170, 172-173
Galium odoratum	223
Gaultheria procumbens	281-283, **281**
Giant reed	187, 190
Gingidium	330-332
Ginkgo biloba	210-211
Ginnie pepper	71
Ginseng	163
Glaucium flavum	245, **247**
Gloriosa superba	117
Glory lily	117
Glycine max	129-131, **129**, 358-359, **359**
Glycyrrhiza glabra	**174**, 175-176, **176**
Goat's rue	167-168, 170
Golden rain acacia	238
Gossypium	337, 340
Grecian foxglove	121-122, 124
Guaiacum officinale	**180**, 181-182
Guaraná	53
Guava	106
Helianthus annuus	355, **355**, 358
Helianthus tuberosus	216
Helleborus albus flore viridi	323
Helleborus niger	323
Hemlock	309
Henbanes	31
Heracleum mantegazzianum	15, 21
Heracleum sphondylium	15
Herba Nicotiana	231
Herb William	16
Hermodactylus	114-116, 119
Hierochloe odorata	223
Hippuris	137, 138
Holly	53
Hordeum jubatum	148, **186**, 187, 189-190
Hordeum vulgare	**194**, 195, **196**
Horned poppy	245, 247
Horseheal	213
Horsetail	138-139, 348, 350
Hydrangea febrifuga	**200**, 201-202
Hydrangea quercifolia	**205**
Hyoscyamus albus	31, 33, **33**
Hyoscyamus niger	33, **33**, 35, **35**
Ilex	52-53
Ilex paraguariensis	53
Illicium anisatum	**206**, 207
Illicium verum	207
Indian anise star	208
Indian or Calechut pepper	70
Indian snakeroot	273
Inula helenium	**212**, 213-215, **214**, 217

Jaborandi   262-263
Japanese star anise   207, 209
Jerusalem artichokes   216
Jesuit's bark   95
Jimson weed   32, 36
Juniperus communis   308

Kale   357, 359
Khella   329, 332
King's claver   219
King's clover   219
King's crown   220
Kola tree   53
Korean angelica   163

Laburnum anagyroides   229, 235, 238, **239**
Latua pubiflora   31, 38
Laus tibi   160
Lemon   99, 103-104, 106-109
Leucacantha   294
Leucoion bulbosum praecox minus Byzantinus   160
Leucojum aestivum   163
lignum-vitae   181
Limes   106, 108-109
Liquorice   175-179, 183, 215
Lobelia   38-39, 229, 235-236, **236**
Lobelia inflata   235-236, **237**
Lobelia siphilitica   236
Lycoris   163

Madagascar periwinkle   77, 80
Ma-huang   138, 140, 143
Maize   337, 341, 352-353
Malus domestica   350, **350**, 353, 355-356
Mandragora officinarum   31-32, **32**, **36**, 163
Mandrake   31, 34-35, 163, 202, 309
Mango beans   349
Matthiola   160
Mayapple   266
Meadow saffron   113-114, 116
Meadowsweet   282
Melilot   219, 221, 359
Melilotus officinalis   **218**, 219-221, **220**, 359
Mexican poppy   245-246
Milk thistle   293-295

Moly   163-164
Mormon tea   139
Morus alba   225-226
Morus nigra   **224**, 225

Naked ladies   113
Narcisso-Leucoium   160
Narcissus   159-160, 163, **165**
Nardus sylvestris   316
Nascata   87
Nicotiana benthamiana   234
Nicotiana minor   232
Nicotiana rustica   **228**, 229, 235, **235**
Nicotiana tabacum   29, 229-231, **229**, **231**,
                                            241, 250

Old maid   77
Olea europaea   337, 358, **358**, **361**
Oleander   309
Olive   337, 358
Onion   354
Opium poppy   243, 249, 251
Oranges   106-110
Origanum majorana   117
Oryza sativa   337, 348, 351, 353-354

Pacific yew   307-308, 310, 312
Palm   337
Papaver bracteatum   245
Papaver orientale   245
Papaver rhoeas   **242**, **244**, 243-244, 251
Papaver somniferum   **248-249**, 249-251, **251**
Paullinia   52-53
Pellitory of Spain   251, 300
Phaseolus minimus   349
Phaseolus radiatus   349
Phaseolus vulgaris   348, 350-351, 356
Physostigma venenosum   41, 87, 89, **256**, 257,
                                            259, 261
Pilocarpus microphyllus   262-263, **262**
Pilocarpus pennatifolius   **262**
Pimpinella anisum   38, 209
Piper Americanum   71
Piper nigrum   69
Plane tree   138

Plum yew 83-84
Podophyllum hexandrum 265-266, **270**
Podophyllum peltatum 63, 67, **264**, 265-266
Podophyllum 'Spotty Dotty' **265**
Pomegranate 251
Potato 31, 106, 337, 341, 351, 353-354
Pteridium aquilinum 350
Purple foxglove 121, 123
Pyrethrum 299-302, 304
Pyrethrum cinerariifolium 299
Pyrethrum officinale 300
Pyrethrum vulgare officinare 301

Qing-Hao 25-26
Qinghaosu 26
Queen Anne's lace 16
Quillaja 344
Quina-quina 94-95
Quinquina tree 94

Radium weed 151
Rauvolfia flexuosum 275
Rauvolfia serpentina **272**, 273-274, 276, 278-279, **277**
Rauvolfia tetraphylla 275
Rauvolfia vomitoria 273
Red pepper 110
Ricinus communis 311, 337, **338**, 339, **339**, 344
Rocket 350, 356
Rosy periwinkle 77
Roughbark lignum-vitae 181
Ruta capraria 167-168
Ruta graveolens 21, 168
Ruta montana 168
Ruta sylvatica 168
Ruta sylvestris 168

Saffron crocus 113
Salix alba **280**, 281-282, **283**, **291**
saponaria 343-344
Sarpagandha 274
Scabwort 213
Sclerotina sclerotiorium 21
Scopolia carniolica 31-32, **32**, 34-35, **35**
Scythian root 176

Secamone 71-72
Serpentaria virginianae 274
Serpentine wood 273
Setwal 316-317
Siliquastrum 70
Silver birch 45
Silybum marianum **292**, 293-294, **294**, **297**
Smilax 309
Snakeroot 273
Snowdrop 160-164
Snowflake 159, 163
Solanum lycopersicum 353, **353**
Solanum tuberosum 337, 351, **351**, 353-354
Solatro maius 38
Sorcerers' tree 31
Soya bean 359
Spina alba 294
Spinach 357, 359
Spiraea ulmaria 283
Spruce 107
Squash 348, 354
St John's wort 223
Strychnos nux-vomica 36, 89
Strychnos toxifera 87, 89
Summer fools 160
Summer snowflake 163
Sunflower 213, 355, 358
Sweet clover 219, 221-223
Sweet grass 223
Sweet marjoram 117
Sweet root 176
Sweet woodruff 223
Sweet wormwood 25

Tamarinds 108
Tanacetum cinerariifolium **298**, 299-300, **300**
Tanacetum coccineum 299, **299**
Taraxacum officinale 153
Taxus baccata 63, 67, **306**, 307-309, **309**, 340
Taxus brevifolia 307-308, **307**
Tea 25, 35, 51-53, 55-56, 58, 60, 78, 125, 139, 145-146, 148, 209, 251, 373
Theobroma cacao 51-52, 54, **54**, **61**
Thorn apple 31-32, 36
Tomato 353

Tonka beans	221, 223
Tragacanth	152, 337, 342
Tree of joy	63
Tree of life	63, 337
Triticum aestivum	337, 348, 351, 353-356, 358
Trychnon	37
Tussilago farfara	38
Valerian	315-319
Valeriana officinalis	315-316, **314-316**
Velvet leaf	87
Veratrum album	**320**, 321-323, **325**, 327
Veratrum californicum	321-322, **321**, **323**, 326
Veratrum caule ramosa	323
Veratrum nigrum	321-323, **321**, **323**
Veratrum viride	325, 327
Vinca parviflora	78
Vinca rosea	78
Viola alba	160
Virginian snake root	274
Visnaga daucoides	16, **328**, 329-332, **330-331**, 335
Wallflowers	160
Water cress	107
Water chestnut	63
Water tung	63
Wheat	197, 337, 348, 351, 353-356, 358-359
White false hellebore	322, 324
Wild carrot	16
Wild sunflower	213
Willow	281-286, 290-291
Wintergreen	282
Winter's bark	103, 107
Woolly foxglove	121-122, 124
Yam	129, 132-133, 135
Yellow melilot	219
Yellow sweet clover	219
Yerba Mate	53
Yew	63, 307-312
Zea mays	337, 352, **352**

OTHER IMAGES

Alexander of Tralles	**367**
Avicenna *Liber Canonis*	**23**
Belladonna extract bottle	**37**
Blane, Gilbert	**108**
Boisduval scale	**305**
Carter's Little liver pills	**271**
Colchicine wine	**116**
Culpeper, Nicholas	**368**
de L'Obel, Matthias	**369**
Digitalis tincture	**125**
Dioscorides	**130**
Dodoens, Rembert	**368**
Ergot extract bottle	**198**
Fuchs, Leonart	**369**
Liquorice	**179**
Mattioli, Pietro	**371**
Parkinson, John	**370**
Pockwood (Guiacum)	**185**
Podophyllin resin	**267**
Quinine bottle	**97**
Withering, William	**127**

The authors

Dr Susan Burge OBE DM FRCP

Susan Burge is an Honorary Consultant Dermatologist at Oxford University Hospitals and a Garden Fellow of the Royal College of Physicians. Her clinical interests and publications have focussed on medical dermatology and medical education. She was Director of Clinical Studies in Oxford University and has been President of the British Association of Dermatologists. She continues to teach dermatology although retired from clinical practice. She is also a keen gardener (with a passion for snowdrops) and leads tours in both the Oxford Botanic Garden and the Medicinal Garden of the Royal College of Physicians.

Dr Timothy Cutler MA FRCP FLS

Timothy Cutler is a retired Consultant Dermatologist who had a special interest in paediatric dermatology. For twenty years (1996-2016) he was the Royal College of Physicians' representative on the Advisory Board of Management at the Chelsea Physic Garden. He is a Past Master of the Worshipful Company of Barbers in the City of London and currently the Honorary Curator of the Physic Garden at Barber-Surgeons' Hall. He is a Garden Fellow of the Royal College of Physicians and a Liveryman of the Worshipful Company of Gardeners. He is a keen gardener with a special interest in the *Pelargonium* species, inspired by the historic collection at the Chelsea Physic Garden.

Professor Anthony Dayan LLB, MD FRCP FRCPath FRSBiol FFPM FFOM

Anthony Dayan was Professor of Toxicology in the University of London at Queen Mary University, London. He has been involved with the development and regulation of drugs and the safety of consumer products for more than 40 years in universities, official agencies in many countries, and in the pharmaceutical industry. He has been Chairman of the British Toxicology Society, and in 2014 the American College of Toxicology elected him Distinguished Scientist of the Year. He has been a Garden Fellow of the Royal College of Physicians since 2014, a co-author of *A Garden of Medicinal Plants*. He catalogued the Pharmaceutical Society herbarium at the College with Professor Michael de Swiet. He has a particular interest in the historical aspects of the dual use of certain plants as foods and medicines. He has lectured to many University of the Third Age and other groups on toxic risks, and on plants and medicines.

Dr Arjun Devanesan MSc MCRP FFICM

Arjun Devanesan was born in India, grew up moving around Southeast Asia and now lives and works as an intensive care physician in London. He is a Garden Fellow of the Royal College of Physicians with a particular interest in plants from the history of anaesthesia. He trained at Barts and the Royal London School of Medicine and Dentistry and specialised in anaesthesia and intensive care medicine. He is a current Fellow of the Faculty of Intensive Care Medicine at the Royal College of Anaesthetists as well as a Member of the Royal College of Physicians. He has a Master of Science degree awarded by the London School of Economics and is currently studying for a Doctorate in Philosophy at King's College, London. He has published in medicine, philosophy and poetry, including a song cycle libretto.

Professor Graham Foster FRCP PhD

Graham Foster is Professor of Hepatology at Queen Mary University of London and has a PhD in Molecular Biology. He runs a research programme studying the natural history of viral hepatitis, particularly hepatotropic viruses and hepatitis C; is chair of the NHS England Hepatobiliary Specialised Commissioning Clinical Reference Group; and past president of the British Association for the Study of the Liver. He is editor of the *Journal of Viral Hepatitis* and has published widely in this field. He is a Garden Fellow at the Royal College of Physicians and teaches on the plants there. He is the lead for the College's link with York University and the Sanger Institute for assessing the full genome of its medicinal plants. He has a special interest in plants which are toxic to the liver.

Jane Knowles MA, Head Gardener and Horticultural Curator, Royal College of Physicians

A graduate of Cambridge University, Jane started her horticultural career at the Chelsea Physic Garden in London where she spent 12 years as Head of Propagation and Medicinal Plants. She was appointed Head Gardener to the Royal College of Physicians in 2005 at the time when the Garden was being re-planted as a collection of medicinal plants. She has gone on to develop and manage it as a botanical collection, introducing several hundred more species and a labelling system, researching and redesigning many areas including most notably the *Pharmacopoea Londinensis* gardens in the College precinct of St Andrews Place. She was instrumental in the setting up of the Garden database http://garden.rcp.ac.uk. She co-authored *A Garden of Medicinal Plants* and *An Illustrated College Herbal* and has written numerous leaflets about the Medicinal Garden.

Professor John Newton FRCP FFPH FRSPH

John Newton is a public health physician and epidemiologist currently working as Director of Public Health Analysis at the UK Government's Department of Health and Social Care. He is also Professor of Public Health and Epidemiology at the University of Exeter and at the University of Manchester (Honorary) and was previously an academic epidemiologist in the University of Oxford, Chief Executive Officer of UK Biobank, Regional Director of Public Health for NHS South Central and Director of Health Improvement at Public Health England. He has been vice-president of the Faculty of Public Health of the Royal College of Physicians and is currently President of the Scientific Council of Santé Publique France.

He has been a Garden Fellow at the Royal College of Physicians since 2012 and teaches in the Medicinal Garden with an especial interest in the use of evidence in relation to plants and medicine, and how it has been understood and used over the centuries. He has a related interest in designing and cultivating a garden that is resistant to large grazing herbivores, namely fallow deer.

Dr Henry Oakeley FRCP VMH FLS

Henry Oakeley is a retired consultant psychiatrist who has been interested in plants since the age of eight, and an international authority on a group of South American orchids, on which he has written the definitive monograph and held the UK National Collections. Sometime adviser to the Chelsea Physic Garden, Honorary Research Associate at Kew and Singapore Botanic Gardens, chairman of the RHS Orchid Committee, RHS Council Member and currently RHS vice-president. He has lectured on orchids and exhibited them around the world; written over 250 articles on orchids and written (or co-authored) ten books relating to plants and their uses, and others on the English Civil War, the Anglo Boer war, and medical biographies. Since 2005 he has been Garden Fellow at the Royal College of Physicians, where he lectures on the plants in the Medicinal Garden. His orchid herbarium and drawings have been deposited at Kew, and his medicinal plant and orchid photographic archives at Kew and elsewhere. His current interest is in documenting the change of use of medicinal plants over the past two millennia.

Dr Noel Snell FRCP FFPM FRSB FLS

Noel Snell is a physician and chartered biologist; he specialised in respiratory medicine and clinical research. Now partly retired, he is still active as an honorary senior lecturer at the National Heart & Lung Institute, Imperial College London and as a vice-president of the British Lung Foundation. He has longstanding interests in the history of medicine and medical botany; he was awarded a diploma in the history of medicine by the Society of Apothecaries of London and is a Garden Fellow of the Royal College of Physicians, and a Fellow of the Linnean Society, the Faculty of Pharmaceutical Medicine, and the Royal Society of Biology. He is a former editorial board member of the journal *Phytotherapy Research*. He has published over 120 research papers, review articles, editorials, chapters in books, abstracts and scientific letters. He lectures regularly on 'plants in respiratory medicine' and allied topics.

Professor Michael de Swiet MD FRCP

Michael de Swiet was a Consultant Physician in Obstetric Medicine at Queen Charlotte's Hospital in London and is now Emeritus Professor in Obstetric Medicine at Imperial College, London. He has been the author or co-author of over 130 refereed papers, 110 chapters in books, and more than 200 abstracts, reviews and other publications. He established the subspecialty of obstetric medicine after graduating from Cambridge University in 1966. *Medical Disorders in Obstetric Practice*, edited and produced by him in 1984, went to 4 editions and was one of the first international textbooks to focus exclusively on providing expert care in this field.

Since 2004 he has been a volunteer at the Royal Botanic Gardens Kew, working with orchid displays in the Princess of Wales Conservatory, micropropagation, and assisting overseas workers at the Jodrell Laboratory with manuscript preparation for the *Botanical Journal of the Linnaean Society*. He has been a Garden Fellow at the Royal College of Physicians since 2012, with a special interest in medicinal plants for women's health. He has arranged public lecture courses on medicinal plants at the College; with Professor Dayan he catalogued the College's Pharmaceutical Society herbarium. He taught on the botany course at Highgate Literary and Scientific Institute. He has taken numerous tours of the College's Garden, taught on its medicinal plants and co-authored *A Garden of Medicinal Plants* (Little Brown, 2015). Since 2016 he has been the Royal College of Physicians' representative on the Advisory Committee at the Chelsea Physic Garden another famous London garden specialising in medicinal plants where, additionally, he teaches the student gardeners and volunteers about medicinal plants.

Angela Tunstall

Assistant Gardener and Garden Database Manager, Royal College of Physicians.